Geological Line Selection for the Qinghai-Tibet Railway Engineering

Jincheng Li · Wenwu Chen
Zhengping Liu

Geological Line Selection for the Qinghai-Tibet Railway Engineering

兰州大学出版社
LANZHOU UNIVERSITY PRESS

Springer

Jincheng Li
China Railway First Survey and Design Institute Group Co., Ltd.
Xi'an, Shaanxi
China

Zhengping Liu
China Railway First Survey and Design Institute Group Co., Ltd.
Xi'an, Shaanxi
China

Wenwu Chen
School of Civil Engineering and Mechanics
Lanzhou University
Lanzhou, Gansu
China

and

Key Laboratory of Mechanics on Disaster and Environment in Western China
Ministry of Education
Lanzhou, Gansu
China

ISBN 978-3-662-55570-5 ISBN 978-3-662-55572-9 (eBook)
https://doi.org/10.1007/978-3-662-55572-9

Jointly published with Lanzhou University Press, Lanzhou, China
ISBN: 978-7-311-05029-0 Lanzhou University Press, Lanzhou, Gansu, China

The printed edition is not for sale in China Mainland. Customers from China Mainland please order the print book from Lanzhou University Press, Lanzhou, China.

Library of Congress Control Number: 2017951412

Translation from the Chinese language edition: 青藏铁路工程地址选线 by Jincheng Li, Wenwu Chen, Zhengping Liu

Printed on acid-free paper

This Springer imprint is published by Springer Nature
The registered company is Springer-Verlag GmbH Germany
The registered company address is: Heidelberger Platz 3, 14197 Berlin, Germany

Preface

The construction of the Qinghai–Tibet Railway was a major strategic decision of the Party Central Committee and State Council. It was also a landmark project in Western development. People in the Tibet autonomous region and Qinghai Province looked forward to this project for a very long time, it was an aspiration of the former leaders of the new China and several generations of railway builders.

The Qinghai–Tibet railway started from Xining City, the capital of Qinghai Province, and ended in Lhasa City, the capital of the Tibet autonomous region, with a fulllength of 1956 km. The first stage of the project from Xining to Golmud was 814 km long. It was constructed in 1979 and began operating in 1984. The second stage of the project from Golmud to Lhasa was 1142 km long. It is a high-plateau permafrost railway and is known to be the longest line and run at the highest altitude above sea level of any railway in the world. This railway section crosses a 960 km long area with altitudes greater than 4000 m and a 550 km long area with continuous permafrost, with the highest point at 5072 m in the Tonglha Mountain Pass.

The time limit for the second phase construction of the Qinghai–Tibet Railway was 5 years. Construction began on June 29, 2001, and it officially opened on July 1, 2006 with a designed transport capacity of bus 8, and a freight volume of 5 million tons. The rear channel connecting Lanzhou–Beijing and the Asian–European land bridge is an important main line of our country's railway network.

During the construction of the Gela section of the Qinghai–Tibet Railway (hereinafter referred to as the Qinghai–Tibet Railway), it was necessary to consider problems that included alpine hypoxia, a complex permafrost condition, a fragile natural environment, long lines, and very difficult engineering tasks. We have some deep feelings to review of the Qinghai-Tibet railway survey and design different types of geological conditions during the line scheme selection and optimization process. Over a period of nearly two years, we summarized the main geological line selection process along the Qinghai–Tibet Railway, analyzed the influence of the geological conditions on the engineering of the line, cited numerous engineering examples, and finally finished the book.

This book reviews the complex geological conditions of the Qinghai–Tibet Railway. In addition, we highlight the characteristics of the Qinghai–Tibet Railway line selection, which had to consider permafrost, earthquakes and active faults, slope geological disasters, environmental protection, geothermal characteristics, wind, and windblown snow.

The book contains seven chapters.

Chapter 1 is an introduction. This chapter provides an overview of the main geological problems and their influence on the Qinghai–Tibet Railway construction, along with the basic principles of railway line selection under various geological conditions.

Chapter 2 reviews the Qinghai–Tibet Railway's geological environment. In this part, the authors analyze the engineering geological environment of the Qinghai–Tibet Railway, including the Tibetan Plateau's unique geographical location, topography, lithology, hydrogeological conditions, and permafrost distribution. A variety of other adverse and special geological conditions and climate characteristics, and the ecological environment, are also introduced.

Chapter 3 considers geological line selection in permafrost regions. In this chapter, permafrost is described in detail, including the classification of frozen soil and engineering geological properties. Permafrost does great harm to railway projects, and is likely to cause major disasters such as roadbed frost heaving and thawing settlement, cutting slope collapse, mudflows, frost heaving and thawing settlement of bridge and culvert foundations, deformation and cracking, tunnel heaving and thawing cracking, and icing inside tunnels. This chapter also discusses the principles of line selection in permafrost regions, and lists the main engineering geology instances in the permafrost regions along the Qinghai–Tibet Railway.

Chapter 4 discusses the active fault zone for strong earthquakes and the geological line selection. This chapter explains the spatial distribution of the seismotectonic zones of Kunlun Mountain, Bengcuo, Gulu–Sangshung, and Yambajan–Damxung. It also expounds on the activity rate of faults and major earthquake occurrences. The active fault belts not only affect the local life, but also do great harm to railway engineers. The main problems include landslides, debris flows, a tectonic fracture zone, thawing mud flows, thaw slumping, thermal lakes and ponds, frost heaving, ice pyramids, hummocks, the ice mantle, and permafrost swamps. From the perspective of the threats caused by an earthquake, this chapter mainly explains some of the damages from the Ms 8.0 Yambajan earthquake, Ms 8.0 Bengcuo earthquake, Ms 7.5 Jiuzina earthquake, and Ms 8.1 Kunlun Mountain earthquake. Therefore, when railway engineering is conducted in active fault zones and strong earthquake areas, a serious investigation is needed to determine the active faults along the railway and perform seismic intensity zoning. An economical and reasonable railway can only be constructed after the selection principle and engineering protection measures are developed. This chapter finally introduces some engineering examples of geological line selection in high earthquake-intensity areas and active fault zones.

Chapter 5 discusses railway line selection in a slope area with geological hazards. In this chapter, slope classification and its distribution law are first introduced.

Then, the negative influences on the construction of the Qinghai–Tibet railway are analyzed, including those from dangerous rocks, rock falls, landslides, collapse, rock heaps, debris flow, thaw slumping, and slope wetland. The route selection principle for slope geological hazards and prevention measures for these hazards are also included. In addition, it elaborates on the analysis procedure in a scheme comparison, and the reason that we chose one scheme rather than another.

Chapter 6 examines railway line selection in nature reserves. This part introduces the general situation of nature reserves, with a detailed description of the relationship between nature reserves and railway locations. It introduces the principles used when selecting the optimum route, emphasizes the importance of railway line selection in the Sanjiangyuan and Kekexili nature reserves, and points out the influence of a railway on an ecological environment and solutions to mitigate this influence.

Chapter 7 focuses on geothermal geological line selection in sand and windblown snow areas. This chapter is divided into three parts. The first section considers geological line selection in a geothermal area, introducing the distribution and classification of geothermal zones, and the engineering characteristics in our country, along with analyzing the geothermal influence on railway engineering, and formulating the principles of line selection in a geothermal area. The second section examines geological line selection in a wind-sand area. It expounds on the causes for the formation of sand and the motion law of the Qinghai–Tibet plateau, and analyzes the effects of sand on railway engineering and the principle for selecting a sand area and protective measures. The third section discusses geological line selection in as now drifting area, illuminates the reason for the formation of windblown snow disasters on the Qinghai–Tibet plateau and the motion law, analyzes the influences of windblown snow and avalanches on railway engineering, and introduces the principles for selecting wind snow areas and protective measures.

Because of the special climatic environment, complicated permafrost geological conditions, high seismic intensity, large number of active fault zones, and especially fragile natural ecological environment, great attention was given to railway line selection from the very beginning of reconnaissance on the Qinghai–Tibet Railway. The unremitting efforts of explorers and scientific and technical personnel made it possible to determine the main geological problems along the railways, especially the distribution characteristics and development law of permafrost, which provided a powerful guarantee for the success of the final railway line selection and decreased the capitalized cost to the minimum margin.

Railway line selection is a comprehensive process, which requires a comprehensive consideration of outside factors. One of the characteristics of this book is a discussion of the Qinghai–Tibet Railway route selection exclusively from the aspect of the influence of geological conditions. In this book, we provide our consideration and experiences when we made the decision under the involved main geological conditions at the end of every chapter to remain space for readers to think and discuss.

This book was based on the large amount of data collected and a summary of survey and design reports, information, design documentation, and published articles about the Qinghai–Tibet Railway. Adhering to the principle of taking our side as the dominant factor, the engineering projects discussed in the book and the final route selection decisions were conducted under practical survey and design service professionals working in the circuit, subgrade, geology, bridge and tunnel, and environmental protection fields, which sufficiently shows the collective intelligence.

Xi'an, China Jincheng Li
Lanzhou, China Wenwu Chen
December 2016

Acknowledgements

I am the Chief Editor and author of the manuscript for this book, and some teachers and graduate students from the School of Civil Engineering and Mechanics of Lanzhou University, along with several professionals from FSDI, assisted in the writing. Based on the massy research, we absorb extensive scientific payoffs, take the scientific concept on development as guidance and finally complete writing this book. Therefore, this book is the crystallization of the collective wisdom and achievements of the entire team. Chen Wenwu, a Professor from the School of Civil Engineering and Mechanics of Lanzhou University, made valuable suggestions on the structure of this book, and Liu Zhenping, a Senior Geological Engineer from FSDI, checked the entire book and provided valuable information. We also owe thanks to the personnel who took part in the writing and mapping, including Yuan Binxiang, Lin Lansheng, Li Weiqi, Ren Junhui, Zhou Wenjun, Qian Yaofeng, Wang Duoqing, Liu Siwen, Wang Qiuming, Chen Jia, Xie Jing, Mang Jun, Hang Lin, Sun Guangji, Kang Chao, Huo Zhangli, and Ma Xiangxian. Lecturer Pengbo Yuan; Ph.D. students Yanrong Xv, Yumin Du, Guopeng Wu and Weipeng Zhang and graduate students Hongzhi Peng, Haozhou He, and Pengpeng Yang also assisted in the translation work.

To comply with idiomatical engineering terms, we used some nonstandard units such as kilo meter, unit area, and mb.

During the process of writing, we referred to many related treatises, papers, and survey reports published in the press. We also quote numerous related information, and even utilize theories, viewpoints, and photographs. Because of the long length of the book, the references are not listed individually. I sincerely apologize for this.

Because of our limitations, mistakes and shortcomings are inevitable in the book. We look forward to receiving criticism and revisions from the public.

Thanks for following friends from Lanzhou University and China Railway First Survey and Design Institute Group Co., Ltd.

Pengbo Yuan, Hongzhi Peng, Gaochao Lin, Guopeng Wu, Weipeng Zhang, Haozhou He, Pengpeng Yang, Bingxiang Yuan, Jia Cheng, Jing Xie, Jun Man, Lin Han, Guangji Sun, Chao Kang, Zhangli Huo, Xiangxian Ma, Lansheng Lin, Weiqi Li, Junhui Ren, Wenjun Zhou, Yaofeng Qian, Duoqing Wang, Siwen Liu, Qiuming Wang, etc.

Contents

Chapter 1
Introduction

Abstract With unique geographical position and severe natural condition, Qinghai–Tibet Plateau has also developed various unfavorable geological phenomena, and these geological problems bring many impacts to the Qinghai–Tibet Railway line selection. In this chapter, main engineering geological problems along Qinghai–Tibet Railway and principles of railway line selection under different geological conditions are summarized.

Keywords Summaries · Engineering geological problems · Principles of railway line selection

The Qinghai–Tibet Plateau, the largest and highest plateau in the world, has complex geological conditions and a harsh environment characterized by high elevation, low temperature, droughts, strong winds, dust storms, and snowdrifts in addition to permafrost, seismic and geological hazards in slope areas, and geothermal activity. These conditions should all be considered in the route planning of the Qinghai–Tibet Railway.

1.1 Engineering Geological Problems Along the Qinghai–Tibet Railway Line

1.1.1 Geological Problems in Permafrost Areas

The northern boundary of the permafrost zone along the Qinghai–Tibet Railway lies in the Xidatan Fault Basin on the northern side of Kunlun mountains, at an altitude of 4350 m at railway boundary marker DK957+640, and the southern boundary lies in the Amdo Valley on the northern side of the Tonglha Mountains, at an altitude of 4780 m at railway boundary marker DK1513+770. This zone is 546.43 km long, with the talik and permafrost regions comprising 101.68 and 444.75 km, respectively.

J. Li et al., *Geological Line Selection for the Qinghai-Tibet Railway Engineering*,
https://doi.org/10.1007/978-3-662-55572-9_1

Permafrost is highly sensitive to temperature. Under the effect of external loads (e.g., roadbed fills), pressure is induced in the permafrost because of the distribution of moisture in the soil during the freezing process, which in turn causes frost heaving because of changes in the soil particle structure and density. As the permafrost melts, the consequent drainage consolidation under the effects of self-weight and external loads leads to soil deformation-induced settlement. Furthermore, because of the effects of climatic changes and human factors, the active layer in the permafrost regions undergo seasonal thawing and freezing every year, inducing periodic freezing and thawing deformation in the roadbed and foundation soil as well as various adverse geological phenomena associated with frozen soil. Thus, railway construction in permafrost regions entails a series of special engineering geological problems, all of which greatly influence the stability of the railway lines.

Railway construction affects the natural water–thermal regime of frozen soil and breaks its water–heat balance, which consequently reduces the mechanical stability of frozen soil; this causes many problems that severely affect railway engineering construction, such as roadbed frost heaving and thawing settlement, cutting slope collapse, mudflow, frost heaving and thawing settlement of bridge and culvert foundations, deformation and cracking, tunnel heaving and thawing cracking, and icing inside tunnels.

1.1.2 Geological Problems in Active Fault Areas

The Qinghai–Tibet Plateau has the most active crustal tectonic movement in the mainland of China. This region has high earthquake risk, and engineering construction here requires a strong consideration of the antiseismic problem. The Wonkhu-to-Tonglha mountains segment of the Qinghai–Tibet Railway passes through the head of the Qinghai–Tibet Plateau eta-type tectonic system. Most of this area features intense uplift geosynclines, and linear fractures appear in bundles. The strata include many folds and faults from the Sinian Suberathem, and the new tectonics (mainly fractures) show obvious inheritance from and reconstruction of the old fractures with strike faults. In each generation, different tectonic movements generate different fracture systems.

The main folds along the Qinghai–Tibet Railway are the Kekexili Synclinorium, Fenghuo Mountain Syncline, Kaixinling Anticline, and the Tonglha Mountain Synclinorium, and the main faults include the Xidatan Concealing Compressive Fracture, Kunlun Mountains Bealock Compressive Shear Faults, Fenghuo Mountain Compressive Fracture at the northern side, Wuli Fracture Zone, Yanshiping Regional Compressive Shear Faults, and the Tonglha Mountain Fault Zone.

Typically, the fault rock quality along both sides is poor; thus, fault cliffs and fault triangular facets occur often, forming a volley surface. Consequently, geological disasters such as landslides and debris flow can be triggered easily. A strong earthquake can produce significant surface ruptures, cause severe road deformation and ground surface subsidence, and damage bridges. In addition, it can induce geological phenomena such as sand liquefaction, collapse, and landslides, aggravating the disastrous effects of the initial earthquake. In the north of the Qinghai–Tibet Plateau permafrost area, fault activities not only cause roadbed deformation, pavement cracking, and engineering failure but also induce uneven frost, structural fractures, and migrating ice mounds, among other geological disasters.

1.1.3 Geological Problems in Slope Areas

Various geological diseases often occur on slopes, both natural (e.g., hillsides, banks of rivers, ditches, and reservoirs) and artificial. The major geological disasters along the Qinghai–Tibet Railway include dangerous rocks, rockfall, landslide, collapse, talus, and debris flow; furthermore, solifluction, thaw slumping, slope wetlands occur in permafrost slope areas. Dangerous rocks and rockfall occur mainly in the front pebble layer and rock slopes of Doilung Qu Gorge Terrace area of the Qinghai–Tibet Railway as well as near Nachitai, the south bank of Kunlun River, and Yambajan tunnels. Rock heaps are mainly distributed on both sides of Kunlun Mountain Luanshi Ditch and on both sides of the Doilung Qu slope toe area. Debris flow that affects the Qinghai–Tibet Railway mainly occurs between Gangou and the inlet section of Kunlun Mountain Tunnel as well as on the two sides of the cross-strait ditch in the Doilung Qu Gorge area. Solifluction and thaw slumping occur mainly in the Kunlun Mountain are and the Fenghuo Mountain area, and the slope wetlands are mainly in the Tonglha Mountain area.

1.1.4 Geological Problems in Geothermal, Aeolian, and Snowdrift Areas

The area along the Qinghai–Tibet Railway is rich in geothermal and wind resources; these resources also induce damaging geological phenomena such as geothermal heat, aeolian drift, and snowdrift, all of which adversely affect the construction and maintenance of the Qinghai–Tibet Railway.

The Yambajan–Nagqu tectonic belt powers boiling springs, including more than 200 hot springs and warm springs, and fountains whose temperature exceeds 90 °C. Geothermal activities usually change the geological environment, inducing various geological disasters, such as collapse and landslides due to changes to the structural features of a geological body, aquifer dewatering-induced land subsidence, and soil salinization-induced geothermal fluid immersion. Similarly, geothermal activity may accelerate, decelerate, or stop ground-material movement, which may lead to hydrothermal landslides, soil salinization, and the formation of cemented hydrothermal sediments. The potential for aeolian disasters on the Qinghai–Tibet Plateau is dependent on natural factors as well as on the interaction of natural and anthropogenic processes: Aeolian action increases the risk of sandstorm disasters, a risk that is further strengthened by the current climate warming and drying trend across the Qinghai–Tibet Plateau. In addition, the rapid increase in the number of livestock and the intensity of human activity has resulted in the excessive exploitation of land resources in the area, further worsening the sand activity.

The zones adversely affected by snowdrift on the Qinghai–Tibet Plateau are in the northeast and southeast (e.g., Hengduan Mountains and West Sichuan Plateau) and the surrounding mountains. The southern and western mountainous areas higher than 5000 m see strong annual snowdrift. The inner plateau has a dry and cold climate and receives low rainfall, which affects snowdrift.

Approximately 90% (by affected length) of the snow damage to the Qinghai–Tibet Railway is caused by snowdrift, mainly in Kunlun Bealock–Nagqu and especially in Tonglha Bealock–Nagqu. Avalanches do occur in Kunlun Mountains, Tonglha Mountains, and Nyainqentanglha Mountains, but their number and severity are less than those of snowdrift.

1.2 Principles of Railway-Line Selection

1.2.1 Line Selection in Permafrost Regions

In addition to the general principles, railway-line selection in permafrost regions should consider the following:

1. Avoid unstable permafrost regions and warm areas where the annual average ground temperature exceeds −0.5°.
2. To ensure foundation stability, prefer embankments over cutting so that the frozen soil remains undisturbed.
3. In hilly areas, lay the line in the upper part of gentle slopes. Excavating tunnels is preferred over going around the hill. When the formation conditions are similar, a single long tunnel is safer than are multiple short tunnels, even if this necessitates merging of short tunnels. Avoid locating tunnels in strata with high ground water and thick underground ice, and avoid placing the tunnel entrances and exits in permafrost zones with unfavorable geological phenomena.
4. In valleys, locate the line in melted regions or in high river terraces, which tend to be stable. This principle must also be applied when a long- or medium-length bridge is required to cross a river. Moreover, avoid constructing the bridge both on both melted soils and frozen soils, which leads to differential settlement.
5. Avoid thick underground ice, thaw slumping, lakes, and ponds caused by thawing, icing, frost mounds, and freezing swamps. In other words, locate the line in dry regions, permafrost areas with less ice and composed of rock, macadam, and coarse sand.
6. In regions with a thick ice layer at a shallow depth, regions with warm and ice-rich frozen soils, and regions with cross-stratification of fine and coarse

particles, replace the embankment with a bridge to prevent nonuniform deformation. Such bridges, which account for more than 130 km of the Qinghai–Tibet Railway, improve reliability and protect the environment.

7. Gravel-soil-taken conditions should be considered. Designate the gravel-soil-taken fields and bind the vehicle routing. To avoid forming new lakes and ponds caused by thawing, do not remove soil from the roadsides and do not transport it outside these designated routes.
8. Wherever the railway line is parallel to the Qinghai–Tibet Highway, locate the line close to the highway; this avoids additional adverse effects on the environment and the ecology. However, do consider a buffer zone of more than 100 m between the railway and the highway to account for future reconstruction and extension.
9. Wherever possible, use large-radius curves to facilitate high train speed.

1.2.2 Line Selection in Active Fault Zones

As per the principles of engineering geological line selection in high-earthquake-intensity regions, railway lines should be located away from active fault zones unless unavoidable, in which case the line should be designed as a simple structure, with a large angle at its narrowest section. In addition, large and high bridges and tunnels should be avoided in active fault zones, because these structures are difficult to repair when damaged.

The strike of most active fault zones along the Qinghai–Tibet Railway either is perpendicular to the railway line or intersects at a large angle, depending on the neotectonic movement characteristics of the Qinghai–Tibet Plateau. In fact, most faults along the planned route could not be avoided. During the engineering investigation and design stage, all involved active faults were comprehensively evaluated and the different design schemes compared. In the final construction-drawing design stage, the aforementioned line selection principles were followed along the main active fault zones, whereas for relatively unimportant active fault zones, efforts were made to avoid high bridges, to shorten bridge spans and lengths, and to replace bridges with culverts. At certain construction sites that influenced the route scheme and engineering setting, additional geological investigations were performed to ensure that the final line scheme and engineering measures were optimal.

1.2.3 Line Selection in Slope Areas

Slope areas are characterized by complex geological and topographical conditions and large variations in elevation. These conditions strongly influence design parameters such as the railway grade, minimum curve radius, railway strike, and line standard.

Before line selection in slope areas, the engineering geological conditions and geological hazards along the slopes should be evaluated. In such areas, the line selection of the Qinghai–Tibet Railway adhered to the following principles:

1. Adopt precautionary measures in tectonically active areas, namely, areas with rock-mass accumulation, avalanches, dangerous rocks, falling rocks, landslides, and unstable permafrost.
2. In elevation design, ensure that the line is higher than the design flood level, and reduce the excavation volume by not locating the line in the upper slopes. To this end, it is essential to fully utilize the favorable topographical conditions and to consider the developmental characteristics of the geological hazards. To minimize the bridge elevation and valley embankment, the railway exhibition line should be started early. By considering the topographical, hydrological, and elevation characteristics in the design, a line with an appropriate undulation can be realized.
3. In areas with a low slope hazard, compare the costs of the slope-avoidance and straight-line schemes by considering the costs of slope treatment. The merits and demerits of these two design schemes are as follows: (a) The slope-avoidance scheme has poor line reliability because of the unfavorable geological conditions and river erosion. (b) In the straight-line scheme, the line uses cutting slopes, tunnels, and extension lines to pass through the slope hazard areas. This approach has high reliability because the line is short and straight; however, it entails higher construction costs and time than does the slope-avoidance scheme. The straight-line scheme is preferred when the construction costs and time are similar.
4. When embankment stability is uncertain, use tunnels to pass through slope areas with a severe slope hazard. For example, the Kunlun Mountain area and Yambajan area have many long tunnels at a shallow depth. Use bridges in areas where the slopes are so steep that adequate tunnel safety cannot be ensured. For example, bridges were designed to navigate the slopes near the exit of the Kunlun Mountain Tunnel because of the large development and wide distribution of geological hazards (e.g., thaw slumping, thaw-frozen mudflow, and rockfall) around this area.

1.2.4 Line Selection in Geothermal Areas

In geothermal areas, the following line selection principles should be adhered to

1. Locate the line as far away from the surface of hot springs as possible.
2. Avoid areas where landslides may be triggered by geothermal activity, especially those areas where large landslides have occurred historically. If such areas cannot be avoided, they must be subjected to landslide treatments, following which the stability and characteristics of the slopes and landslides must be reevaluated.

3. Surface hot springs are valuable tourism resources. During railway construction, ensure that depression cones, which reduce the water quantity in the springs, are not formed. However, tunnel construction would inevitably affect these springs; hence, preventive engineering measures, such as controlling the excavation temperature and employing a reasonable tunnel lining, a heat-insulation layer, and a water barrier, must be taken to reduce the negative effects. Note that recession in the regional groundwater level depends on the local stratigraphic and tectonic conditions.

1.2.5 Line Selection in Aeolian Areas

Line selection in dust-bowl (aeolian) areas should not only account for the activities during the investment and construction periods but also those during the maintenance and operation periods. In particular, the following principles should be considered

1. Avoid areas with severe sand drift, especially if the additional line length and construction cost is not too high.
2. When sand-drift areas must be crossed, locate the line over regions such as lake and beach plains, valley terraces, old riverbeds, and the edges of proluvial fans.
3. Cross fluid dunes at their narrowest point, and locate the line along the headwind side of the wind erosion of the dune. Where possible, pass through dune groups in the low-lying section between dunes.
4. To decrease wind erosion of the embankment and the sand-burying hazard, ensure that the line is either parallel to the dominant wind direction or intersecting at a small angle.
5. Design reasonable horizontal curves and vertical sections considering the wind direction and topography. The convex surface of horizontal curves should face the dominant wind direction. To ensure smooth transit, the vertical section should not go all the way. Avoid cutting the terrain, and locate the line such that it follows the natural topographical relief. Moreover, the embankment height should not be too high.

1.2.6 Line Selection in Snowdrift Areas

The main snow disaster along the Qinghai–Tibet Railway is snowdrift. Snowdrift along the Qinghai–Tibet Railway depends not only on the climatic conditions and elevation but also on such factors as the regional geographic and geomorphic conditions, terrain, longitudinal section, and embankment cross-section. In general, areas prone to severe snow disasters must be avoided to shorten the construction period and reduce the engineering investment. Use tunnels and shed-tunnels when such areas cannot be avoided. The following points should be considered during line selection, especially when the additional line length and cost required to avoid such areas is not too high

1. Line direction—The line should either be parallel to the direction of the prevailing winds or intersect this direction at a small angle, especially when the line is adjacent to a mountain. Generally, the smaller this intersection angle is, the weaker is the effects of the snow disaster.
2. Line location—The lower is the elevation of the line, the weaker is the effects of the snow disaster, because the snow retention period is longer and the snowmelt time later at lower elevations. Thus, when the line must cross a mountain, prefer lower elevations. During insolation, sunny slopes receive stronger solar radiation for longer periods than do shaded slopes, meaning that the snow on the former melts earlier than that on the latter. Accordingly, sunny slopes should be preferred for line location, except when a sunny slope is too undulatory or when a shaded slope has abundant vegetation. Snow disasters are more severe on the leeward slope than on the windward slope. In particular, along a leeward slope adjacent to a steep slope, snow disasters are severe at hairpin bends; thus, the number of hairpin bends should be kept minimal in areas with heavy snowfall. Furthermore, locate the line an open valley rather than a narrow valley and in valleys with less undulation. The mountain ridge should be preferred over the mountain side, and avoid locating the line in the gap between mountains.
3. Embankment design—(a) In areas with heavy snowfall, the effects of snow disasters are strong irrespective of the angle between the line and the prevailing wind direction, especially in shallow and long cutting slopes. Thus, embankments should be preferred over cutting slopes, and where cutting slopes are unavoidable, use very deep and wide side ditches on the windward side. Occasionally, the railway road in a cutting slope appears as an embankment. Moreover, snow storage fields and snow collection ditches must be built along the line. (b) For leeward road sections with semifilling and semiexcavating as well as upwind road sections, the earthwork excavation may need to be deepened depending on the local conditions. Snow storage fields, wide side ditches, and large gravel-soil-taken fields are essential to mitigate and prevent snow disasters. (c) Large amounts of snow accumulate at hairpin bends. Therefore, to eliminate snow deposition caused by vertical and horizontal vortex, design curves with a large turning radius and remove part or whole of the residual mountain body. (d) Along mountain saddles with high risk of snow disasters, conventional tunnels, cut-and-cover tunnels, and shed-tunnels should be used when the line length in such regions is short. Although these structures are expensive, they are highly efficient in preventing snow damage.
4. Drainage design—In areas with heavy snowfall, particular attention should be paid to drainage design, because snowmelt runoff increases as the weather becomes warmer. Specifically, construct culverts in the valley for adequate drainage of the snowmelt runoff, thus ensuring embankment stability.

1.2.7 Line Selection in Nature Reserves

There are 24 national and provincial nature reserves along the Qinghai–Tibet Railway: five national and four provincial nature reserves in Qinghai Province and nine national and six provincial nature reserves in Tibet. Of these, 11 are close to various railway lines: Hoh Xil Nature Reserve, Sanjiangyuan National Nature Reserve, Yarlung Tsangpo Nature Reserve, Qiangtang Nature Reserve, Nam Co Lake Nature Reserve, Damxung–Yambajan–Geda Thermophis Baileyi Nature Reserve, Lhasa Cypress Forest Nature Reserve, Linzhou–Pengbo Black-Necked Crane Nature Reserve, Dagze–Yeba Karst Landform Nature Reserve, Doilungdêgên Ma Unconformable Contact Nature Reserve, and Lhalu Wetlands Nature Reserve. To minimize the effects of railway construction on these nature reserves and the wild animals living along the Qinghai–Tibet Railway, the following principles should be considered in line selection:

1. Avoidance—Design the line to go around these nature reserves, and where this is not feasible, avoid locating the line in the established buffer zones; similarly, avoid locating the line in animal habitats, water conservation areas, biodiversity conservation areas, and alpine ecosystem conservation areas (e.g., wetlands and headwaters). If passing through wetlands is unavoidable, measures must be taken to avoid changing the flow direction of the surface water and groundwater in order to ensure wetland continuity. Bridges are therefore ideal in such areas.
2. Wildlife passageway—To prevent disruption of wildlife reproduction and migration activities, scientifically design adequate passageways of adequate width and height, considering animal population growth and according to the final environmental impact assessment. The major types of wildlife passageways are those under bridges, above tunnels, and on the gentle slopes of embankments. In addition, implement ecological projects to reduce or neutralize the effects of railway construction on these passageways and the animal habitats; in the long term, these efforts would support tourism along the Qinghai–Tibet Railway.

Given the geographically unique, pristine, fragile, and sensitive ecosystem of the Tibet Plateau, the effects of the Qinghai–Tibet Railway on the geographical ecosystem of the Tibet Plateau must be monitored and minimized.

Chapter 2
The Qinghai–Tibet Railway Geological Environment

Abstract The environmental conditions along the Qinghai–Tibet Railway are extremely harsh, the ecology fragile, and the geological conditions complex, complicating the design and construction of the railway line. Therefore, a comprehensive analysis of the various engineering geological conditions along the line is essential for studying the line scheme. In this chapter, the geographical conditions—namely the geography, climatic characteristics, soil and vegetation, wildlife, river systems, and ecological environment—along the Qinghai–Tibet Railway are discussed. Moreover, the geological environment—the topography, formation lithology, geological structure, hydrogeology, seismic and active faults, permafrost distribution, and unfavorable and unique geology—along the railway is described in detail.

Keywords Qinghai–Tibet Railway · Physical geography · Geological environment

The anoxic atmosphere, low temperature, and complex geological and engineering geological conditions, in addition to the permafrost, landslides, debris flow, earthquakes, dust storms, snowdrifts, and other geological hazards, render harsh the landscape that the Qinghai–Tibet Railway passes through. Moreover, the ecological environment along the railway line is fragile and sensitive: if the vegetation in the permafrost region is damaged, the recovery would be very slow. Thus, three major problems were encountered in the construction of the Qinghai–Tibet Railway: permafrost, alpine hypoxia, and fragile ecology.

2.1 Physical Geography

2.1.1 Geographical Location

Located in central Asia at an average altitude of 4500 m and spanning more than 10° latitude, the Qinghai–Tibet Plateau is the largest and highest plateau in the world and is thus known as "the Roof of the World" and "the Third Pole of the

J. Li et al., *Geological Line Selection for the Qinghai-Tibet Railway Engineering*,
https://doi.org/10.1007/978-3-662-55572-9_2

Earth." This 2.5×10^6 km^2 plateau, the highest Chinese terrain, is bordered by the Hengduan Mountains on the east, the Himalayas on the south and the west, and Kunlun Mountains on the north. The plateau comprises the Xizang Autonomous Region, Qinghai Province; parts of the Xinjiang Uygur Autonomous Region, Gansu Province, Yunnan Province, Sichuan Province; and part or all of the countries of Bhutan, Nepal, India, Pakistan, Afghanistan, and Tajikistan.

Located along north latitude 29°30′–36°25′ and east longitude 90°30′–94°55′, the Qinghai–Tibet Railway passes through the hinterland of the Qinghai–Tibet Plateau, specifically through Qinghai and two of Tibet's provinces. This line starts from Golmud city in Qinghai Province and extends 1142 km southward to Lhasa, primarily along the Qinghai–Tibet Highway and across Naij Tal, Kunlun Pass, Wudaoliang, Tuotuoheyan, Yanshiping, Tonglha Mountains, Amdo, Nagqu, Damxung, and Yambajan (Fig. 2.1).

The average altitude of the line is approximately 4380 m, with approximately 960 km of the line being 4000 m above sea level and the nearly 550 km along permafrost regions. Figure 2.2 shows the plan and profile sections of the Qinghai–Tibet Railway. With the maximum altitude of 5072 m, it is the highest and longest plateau permafrost railway in the world, and it passes through a few national and provincial nature reserves, such as the Hoh Xil Nature Reserve, Sanjiangyuan National Nature Reserve, and Chang Tang Nature Reserve.

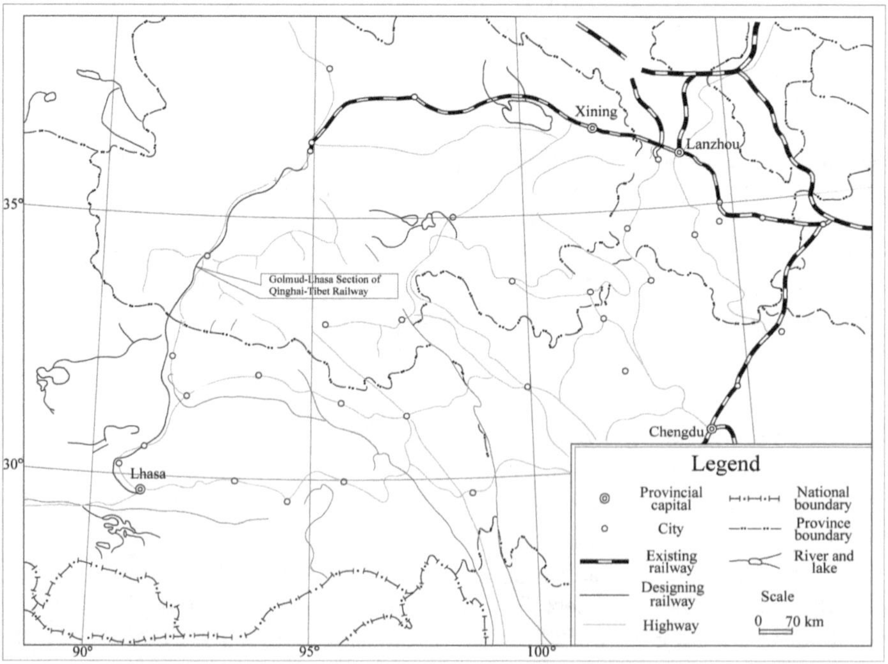

Fig. 2.1 Geographic map of the Qinghai–Tibet Railway

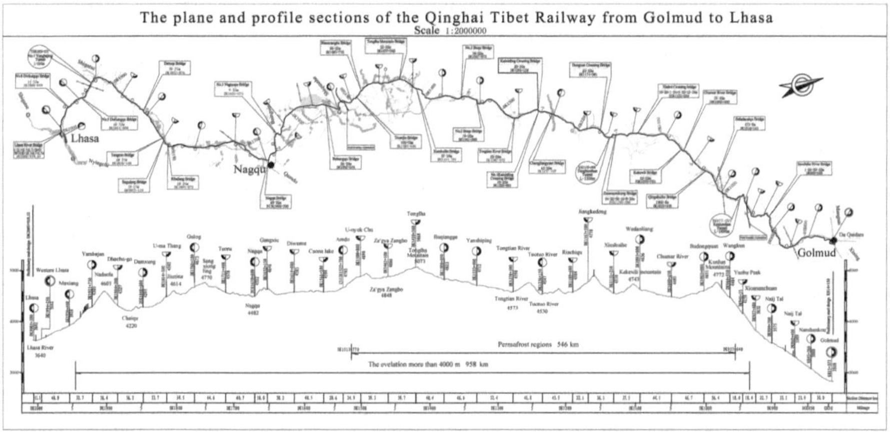

Fig. 2.2 Plane and profile sections of the Qinghai–Tibet Railway

2.1.2 Climatic Characteristics

The Qinghai–Tibet Railway is located in the Chinese mainland, far from oceans and seas. Apart from the southern Tonglha Mountains area, which has a marine climate, and the northern Qaidam area, which has an arid climate, most of the plateau hinterland has a unique climate of periglacial drought, with distinct vertical climatic zoning with increase in altitude. This region is characterized by a cold, dry, and changeable climate, without any clear seasonal distinctions; furthermore, the air rarefied, and the air pressure is low. The plateau is frozen for 7–8 months each year, typically from September to April–May, and the evaporation rate far exceeds the precipitation. In the alpine region, most of the precipitation occurs as snow, graupel, and hail; by contrast, the wide high plains see less snowfall even in winter, and rainfall accounts for 60–90% of the precipitation in the warm season. Except in certain mountain regions, snow coverage is generally unstable and thin. The prevailing winds blow toward the northwest and west, with strong winds (grade $\geq$ 8) mainly occurring from October to April. Table 2.1 summarizes the main meteorological conditions along the line.

The annual average temperature along the line is −6.9 to −2 °C, with the highest and lowest temperatures typically occurring in July (~6.5–8.1 °C) and January (occasionally, December; −17.4 to −14.5 °C). The annual average temperature fluctuation is <15–26 °C and never exceeds 50 °C. The daily average temperature fluctuation over a year is <10–19 °C is always less than 35 °C. In contrast to the northeast permafrost region in China, the temperature characteristics in the Qinghai–Tibet Plateau vary slightly over the year but greatly within a day.

Atmospheric transparency along the line is high (low cloud, strong sunshine), and the global solar radiation and sunshine duration are high (2600–3000 h a^{-1}). The radiation incident on regions at altitudes below 5000 m is the highest in all of

Table 2.1 Summary of the main meteorological data along the Qinghai–Tibet Railway

Station name and construction time			Golmud (1955)		
Location and elevation			N: 36°25′ E: 94°54′	Elevation (m)	2807.6
Representative mileage and site			K814+150–DK957+640		
Values and data collection period			Values	Data collection period (years) and date	
Average air pressure (mb)			724.9	10	1990–1999
Temperature (°C)	Annual average		6.7	10	1990–1999
	Extreme	Maximum	35.5	45	August 2, 1999
		Minimum	−33.6	45	January 12, 1959
	Hottest monthly average		17.8	10	July
	Coldest monthly average		−8.9	10	January
	Maximum monthly averaged diurnal temperature range		19.5	10	January 1966
Humidity	Absolute (g m^{-3})	Average	3.3	10	1990–1999
		Maximum	15	45	July 21, 1971
		Minimum	0	10	Nine times
	Relative (%)	Average	32	10	1990–1999
		Minimum	0	10	More than three times
Precipitation (mm)	Annual average		41.8	10	1990–1999
	Annual maximum		101.8	45	1967
	Annual minimum		11.4	45	1965
	Monthly maximum		44.0	45	July 1971
	Daily maximum		32.0	45	July 22, 1971
	Maximum rainfall and duration		32.0	10	13 h
	Annual average precipitation days		28.5	10	1990–1999
Evaporation (mm)	Annual average		2392.6	10	1990–1999
	Annual maximum		3232.3	45	1956
Wind speed (m s^{-1})	Average wind speed and dominant wind direction		2.6W	10	1990–1999
	Seasonal average wind speed	Spring	3.1W	10	March–May
		Summer	2.9E	10	June–August
		Autumn	2.2W, SW	10	September–November
		Winter	2.2SW	10	December–February
	Annual average gale days (grade ≥ 8)		9.8	10	1990–1999
	Maximum wind speed and direction	Timing	24W	45	July 6, 1957
		Instantaneous	43W	10	1969

(continued)

Table 2.1 (continued)

Station name and construction time			Golmud (1955)		
Frozen snow (cm)	First and last snowfall		October 20 and May 14	45	1955–1999
	Largest maximum thickness		6.0	45	March 1985; July 1987
	Maximum seasonal frozen-soil depth and duration		88.0	45	1955–1999
Other	Annual average fog days		None	10	None
	Annual average thunderstorm days		3.3	10	1990–1999
	Annual average sandstorm days		–	–	–
Comment	Averages are calculated using data from the past 10 years; extremes are derived from the station's historical data				
Station name and construction time			Wudaoliang (October 1955)		
Location and elevation			N: 35°13′ E: 93°05′	Elevation (m)	4612.2
Representative mileage and site			DK957+640–DK1160+344		
Values and data collection period			Values	Data collection period (years) and date	
Average air pressure (mb)			578.9	10	1990–1999
Temperature (°C)	Annual average		−5.2	10	1990–1999
	Extreme	Maximum	23.2	44	June 11, 1961
		Minimum	−37.3	44	January 12, 2012
	Hottest monthly average		5.6	10	1990–1999
	Coldest monthly average		16.7	10	1990–1999
	Maximum monthly averaged diurnal temperature range		18.6	44	March 1966
Humidity	Absolute (g m^{-3})	Average	2.8	10	1990–1999
		Maximum	11	44	July 12, 1993
		Minimum	0	44	Three times
	Relative (%)	Average	57	10	1990–1999
		Minimum	0	44	Five times
Precipitation (mm)	Annual average		290.9	10	1990–1999
	Annual maximum		407.0	44	1989
	Annual minimum		136.3	44	1984
	Monthly maximum		133.1	44	July 1991
	Daily maximum		37.1	44	August 7, 1977
	Maximum rainfall and duration		37.1	44	21 h
	Annual average precipitation days		180.0	10	1990–1999
Evaporation (mm)	Annual average		1316.9	10	1990–1999
	Annual maximum		1531.6	44	1973

(continued)

Table 2.1 (continued)

Station name and construction time			Wudaoliang (October 1955)		
Wind speed (m s^{-1})	Average wind speed and dominant wind direction		4.1 W	10	1990–1999
	Seasonal average wind speed	Spring	4.8 W	10	1990–1999
		Summer	3.5 E	10	1990–1999
		Autumn	3.4 W, NE	10	1990–1999
		Winter	5.1 W	10	1990–1999
	Annual average gale days (grade $\geq$ 8)		130.1	10	1990–1999
	Maximum wind speed and direction	Timing	31 W	44	April 2, 1986
		Instantaneous	40 W	44	More than five times
Frozen snow (cm)	First and last snowfall		August 6 and July 27	44	1956–1999
	Largest maximum thickness		14.0	44	October 14, 1967
	Maximum seasonal frozen-soil depth and duration		No observation data		Permafrost area
Other	Annual average fog days		10.9	10	1990–1999
	Annual average thunderstorm days		36.7	10	1990–1999
	Annual average sandstorm days		–	–	–
Comment					
Station name and construction time			Tuotuo River (October 1955)		
Location and elevation			N: 34°13′ E: 92°26′	Elevation (m)	4533.1
Representative mileage and site			DK1160+344–DK1419+300		
Values and data collection period			Values	Data collection period (years) and date	
Average air pressure (mb)			585	10	1990–1999
Temperature (°C)	Annual average		−4	10	1990–1999
	Extreme	Maximum	24.7	44	June 29, 1988
		Minimum	−45.2	44	January 6, 1986
	H hottest monthly average		7.6	10	1990–1999
	Coldest monthly average		−16.2	10	1990–1999
	Maximum monthly averaged diurnal temperature range		21.8	10	January 1986
Humidity	Absolute (g m^{-3})	Average	3	10	1990–1999
		Maximum	11.5	44	June 14, 1999
		Minimum	0	44	Nine times
	Relative (%)	Average	53	10	1990–1999
		Minimum	0	44	More than five times

(continued)

Table 2.1 (continued)

Station name and construction time			Tuotuo River (October 1955)		
Precipitation (mm)	Annual average		248.5	10	1990–1999
	Annual maximum		459.4	44	1985
	Annual minimum		164.2	44	1984
	Monthly maximum		174.0	44	July 1972
	Daily maximum		50.2	44	October 18, 1985
	Maximum rainfall and duration		50.2	44	24 h
	Annual average precipitation days		172.6	10	1990–1999
Evaporation (mm)	Annual average		1638.9	10	1990–1999
	Annual maximum		1945.5	44	1960
Wind speed (m s^{-1})	Average wind speed and dominant wind direction		3.9 W	10	1990–1999
	Seasonal average wind speed	Spring	4.8 W	10	1990–1999
		Summer	3.5 NE, E	10	1990–1999
		Autumn	3.1 W	10	1990–1999
		Winter	4.3 W	10	1990–1999
	Annual average gale days (grade $\geq$ 8)		178	10	1990–1999
	Maximum wind speed and direction	Timing	30 W, SW	44	December 17, 1970
		Instantaneous	40 W	44	More than four times
Frozen snow (cm)	First and last snowfall		August 15 and July 28	44	1956–1999
	Largest maximum thickness		39	44	October 18, 1985
	Maximum seasonal frozen-soil depth and duration		No observation data	44	Permafrost area
Other	Annual average fog days		8.8	10	1990–1999
	Annual average thunderstorm days		47.8	10	1990–1999
	Annual average sandstorm days		–	–	–
Comment					
Station name and construction time			Amdo (November 1965)		
Location and elevation			N: 32°21′ E: 91°06′	Elevation (m)	4800
Representative mileage and site			DK1419+300–DK1633+500		
Values and data collection period			Values	Data collection period (years) and date	
Average air pressure (mb)			574.0	34	–
Temperature (°C)	Annual average		−2.9	34	–
	Extreme	Maximum	23.3	34	June 19, 1972
		Minimum	−36.7	34	January 17, 1968
	Hottest monthly average		7.5	34	July
	Coldest monthly average		−14.7	34	January
	Maximum monthly averaged diurnal temperature range		16.7	34	December

(continued)

Table 2.1 (continued)

Station name and construction time			Amdo (November 1965)		
Humidity	Absolute (g m^{-3})	Average	3.1	34	–
		Maximum	12	34	February 25, 1972
		Minimum	0	34	Multiple times
	Relative (%)	Average	51	34	–
		Minimum	0	34	–
Precipitation (mm)	Annual average		428.4	34	–
	Annual maximum		604.6	34	1971
	Annual minimum		292.3	34	1972
	Monthly maximum		199.8	34	July 1970
	Daily maximum		35.0	34	July 18, 1983
	Maximum rainfall and duration		149.6	34	August 9, 1974; 19 days
	Annual average precipitation days		87.0	34	–
Evaporation (mm)	Annual average		1782.9	34	–
	Annual maximum		1941.1	34	1984
Wind speed (m s^{-1})	Average wind speed and dominant wind direction		4.3N, NE	34	–
	Seasonal average wind speed	Spring	5.8 W, SW	34	–
		Summer	3.6 NE	34	–
		Autumn	3.9 N, NE	34	–
		Winter	5.3 W	34	–
	Annual average gale days (grade $\geq$ 8)		147.1	34	–
	Maximum wind speed and direction	Timing	35 W, SW	34	February 6, 1976
		Instantaneous	38 W	34	January 11, 1977
Frozen snow (cm)	First and last snowfall		August 7 to July 27	34	1988
	Largest maximum thickness		20.0	34	October 4, 1968
	Largest seasonal frozen-soil depth and duration		350.0	34	–
Other	Annual average fog days		1.2	34	–
	Annual average thunderstorm days		74.8	34	–
	Annual average sandstorm days		–	–	–
Comment					
Station name and construction time			Nagqu (July 1954)		
Location and elevation			N: 31°29′ E: 92°04′	Elevation	4507.0
Representative mileage and site			DK1633+500–DK1797+300		
Values and data collection period			Values	Data collection period (years) and date	

(continued)

Table 2.1 (continued)

Station name and construction time			Nagqu (July 1954)		
Average air pressure (mb)			587.4	45	–
Temperature (°C)	Annual average		−1.3	45	–
	Extreme	Maximum	24.2	45	June 10, 1995
		Minimum	−41.2	45	January 16, 1968
	Hottest monthly average		8.0	45	July
	Coldest monthly average		−12.9	45	January
	Maximum monthly averaged diurnal temperature range		19.4	45	–
Humidity	Absolute (g m^{-3})	Average	3.6	45	–
		Maximum	14.2	45	July 22, 1971
		Minimum	0	45	Multiple times
	Relative (%)	Average	54	45	–
		Minimum	0	45	–
Precipitation (mm)	Annual average		421.8	45	–
	Annual maximum		590.4	45	1980
	Annual minimum		307.5	45	1986
	Monthly maximum		211.2	45	August 1964
	Daily maximum		333	45	July 16, 1984
	Maximum rainfall and duration		138.7	45	Jul. 19, 1984; 13 days
	Annual average precipitation days		121	45	–
Evaporation (mm)	Annual average		1961.5	45	–
	Annual maximum		2116.8	45	1966
Wind speed (m s^{-1})	Average wind speed and dominant wind direction		4.1 W	45	–
	Seasonal average wind speed	Spring	4.3 W, SW	45	–
		Summer	2.8 S, SW	45	–
		Autumn	2.6 W	45	–
		Winter	3.5 W	45	–
	Annual average gale days (grade $\geq$ 8)		106	45	–
	Maximum wind speed and direction	Timing	37 SW	45	June 2, 1988
		Instantaneous	40 W	45	January 31, 1989
Frozen snow (cm)	First and last snowfall		August 29 to August 1	45	1985
	Largest maximum thickness		21	45	October 16, 1990
	Largest seasonal frozen-soil depth and duration		281	45	Multiple times in 1974

(continued)

Table 2.1 (continued)

Station name and construction time			Nagqu (July 1954)		
Other	Annual average fog days		4.8	45	–
	Annual average thunderstorm days		80.6	45	–
	Annual average sandstorm days		3.1	–	–
Comment					
Station name and construction time			Damxung (1962)		
Location and elevation			N: 30°29′ E: 91°06′	Elevation (m)	4201.1
Representative mileage and site			DK1797+300–DK1953+670		
Values and data collection period			Values	Data collection period (years) and date	
Average air pressure (mb)			604.4	37	–
Temperature (°C)	Annual average		1.6	37	–
	Extreme	Maximum	26.5	37	June 8, 1996
		Minimum	−35.9	37	January 16, 1968
	Hottest monthly average		16.8	37	July
	Coldest monthly average		−9.7	37	January
	Maximum monthly averaged diurnal temperature range		18.6	37	January
Humidity	Absolute (g m^{-3})	Average	4.2	37	–
		Maximum	13	37	September 2, 1989
		Minimum	0	37	Multiple times
	Relative (%)	Average	54	37	–
		Minimum	0	37	Multiple times
Precipitation (mm)	Annual average		468.1	37	–
	Annual maximum		685.8	37	1971
	Annual minimum		293.6	37	1972
	Monthly maximum		253.2	37	July 1974
	Daily maximum		50.4	37	July 23, 1975
	Maximum rainfall and duration		164.8	37	August 10, 1998; 25 days
	Annual average precipitation days		116.4	37	–
Evaporation (mm)	Annual average		1866.1	37	–
	Annual maximum		2295.5	37	1972
Wind speed (m s^{-1})	Average wind speed and dominant wind direction		2.4 SW	37	–
	Seasonal average wind speed	Spring	3.0 SW	37	–
		Summer	2.3 NE	37	–
		Autumn	1.9 N, NE	37	–
		Winter	2.4 SW	37	–
	Annual average gale days (grade $\geq$ 8)		57.1	37	–
	Maximum wind speed and direction	Timing	25 W, SW	37	March 6, 1993
		Instantaneous	–	–	–

(continued)

Table 2.1 (continued)

Station name and construction time			Damxung (1962)		
Frozen snow (cm)	First and last snowfall		September 2 to June 4	37	–
	Largest maximum thickness		14.0	37	October 4, 1966
	Largest seasonal frozen-soil depth and duration		11.3	37	October 4, 1968
Other	Annual average fog days		0.7	37	–
	Annual average thunderstorm days		74.8	37	–
	Annual average sandstorm days		–	–	–
Comment					
Station name and construction time			Lhasa (1955)		
Location and elevation			N: 29°40′	Elevation (m)	3648.70
			E: 91°08′		
Representative mileage and site			DK1953+670–DK2006+700		
Values and data collection period			Values	Data collection period (years) and date	
Average air pressure (mb)			652.2	44	–
Temperature (°C)	Annual average		7.8	45	–
	Extreme	Maximum	29.6	45	June 8, 1995
		Minimum	−16.5	45	January 17, 1968
	Hottest monthly average		14.5	45	July
	Coldest monthly average		−1.8	45	January
	Maximum monthly averaged diurnal temperature range		16.6	45	January
Humidity	Absolute ($g\ m^{-3}$)	Average	4.9	45	–
		Maximum	16.4	45	July 14, 1970
		Minimum	0	45	Multiple times
	Relative (%)	Average	45	45	–
		Minimum	0	45	Multiple times
Precipitation (mm)	Annual average		406.8	45	–
	Annual maximum		796.6	45	1962
	Annual minimum		229.6	45	1983
	Monthly maximum		313.5	45	August 1958
	Daily maximum		41.6	45	July 28, 1969
	Maximum rainfall and duration		262.1	45	July 6, 1984; 27 days
	Annual average precipitation days		90.1	45	–
Evaporation (mm)	Annual average		1975.7	45	–
	Annual maximum		2873.1	45	1989

(continued)

Table 2.1 (continued)

Station name and construction time			Lhasa (1955)		
Wind speed (m s^{-1})	Average wind speed and dominant wind direction		2.0 E, SE	45	–
	Seasonal average wind speed	Spring	1.7 E	45	–
		Summer	2.4 E, SE	45	–
		Autumn	1.8 E, SE	45	–
		Winter	1.6 E, SE	45	–
	Annual average gale days (grade $\geq$ 8)		26	45	–
	Maximum wind speed and direction	Timing	16.3 N, NE	45	October 7, 1980
		Instantaneous	32.3 SW, W	45	1974, 1976
Frozen snow (cm)	First and last snowfall		October 7 to May 4	45	–
	Largest maximum thickness		12	45	December 11, 1981
	Largest seasonal frozen-soil depth and duration		2.6	45	January 29, 1966
Other	Annual average fog days		None	45	–
	Annual average thunderstorm days		68.1	45	–
	Annual average sandstorm days		–	–	–
Comment					

China ($\sim$600–800 kcal cm^{-2} a^{-1}), and the radiation of each month is positive. Because of the strong winds across the plateau, the sensible heat flux and latent heat flux (60–80 and 20–30% per year, respectively) consume most of the surface energy. The vast majority (98.8%) of the radiation incident on regions at altitudes below 5000 m effuses into the atmosphere through turbulent exchange in the form of sense heat or latent heat. The remaining 1.2% of the radiation warms the soil and thaws the frozen ground, minimally increasing the ground-surface temperature in the process; however, the subsurface temperature remains low.

The Qinghai–Tibet Railway crosses three large natural climatic zones: the arid climatic zone north of Kunlun Mountains (e.g., Golmud), the plateau arid climatic zone between Kunlun Mountains and Tonglha Mountains, and the plateau subarid climatic zone south of Tonglha Mountains. From Kunlun Mountains to the hinterland of the Qinghai–Tibet Plateau, the altitude of the arid climatic zone rises up to 4500 m, the temperature and evaporation gradually decreases, precipitation gradually increases, the air pressure decreases to 560–580 mb, and the relative humidity increases to 49–52%; the annual average wind speed in this zone is 3.9–4.1 m s^{-1}. The plateau subarid climate tends to be relatively warm and humid, with an atmospheric pressure of 587–652 mb, relative humidity of 54%, and annual average wind speed of 2.0–4.1 m s^{-1}. Overall, the area along the line can be characterized as having an alpine semiarid to semihumid climate. Nevertheless, given the vastness of the landmass and its large elevation variation, characteristics of the mountainous areas on the plateau clearly differ from those of the high plains.

The annual average temperature in Kunlun Mountains, Hoh Xil Mountains, Fenghuo Mountain, and Tonglha Mountains is less than −6 °C. From October to May, the temperature is less than 0 °C. Rainfall occurs during June–September, and the annual evaporation exceeds 1300 mm. The wind is strong and changeable; in some intermontane valleys, the wind speed is lower and the wind direction changes frequently.

The annual average temperature of the high plains in the hinterland of the Qinghai–Tibet Plateau is −4.5 to −4 °C. From October to April, the air temperature is less than 0 °C, and the average minimum temperature is below −10 °C. The annual precipitation is approximately 300 mm. The prevailing wind during September–May is from the west and that during June–August is from the north; the maximum wind speed is 30–31 m s^{-1} and occurs mostly during November–March.

The overall climatic characteristics of Amdo Valley in the south Tonglha Mountains region tends to be warm and humid despite its elevation of more than 4700 m above sea level. The annual average temperature here is −2.9 °C, with subzero temperatures occurring mainly during October–April. The annual precipitation is 428.4 mm, more than 80% of which occurs during July–September. The annual evaporation is 1782.9 mm, and the annual average wind speed is 4.3 m s^{-1}.

In the region between the Nyainqentanglha Mountains and Damxung, Lhasa, the climate is relatively warm and humid, with an annual average temperature of 1.6 °C, and annual average precipitation of 468.1 mm; the annual average temperature and precipitation in Lhasa alone is 7.8 °C and 406.8 mm, respectively.

2.1.3 Soil and Vegetation

Because of the strong uplift of the Qinghai–Tibet Plateau, the soil and vegetation characteristics differ from those expected at the corresponding latitudes. The soil and vegetation in the plateau can be characterized as alpine meadow, grassland, and alpine desert landscapes. Given the cold and arid climatic conditions, the natural vegetation is short and sparse and exhibits such characteristics as drought resistance, wind resistance, and salt tolerance. Alpine meadow soil and alpine desert soil are widely distributed on the plateau; these soils have the characteristics of coarse texture, thin soil layer, slow pedogenesis, and poor development. The mountain altitudinal belt structure is clear at the outer edge of the plateau.

2.1.3.1 Soil

The Qinghai–Tibet Plateau mainly has meadow soil and alpine soil. Meadow soil develops under meadow vegetation in cold and wet conditions. Its formation is directly influenced by groundwater infiltration and is characterized by the accumulation of humus as well as the alternation of soil oxidation and reduction because of periodic fluctuations in the groundwater level, which in turn results in the formation of a rust-colored stripe layer. Because meadow soil has relatively high moisture content, it supports lush vegetation with deep and dense roots. Furthermore, this vegetation supplies a large amount of organic content to the soil. When the soil freezes, the organic residues decompose slowly and incompletely, leading to gradual humus accumulation in the soil.

Alpine soil in the Qinghai–Tibet Plateau is mainly that soil formed in the vast windbreak-less mountain forest belt surrounding the plateau and the high mountain snow–ice zone. The alpine soil forms under the alpine bioclimate and exhibits distinct vertical distribution regularity: The upper permafrost layer is formed in the highest and coldest periglacial zone, which is closest to the snowline; the second layer comprises the meadow soil, frigid calcic soils, and frost desert soils in the high frigid zone; the third layer comprises the dark felty soils, cold calcic soils, and cold desert soils in the subfrigid zone; and the lowest layer is the gray-cinnamon soil of the temperate high land.

The soil along the railway line in Qinghai Province is mainly alpine meadow soil, which forms under a wet and cold climate and the alpine meadow vegetation. This soil exhibits distinct soddy process, clear humification, and varying degrees of leaching. This alpine soil has strong frost weathering characteristics, shows clear layering in the soil, and is unsuitable for planting. It can be further classified as alpine meadow soil, alpine steppe meadow soil, alpine shrub meadow soil, and alpine moist meadow soil.

On the Qinghai–Tibet Plateau, the soil distribution changes with the climate: From the southeast to the northwest, the climate changes as humid, subhumid, and semiarid, and the soils change as alpine meadow soil, dark felty soil–frigid calcic soil, cold calcic soil, cold desert alpine soil, and cold desert soil.

2.1.3.2 Vegetation

The plants along the Qinghai–Tibet Railway are of the Qinghai–Tibet Plateau subalpine meadow steppe type. The soil of the Qinghai–Tibet Plateau is generally poor because of the harsh topography and climate; consequently, the plateau vegetation is sparse, with meadow steppe and cushion plants forming most of the landscape. Given the vastness of the plateau and its diverse regional variations, the vegetation distribution varies across the plateau, with the vertical distribution pattern being especially evident. Desert steppe plants dominate the deserts and dry valleys, and the subalpine regions are covered by grassland and shrub plants. Meadow shrubs are common in the high mountain regions, whereas cushion plants are present near the summits, under the snowbelt. Only two types of vegetation can be found in the dry and cold desert regions, such as the Qiangtang Plateau: alpine cushion plants and alpine desert plants. The Qinghai–Tibet Plateau does not have forests, except in the southeast of the plateau (i.e., the northern section of Hengduan Mountains), where the warm and humid airflow from the ocean supports mountain coniferous forests. In the cold, wet, and humid high plains, such as the Qinghai

Plateau and Yushu Plateau, the meadow steppe serves as pastures and supports animal husbandry. The Qinghai–Tibet Plateau has two dry and cold desert regions: Qaidam Basin and Kunlun–Altun Mountains in the north and the eastern Pamirs in the west. The vegetation in the Qaidam Basin is mainly semishrub and shrub deserts, similar to the desert in Tarim Basin. The bleak desert climate of the Kunlun–Altun Mountains sees almost no vegetation. In Pamir–Karakoram, desert grasslands dominate the upper regions, where the lower regions contain alpine meadows or deserts. Because of limited rainfall, only a few Schrenk spruce trees grow in southern Yecheng. The lower regions of the Qilian Mountains have desert steppe or steppe vegetation, and in the subalpine belt, the southeast monsoon nourishes spruce forests and secondary deciduous forests and shrubs; the upper regions contain alpine grasslands and a few alpine plants. In the southern regions of the plateau, the valley between the Kailas Range and the Himalayas has a landscape of xerophytic thorn shrubs and steppe, whereas the alpine belt is dominated by alpine meadows and cushion plants.

Vegetation on the Qinghai–Tibet Plateau exhibits clear vertical zonation. In the 3000–4000 m elevation belt, the mountain area in the southeast is influenced by the southwest monsoon; this belt has a relatively humid climate, with vertical distributions of subtropical moist evergreen broad-leaved forests, mixed-wood forests, and cold temperate coniferous forests. In the 4000–4500 m elevation belt, the westerly circulation has a strong influence in the northern plateau, whereas the southwest monsoon has a weaker effect; the climate cold here supports vegetation alpine shrub grasslands and meadows. In northern Tibet, regions with an average altitude of 5000 m are mostly influenced by the westerly circulation; the climate is extremely arid and cold, and alpine desert vegetation is common. The temperate belt at 4200–4500 m elevation sees mountain temperate desert vegetation and steppe desert vegetation.

The main types of vegetation along the railway are alpine meadow, alpine shrub meadow, alpine steppe, and alpine desert steppe. In particular, semishrubs and herbs, such as Gramineae, Chenopodiaceae, Compositae, *Dasiphora fruticosa* of Rosaceae, and Arenaria of Caryophyllaceae, which are capable surviving in the arid climate, are common. On the basis of vegetation ecology, the area from Golmud to Lhasa can be classified as follows: Gobi bare land (Golmud to Nanshankou); mountain desert and mountain grassland (Nanshankou to Kunlun Mountains); alpine meadow, grassland, and desert (Golmud to Tonglha Mountains); alpine meadow (Tonglha Mountains to Damxung); and bushveld (Damxung to Lhasa).

Given the variation in water availability, temperature, topography, and especially altitude, the southern areas along the railway line are species-rich, whereas the high plains and the northern regions have successively fewer vegetation species. Table 2.2 lists the vegetation ecological areas along the line.

There are 305 species and varieties (22 shrubs, 283 herbs) of seed plants, belonging to 40 families and 134 genera, in the area along the railway. Around 40 species, including four shrubs, are dominant, examples of which include *Stipa purpurea*, *Festuca rubra*, *Elymus nutans*, *Orinus*, *Kobresia humilis*, *Himalayan Kobresia humilis*, *Carex moocroftii*, and *Edelweiss*.

Table 2.2 Ecological regionalization of the Qinghai–Tibet Railway

No.	Location	Ecological regionalization	Dominant vegetation
1	Nanshankou to Naij Tal	Desert	Xerophilous shrub
2	Dongxidatan to Fenghuo Mountain Pass	Alpine steppe	Carexmoorcroftii, Compositae, Gramineae
3	Wuli to Amdo	Secondary alpine steppe	Compositae, Gramineae, Cyperus
4	North Tonglha Mountains to Wumatang	Alpine meadow	Kobresia pygmaea
5	Damxung area	Meadow–grassland transition	Kobresia Cyperus
6	Yangbaling to Lhasa	Shrub grassland	Potentilla

East of eastern Nyainqentanglha mountains, the rivers are parallel and the relief great; this region is strongly influenced by the southwest monsoons and the southeast monsoons, which bring substantial rainfall. Consequently, this region has tropical, subtropical, and warm temperate forests dominated by woody species, and the distribution of these forests exhibit clear vertical zonation. Southeastern Qinghai and northwestern Sichuan, located in the transition zone of the alpine valleys and the plateau, have a low and gentle surface relief, and the landscape is dominated by alpine shrubs (e.g., *Caragana*, *Dasiphora fruticose*, and *Salix cupularis*) and alpine meadows (e.g., grain grass (Xiaoman grass), *Kobresia humilis*, Tibetan wormwood, and Carex). The vegetation on the meadows is rich in nutrition and has high palatability and thus is all good pastures. Winters here are cold and dry and the summers warm and humid. West of this belt, the grass is low (e.g., *Ceratoides compacta*, *Ajania tibetica*, Arenaria, and Myricaria) and the vegetation coverage sparse, short, and monotonous, with *Stipa* L. (Gramineae), such as *Stipa bungeuna*, *Stipa breviflora*, *Stipa purpurea*, *Stipa krylovii*, and *Artermisia*, being dominant.

Vegetation coverage on the Qinghai–Tibet Plateau has increased in recent years, with anthropogenic activity having little effect in this region. From 1982 to 1991, vegetation in the plateau exhibited an increasing trend, but in the southeast–northwest region, the rate of increase has gradually reduced, which is consistent with the gradually worsening climatic conditions and meridional and zonal variation from the southeast to the northwest. From 1992 to 2002, vegetation in the central and northwest plateau exhibited a decreasing trend, with areas along the Yangtze River, Yellow River, Lancang River, and the source region of Nujiang River being the worst affected; this also evidences that the climatic conditions in the middle and northwest plateau are not conducive to vegetation growth. The middle and northwest regions of the plateau are the most sensitive to climate change. Because of temperature variation, plateau vegetation changed within 7 and 3.5 years, indicating its sensitivity to temperature variation. From 1982 to 2002, seven of the eight main vegetation types in the plateau exhibited varying but overall increasing growth trends; the vegetation in the cold and arid regions however were vulnerable to climate change and were slow to recover.

2.1.4 Rare Wild Animals

More than ten types of rare wild animals, mostly mammals and birds, are distributed along the Qinghai–Tibet Railway. On the basis of their habitat, these animals can be categorized as intermountain lake basin fauna and broad valley fauna, mountain mammalian fauna, and wetland animal fauna.

1. Intermountain lake basin fauna and broad valley fauna—This type includes species endemic to the Qinghai–Tibet Plateau, such as the Tibetan antelope, *Equus kiang*, *Bos mutus*, and *Procapra picticaudata*, which are distributed in flat terrains, intermountain basins, lakeshores, and occasionally in broad river shoals. They avoid predators through aggregation and by running.
2. Mountain mammalian fauna—Animals of this type (e.g., *Pseudois nayaur*, *Ovis ammon*, White-lipped deer, lynx, snow leopard, and brown bear) prefer the mountains. The Artiodactyla animals, for example, evade predators by clustering and relying on their good climbing abilities.
3. Wetland animal fauna—This type includes wading and swimming birds (e.g., black-necked crane, bar-headed goose, red chicken, and brown-headed gull) mainly live in rivers, lakes, marshes, and swamps.

2.1.5 River System

The Qinghai–Tibet Railway passes through five major river systems: (north to south) the Qaidam Inland River System, Yangtze River, Za'gya Zangbo Inland River System, Nujiang River, and Yalu Tsangpo River. The main rivers are Golmud and Kunlun Rivers of the Qaidam Inland River System; Qingshui, Chumar, Tuotuo, Tongtian, and Buqu rivers of the Yangtze River system; Za'gya Zangbo and Rianazangbu Meanders of the Za'gya Zangbo Inland River System; Lari Meander, Beisang Meander, Liantong River, Nagqu, and Muge Meander of the Nujiang River system; and Dam Qu River, Doilung Meander, and Lhasa River of the Yalu Tsangpo River system. These rivers, all of which have a low temperature, are mainly fed by meltwater and meteoric water, and thus, exhibit seasonal variations in their water level.

The major hydrological characteristics of these rivers are as follows:

1. The Qaidam Inland River System develops north of Kunlun Mountains, and its main river is Golmud River, which is known as Kunlun River upstream. This river has a few branch ditches (e.g., Xiaonanchuan and Wanbaogou) on both its shores. The landform is of the mountains and Gobi type, and the flood level varies suddenly and widely.
2. The Yangtze River system develops in the region south of Kunlun Mountains and to northern Tonglha Mountains, and its major constituents are Chumar River, Tuotuo River, and Tongtian River. The runoff for these rivers primarily

depends on precipitation, with minor contributions from glacier-melt and snowmelt water. The landform is of the low mountains and hills type and is composed of large semiarid grasslands and swamp lakes. Hummocks, ice cones, and icy mantles form in winter. This area has the well-known Hoh Xil Nature Reserve, which contains the source of Changjiang River.
3. The Za'gya Zangbo Inland River System forms between the southern Tonglha Mountains and northern Touerjiu Mountains, and its main river is the Za'gya Zangbo Meander. Its runoff is mainly from precipitation and meltwater. The landform is of the low mountains and hills type, with sparse pastures and numerous swamps and wetlands in the valleys.
4. The Nujiang River system, which comprises Nagqu and Beisang Qu rivers, develops in the region south of Touerjiu Mountains and to northern Nyainqentanglha Mountains. The landform is of the high plains semihumid meadow type and has dense pastures and numerous marsh wetlands. The permafrost and nonpermafrost areas are north and south of Amdo, respectively.
5. The Yalu Tsangpo River system, whose major constituents include Lhasa River and its ditches (namely Doilung Meander, Dam Qu, and Xiongqu) run south of Nyainqentanglha Mountains. The upstream region of Lhasa River starts is of the low mountain meadow type, the middle and lower reaches alpine, and the valley broad.

2.1.6 Ecological Environment

The Qinghai–Tibet Plateau can be characterized as having a high altitude, thin air, a cold arid climate, a short plant-growth period, low biomass, and a simple food chain. Its ecological environment is fragile because the material recycling and energy conversion processes are slow. In addition, because of the long low-temperature periods and the short growing season, the recovery of any destroyed vegetation is very slow; this in turn accelerates the melting of frozen soil, leading to desertification and soil erosion. Therefore, the World Wide Fund for Nature has listed the Qinghai–Tibet Plateau as a top priority for global biodiversity conservation.

2.2 Topography and Geomorphology

Except for the region between Golmud and Nanshankou in the southern edge of Qaidam Basin, the Qinghai–Tibet Railway passes entirely through high plains. The main mountains along the railway are Kunlun Mountains, Hoh Xil Mountains, Fenghuo Mountain, Kaixinling, Tonglha Mountains, Touerjiu Mountains, and Nyainqentanglha Mountains. The strikes of these mountains are almost WN–ES, and the relative elevation is mostly less than 300 m. Similar to an undulating dome, these mountains have gentle slopes, narrow bodies, and broad valleys, forming a high plains scenery, which is actually a false impression of the river. These mountains are the watershed and the main source of Golmud River, Yangtze River, Za'gya Zangbo, Nujiang River, and Yalu Tsangpo River and also contain alpine lakes and alpine swamps.

Mountains, rivers, and tectonic movement have shaped the landscape of the Qinghai–Tibet Plateau, which in turn determine the railroad conditions. High plains approximately 50–200 km long are usual between the watershed areas, which are low and even; The railway mostly passes through the valleys, with only few sections passing through the watershed. The terrain units are as follows.

2.2.1 Qaidam Basin (K814+500–DK845+400)

The area from Golmud to Nanshankou forms the Qaidam Basin southern edge alluvial plain, which is flat and inclined northward with a longitudinal slope of approximately 15‰. This basin, whose altitude is 2800–3000 m, has little vegetation, and is part of the Gobi desert.

2.2.2 Valley Terrace North of Kunlun Mountains (DK845+400–DK940+500)

Between Nanshankou and Xiaonanchuan, the line mainly runs along the Golmud River and the Kunlun River Valley.

The area from Nanshankou to Kunlun Bridge is part of the Golmud River strath terrace of northern Kunlun Mountains. The modern riverbed of Golmud River is a sharply cut U-shaped cross-section channel with precipitous sides; it is 10–8 m wide, the longitudinal slope is 15–20‰, and scarps on both sides are 35–50 m high. Some sections of the riverbed have developed deep valleys that are less than 4 m wide but up to 35 m deep. Beyond the channel scarps, the terrain is open and flat and the vegetation sparse, forming the Gobi desert.

From Kunlun Bridge to Xiaonanchuan, the line runs along the valley terrace of Kunlun River, the upstream section of Golmud River. The modern riverbed is 20–200 m wide. The third terrace of Kunlun River is relatively complete and is 30–50 m above the riverbed. The first terrace is 3–10 m above than riverbed, and most of the second terrace is missing. The upper and middle reaches of Kunlun River exhibit cross-flow. The widest part of these flows is in Xiaonanchuan River area, which has a broad and shallow riverbed.

2.2.3 Xidatan Structural Valley (DK940+500–DK973+700)

The Xidatan valley is a flat terrain 30–50 km long (east to west) and 4–7 km wide (south to north). The line passes through the Xidatan basin from east to west at an altitude of 4120–4600 m. The northern boundary of the permafrost is located in DK957+640 at a high altitude of 4360 m.

2.2.4 Kunlun Medium-High Mountain Area (DK973+700–DK1005+500)

The Kunlun medium-high mountain area (4500–4800 m elevation) comprises Luanshi Ditch, which is on the northern slope of Kunlun Mountains; Kunlun Bealock; Kunlun Pass Basin; and Budongquan Valley. The terrain has large undulations, sparse vegetation, and dominant geomorphological features, such as ancient glaciers, modern glaciers, and frost weathering; alluvial and pluvial terraces are also present in some areas. Luanshigou Ditch is narrow and steep. Glaciofluvial deposition and lake basins are present in Kunlun Mountain Bealock and southern bealock area, and the terrain is relatively flat. Moreover, some harmful frozen-soil phenomena occur in this region.

2.2.5 Chumar River High Plains (DK1005+500–DK1072+500)

With an altitude of 4500–4700 m, the Chumar River high plains comprise the alluvial plains of Baladacai Qu, Qingshui River, and Chumar River. The topographical relief is low, with alterations of low hills and depressions; the vegetation is relatively rich. In addition to sandy lands and sand dunes, this area has thaw lakes and ponds. The valleys are characteristic of braiding and wandering river systems, and the stream trenching is indistinctive.

2.2.6 Hoh Xil Mountains (DK1072+500–DK1124+500)

The Hoh Xil Mountain area comprises the southern shore of Chumar River, Wudaoliang, Hoh Xil Mountains, Hongliang River, and Qushui River. The terrain is dissected by ravines and ridges and has wavelike undulations from southern Chumar River to Hoh Xil Mountains. Hoh Xil Mountains are spread out nearly east–west, with peaks at 4500–4700 m elevation; the ridges are gentle, with a relative elevation of 100–300 m. Furthermore, fluid dunes are distributed in Hongliang Valley.

2.2.7 Beilu River Basin (DK1124+500–DK1145+500)

Located at a high altitude of 4500 m, the Beilu River Basin comprises Xiushuihe and Beilu River beach land and terrace and can be described as alluvial high plains. Topographical relief is low, with interphasing low hills and depressions. This region has gullies, the vegetation is sparse, and some areas have sandy lands and sand dunes.

2.2.8 Fenghuo Mountain Area (DK1145+500–DK1165 +500)

With an altitude of 4500–4700 m, the Fenghuo Mountain area comprises Fenghuo Mountain Piedmont hill and its low mountain areas, with relative elevations of 200–300 m. The bedrock is exposed at the mountain summit; the ridge is gentle, with the upper regions of the hillside being precipitous and the lower regions flat. In addition, this region has intermontane valleys, with deep and precipitous slopes. The mountain area has a flat summit with gentle slopes, wide valleys, and short ditches.

2.2.9 Chiqu Valley (DK1165+500–DK1193+200)

The Chiqu Valley area has an altitude of 4580–4600 m and mainly comprises Chiqu River terraces, which has a flat terrain, developed ditches, and sparse vegetation.

2.2.10 Wuli Basin (DK1193+200–DK1202+500)

The Wuli Basin area mainly comprises alluvial and pluvial basins and has an elevation of 4580–4600 m; the area within the basin is flat and has sparse vegetation and a few ditches.

2.2.11 Wuli Mountain (DK1202+500–DK1217+700)

The Wuli Mountain region mainly comprises Wuli low mountains. It has an elevation of 4500–4700 m and has precipitous slopes and many deep-cut gullies. The bedrock is exposed, the summit flat, and vegetation sparse.

2.2.12 Tuotuo River Basin (DK1217+700–DK1245+000)

The Tuotuo River basin is broad in the east and narrow in the west, with alluvial fans and the Tuotuo River Valley and Terraces distributed in the marginal and central regions, respectively. This basin has sparse vegetation coverage, wide sand distribution, and low terrain undulation.

2.2.13 Kaixinling Mountain Area (DK1245+000–DK1252+800)

The peak of Kaixinling Mountain is flat and round, the slopes precipitous, and the gullies cut deeply. The intermountain basin is relatively flat, with wide vegetation coverage.

2.2.14 Tongtian River Basin (DK1252+800–DK1282+800)

The Tongtian River Basin mainly contains alluvial and diluvial plains, which comprise valleys and terraces, at 4600–4700 m elevation; the open and flat basin has a few sandy lands and sparse vegetation.

2.2.15 Buqu Valley (DK1282+800–DK1360+800)

Buqu Valley, elevation 4700 m, mainly comprises Buqu Valley, its floodplain, its first and second terraces, and an occasional third terrace. Encompassed by southern and northern mountains, this narrow valley runs south–north.

2.2.16 *Wenquan Fault Basin (DK1360+800–DK1394+800)*

Wenquan Basin is an N–S-trending extensional fault basin with a high altitude of 4700–4800 m, length of approximately 30 km (south–north), and width of 5–8 km. Its ambient areas contain greatly developed alluvial and diluvial fans. The Buqu riverbed is wide and shallow. This smooth-terrain area has numerous ditches and sparse vegetation coverage.

2.2.17 *Tonglha Mountains and Intermontane Basin (DK1394+800–DK1485+200)*

Tonglha Mountains, elevation 4700–5200 m, can be divided into three sections: the gorge area of the source of Buqu, the piedmont glaciofluvial deposit plain, and the mountainous and hilly regions.

1. Gorge area of the source of Buqu (DK1394+800–DK1419+300)—This area is 4800–5200 m high. The river channel in this area is curvy and narrow, with steep slopes, incomplete terraces, gullies, and sparse turfs. The topographical relief is slightly high, with exposed bedrock and thin Quaternary coverage.
2. Piedmont glaciofluvial deposit plain (DK1419+300–DK1458+700)—On this wide plain, surface runoff develops as a cross-flow, and no distinct gullies are present. Vegetation is sparse, and the ground glaciofluvial deposit layer is soft and difficult to walk on during the rainy season. Some thermokarst lakes are also present.
3. Mountainous and hilly regions (DK1458+700–DK1485+200)—This region has high topographical relief and sparse vegetation. Za'gya Zangbo River runs windingly in this region, but no river terraces are obvious. Moreover, this region has glacial deposits.

2.2.18 *Touerjiu Mountain Area (DK1485+200–DK1513+770)*

The landform of the Touerjiu Mountain Yueling area (elevation 4800–5100 m) is of the medium alpine landform, which is not very flat; it has numerous gullies and grooves and high vegetation coverage. Furthermore, permafrost swamping wetlands are present in the gully and slope area. Along the Galebu Meander, the landform becomes a V-shaped valley with a shallow riverbed and inconspicuous terraces.

2.2.19 *High Plains Area in Northern Tibet (DK1513+770–DK1801+100)*

The high plains area in northern Tibet, elevation 4600–4800 m, can be divided into two sections: the gorge area of the source of Duopuer Meander and the Anduo Senlis Valley.

1. Gorge area of the source of Duopuer Meander (DK1513+770–DK1801+100)—Gullies have developed in the area between Amdo and the source of Duopuer. Most of the bedrock on the shore is exposed. The riverbed is narrow, and mountains on both sides of the gorge are precipitous. Swamping wetlands are distributed on the gentle slopes.
2. Anduo Senlis Valley (DK1520+100–DK1801+100)—The major landforms along the Qinghai–Tibet Railway are piedmont alluvial plains, half-fixed sandy lands on the east coast of Beisang Qu, barchans on the east coast of Cuona Lake in Nagqu Valley, east coast alluvial plains and hills, and alluvial plains and hummocky moraine in Muge Qu. The general topography can be described as highland hills characterized by a gentle hillside, rare mountain peaks, inconspicuous gorges, and precipitous mountains with relative elevations of 300 m. The flat and open area between the mountains and hills contain alluvial, till, and lacustrine plains; drainage networks; wide and shallow riverbeds; and rivers with receding banks; large lakes; and strip wetlands, with high vegetation coverage of 60–90%.

2.2.20 *Medium-High Mountain Area Between Samli and Yangbaling (DK1801+100–DK1904+500)*

Located at a high altitude of 4200–4700 m, the medium-high mountain area between Samli and Yangbaling contains the Sang Qu medium-high mountain strath terrace, Jiuzina Mountain area, Damxung Basin, and Yangbaling piedmont alluvial plains.

1. Northern Jiuzina area (DK1801+100–DK1808+500)—The Qinghai–Tibet Railway line passes through Samli medium-high mountain strath terrace, which has a low and gentle surface relief, nearly 10‰ lateral ground slope, and sparse vegetation coverage.
2. Jiuzina Mountain area (DK1808+500–DK1810+000)—This area is affected by weathering, erosion, and tectonic movement and has the characteristics of a ridgelike landform, such as low hills, steep lateral slope, and sparse vegetation coverage.
3. Area between Jiuzina and Yangbaling (DK1810+000–DK1904+500)—The main geomorphology in this area are intermontane valleys and clinoplains, which are broad and flat. The valley is 10 km wide, on either sides of which are

high mountains with steep slopes and exposed bedrocks. The valley terrace is smooth, with lateral slopes of 2–6‰. Low terraces and floodplains can also be seen.

2.2.21 Southern Nyainqentanglha Mountain Area from Yangbaling to Lhasa (DK1904+500–DK2006+700)

The southern Nyainqentanglha Mountain area, elevation 630–4600 m, contains Yambajan Basin, Doilung Qu Gorge area, and Lhasa River strath terrace.

1. Yambajan Basin (DK1904+500–DK1921+500)—This basin is 6–10 km wide (south–north) and tens of kilometers long (east–west). The Qinghai–Tibet Railway line passes through a broad and flat section of the basin, which also has floodplains and low terraces.
2. Doilung Qu Gorge area (DK1921+500–DK1993+500)—Also known as Yambajan Canyon, this gorge area has a relative height difference between the stream valley and banks of 500–1000 m and bank slopes of 35°–40°. The present riverbed is 20–50 m wide, with a longitudinal slope of 15–25‰; the valley is narrow and precipitous and sees rapid flow. The valley is 1–3 km wide, gradually becoming wider as it approaches Lhasa. The present riverbed is 100–300 m wide, with a longitudinal slope of 2–10‰. Terraces and ancient alluvial fans are present on each bank, but most of these are currently being utilized as farmlands.
3. Strath terrace in Lhasa River (DK1993+500–DK2006+700)—This terrace is 4–6 km wide (south–north), with the relative height difference of the banks being 500–800 m. The present riverbed is 500–800 m wide, with a longitudinal slope of 1–1.5‰. The riverbed is not very stable, and the mainstream flow varies widely.

2.3 Formation Lithology

The main strata along the Qinghai–Tibet Railway comprise the Quaternary system, Tertiary system, Cretaceous, Jurassic system, Triassic system, Permian system, Paleozoic group, Silurian system, Sinian system, lower Proterozoic, Caledonian granite, and magmatic rocks. The following sections describe the main strata along the line (i.e., from Golmud to Lhasa) from the newest to the oldest layer.

2.3.1 Quaternary System

2.3.1.1 Holocene Series

1. Silty clay ($Q_4^{al2,\ pl2,\ dl2}$)—Silty clay, mainly distributed in river terraces, piedmont alluvial plains, and the gentle surface layer, is 0.5 m thick on average and up to 10 m thick in the lacustrine plains between Amdo and Samli. This light yellow or tan clay is a flow plastic–to–hard-plastic grade II soft rock, with a bearing capacity (σ_0) of 80–150 kPa. On freezing, it becomes a grade IV soft rock with σ_0 = 250–350 kPa.
2. Silt soil ($Q_4^{al3,\ pl3,\ dl3}$)—Silt soil is mainly distributed in river terraces, piedmont alluvial plains, and the gentle surface layer. The average thickness of the predominantly light yellow soil layer is 0.5–5 m. This prevalent grade II soil is in a slightly wet-to-saturated state, with σ_0 = 80–150 kPa. On freezing, it becomes a grade IV soft rock with σ_0 = 250–400 kPa.
3. Silt and fine sand ($Q_4^{al4,\ pl4,\ dl4}$)—occurring as lens-shaped, silt and fine sand are mainly distributed in the river floodplains and terraces and piedmont alluvial plains. On hillsides, these sands are primarily distributed on the surface, with a typical thickness of less than 5 m. This grade I loose soil is light yellow or steel-gray, loose-to-medium dense, and wet-to-saturate, with σ_0 = 100–200 kPa. On freezing, it becomes a grade IV soft rock with σ_0 = 250–450 kPa.
4. Silt and fine sand (Q_4^{eol4})—Silt and fine sand distributed are mainly distributed in the following areas: the gentle slope zone from Kunlun River to Naij Tal, some areas between Budongquan and Hoh Xil area as well as the local segment between Wuli and Kaixinling, the surface of piedmont alluvial plains from Tongtian River to Yanshiping, Cuona Lake area, and Lhasa River Valley terrace. In addition to the local dune chains and meniscus dunes, the vast majority of landform comprises half-fixed sandy lands and semifixed sand dunes. This prevalent grade II sand is loose, less than 5 m thick, light yellow or steel-gray, and slightly wet-to-wet.
5. Coarse sand ($Q_4^{al5,\ pl5,\ dl5}$)—Occurring as lens-shaped intercalates, coarse sand is mainly distributed in floodplains, terraces, and alluvial and diluvial plains. This stratum is less than 5 m thick and is light yellow or steel-gray, loose-to-medium dense, and slightly wet–to-saturated. It is a grade I loose soil, with σ_0 of medium sand being 150–300 kPa, and that of coarse and gravelly sand being 200–300 kPa. On freezing, it becomes a grade IV soft rock with σ_0 = 300–600 kPa.
6. Thick pellet soil and brecciated soil ($Q_4^{al6,\ pl6,\ fgl6,\ dl6}$)—Thick pellet soil and brecciated soil are mainly distributed in river terraces, Xidatan and Chumar River high plains, alluvial plains, glaciofluvial deposit plains, and the surface layer of gentle slopes. The thickness of the strata with alluvial features can exceed 10 m, whereas that of the slope-made strata is less than 5 m. This prevalent grade II light yellow or dust-colored soil is loose-to-medium dense

and slightly wet-to-saturated, with σ_0 = 300–400 kPa. On freezing, it becomes grade IV soft rock with σ_0 = 400–600 kPa.

7. Cailloutis soil ($Q_4^{al7,\ pl7,\ fgl7,\ dl7}$)—Cailloutis soil is mainly distributed in riverbeds, river terraces, Xidatan Basin, glaciofluvial deposit plains, and Muge Qu and Sang Qu alluvial plains. This loose-to–medium dense wet-to-saturated soil stratum is >10 m thick. It is a grade III stiff soil with σ_0 = 500–600 kPa. On freezing, it becomes a grade IV soft rock with σ_0 = 800 kPa.
8. Gravelly soil ($Q_4^{col7,\ dl7}$)—Gravelly soil is mainly distributed in the toe area of Golmud River valley slopes, ambient areas in the rock heap foreland and collapse bodies in Naij Tal, and other local piedmont slope toes in alluvial plains. This grade III stiff soil is medium dense and slightly wet–to-saturated, with σ_0 = 500 kPa. On freezing, it becomes a grade IV soft rock with σ_0 = 800 kPa.
9. Granular soil and boulder clay ($Q_4^{al8,\ pl8,\ col8,\ dl8}$)—Granular soil and boulder clay are mainly distributed in piedmont alluvial terraces, moraine hills, and the toe of slopes. The deposition layer is >10 m thick, and this steel-gray grade IV soft rock is medium dense and slightly wet-to-saturated, with σ_0 = 600 kPa. On freezing, it remains a grade IV soft rock but with σ_0 = 800 kPa.
10. Sinter (Q_4^{ch})—Sinter mainly appears in hot springs, the summit area of Tonglha Mountains, and ambient area in Amdo; sinter has a granular structure and is mainly composed of travertine; it is white or off-white, light, and semihard.
11. Humic soil (Q_4^{l})—Humic soil is mainly distributed in the wetland surface layer between Amdo and Lhasa. Comprised of turf, rotten roots, and silty clay, this dust-colored or gray flow plastic-to-soft plastic, grade I loose soil layer is 0.5–2 m thick. This soil makes for a poor foundation and thus must be avoided during line selection.

2.3.1.2 Upper Pleistocene

1. Silty clay ($Q_3^{al2,\ pl2}$)—Silty clay is distributed in the high terraces and the ancient piedmont alluvial fans on both sides of Dam Qu and Doilung Qu. This prevalent grade II hard-plastic clay layer is 0.5–5 m thick, is brownish yellow or light gray, has homogeneous mass, and has more clay and less gravel, with σ_0 = 150 kPa.
2. Silt and fine sand ($Q_3^{al4,\ pl4,\ fgl4}$)—Silt and fine sand are mainly distributed in the lower sections of Golmud River terraces, moraine hills from Nanshankou to Naij Tal, and some sections on the northern slope of Kunlun Mountains. This grade I loose soil layer is light yellow and less than 5 m thick; it is medium dense and wet-to-saturated, with σ_0 = 150–200 kPa. On freezing, it becomes a grade IV soft rock with σ_0 = 500 kPa.
3. Thick pellet soil ($Q_3^{al6,\ pl6,\ fgl6}$)—Thick pellet soil is mainly distributed in the lower sections of Golmud River terraces, the moraine hill from Nanshankou to Naij Tal, and the surface of till plain in southwest Amdo. This prevalent grade II

soil layer is >10 m thick, light yellow or steel-gray, medium dense, slight wet–to-wet, and occasionally semibonded, with $\sigma_0 = 400$–600 kPa.

4. Gravelly soil and cailloutis soil ($Q_3^{al7,\ pl7,\ fgl7,\ dl7}$)—Gravelly soil and cailloutis soil is mainly distributed in piedmont alluvial terraces, moraine hills between Golmud and Wangkun, high terraces, and the ancient piedmont alluvial fans from Damxung to Lhasa; this grade III stiff soil layer is steel-gray, >10 m thick, medium dense, and occasionally semibonded, with $\sigma_0 = 800$ kPa.
5. Granular soil and boulder clay ($Q_3^{col8,\ dl8}$)—Granular soil and boulder clay is mainly distributed in piedmont alluvial terraces, moraine hills, and the toe of slopes in Golmud to Wangkun; this steel-gray grade IV soft rock layer is >10 m thick and loose-to-medium dense, with occasional shale or calcareous cementation and $\sigma_0 = 800$ kPa.

2.3.1.3 Middle Pleistocene

1. Thick pellet soil (Q_2^{fgl6})—Thick pellet soil is mainly distributed on both sides of Luanshi Ditch. This prevalent grade II, steel-gray, medium dense layer is >10 m thick, with $\sigma_0 = 300$ kPa. On freezing, it becomes a grade IV soft rock with $\sigma_0 = 600$ kPa.
2. Cailloutis soil (Q_2^{fgl7})—Cailloutis soil is mainly distributed on both sides of Luanshi Ditch. This prevalent grade III soil layer, steel-gray, medium dense sediment layer is >10 m thick, with $\sigma_0 = 500$ kPa. On freezing, it becomes a grade IV soft rock with $\sigma_0 = 800$ kPa.
3. Boulder clay (Q_2^{fgl8})—Boulder clay is mainly distributed on both sides of Luanshi Ditch and on the surface of the moraine hill from Amdo to Samli. This grade IV, steel-gray, medium dense sediment layer is >10 m thick, with $\sigma_0 = 700$ kPa. On freezing, it becomes a grade IV soft rock with $\sigma_0 = 800$ kPa.

2.3.1.4 Lower Pleistocene

Silty clay (Q_1^{l2})—Silty clay is mainly distributed in Kunlun Pass Basin and Chumar River high plains. With a thickness of >5 m, this prevalent grade II clay is dust-colored or charcoal gray and has homogeneous mass, with $\sigma_0 = 100$ kPa. On freezing, it becomes a grade IV soft rock with $\sigma_0 = 250$–350 kPa.

2.3.2 *Tertiary System*

2.3.2.1 Neogene

Sandstone, mudstone, marl ($N^{Ss+Ms+Am}$)—Neogene, which is a light red or multicolored continental sediment composed of sandstone, mudstone, and marl as well as gypsum clastic rocks, are mainly distributed in Chumar River high plains, Togton River Basin, Tongtian River Basin, and Yanshiping and Buqu high terraces. The bottom section of this sediment is a thin layer of light-red quartz sandstone and sandstone, the middle section is composed of alternating layers of tangerine and gray-green sandstone and mudstone, and the top sections are an interbed of light-red sandstone with dust-colored or gray siltstone, atop which lies an interbed of gray marl and gray-yellow mudstone, respectively. The rocks in this sandy sedimentary soil are in a transition state between being fully weathered and strongly weathered. Neogene is a grade III stiff soil with σ_0 = 150–200 kPa, which on freezing becomes a grade IV soft stone with σ_0 = 250–350 kPa.

2.3.2.2 Lower Tertiary

Mudstone and sandstone (E^{Ms+Ss})—Mudstone and sandstone rocks are mainly distributed in Fenghuo Mountain area, where they appear as extra-thick alternating beds of heliotrope, dark red sandstone, and mudstone that been undergone different degrees of weathering. They are grade IV soft rocks surrounded by grade IV rocks, with σ_0 = 250–350 kPa.

2.3.3 Cretaceous

1. Sandstone (K^{Ss})—Sandstone is scattered in the area between Touerjiu and Amdo, the ambient area of Liangdaohe Army Service Station, the southern part of Jiuzina pass, and the area between Yambajan and Sai Qu. The sandstone here is purplish-red and is interbedded with shale and conglomerate. The regolith is 3–5 m thick, and these weathered rocks are grade IV soft rocks with σ_0 = 400–600 kPa, whereas the whole bedrock is grade V secondary hard rock with σ_0 = 800 kPa.
2. Conglomerate (K^{eg})—Conglomerate is distributed along the left bank of Nagqu and has the following characteristics: bedded structure, fragmental texture, calcium and mud cementation, numerous joints, crushed rocks, and strongly weathered-to-weakly weathered. The regolith is 2–4 m thick; this rock is grade IV soft rock with σ_0 = 500 kPa. The bedrock is grade V secondary hard rock with σ_0 = 800 kPa.
3. Limestone (K^{ls})—Off-white and hard limestone is scattered in the area between Touerjiu and Amdo, with regolith of thickness 1–3 m; this grade IV soft rock has σ_0 = 800 kPa. The whole bedrock is grade V secondary hard rock with σ_0 = 1000 kPa.

2.3.4 Jurassic System

1. Mudstone interbedded with shale (J^{Ms+Sh})—Mudstone interbedded with shale is mainly distributed in Yanshiping to Amdo. These stratified rocks have alternating light gray and brownish-red layers and exhibit calcium and mud cementation. Varying from fully weathered-to-strongly weathered, the regolith is 3–5 m thick and is grade IV soft rock with σ_0 = 400–600 kPa. The whole bedrock is grade V secondary hard rock with σ_0 = 600–800 kPa.
2. Sandstone interbedded with shale (J^{Ss+Sh})—Sandstone interbedded with shale is distributed between Yanshiping and Amdo, the upper low-alpine sections between Amdo and Samli, between Samli and north Damxung, and in the northern Doilungdêqên. These purplish-red and gray-green sediments have thin layer construction and an arenopelitic structure. The regolith is 3–5 m thick grade IV soft rock with σ_0 = 400–600 kPa. The whole bedrock is grade V secondary hard rock with σ_0 = 800 kPa.
3. Limestone (J^{ls})—Off-white and hard limestone is mainly spread between Yanshiping and Amdo and in the ambient area of Doilungdêqên. The regolith is 1–3 m thick and is grade IV soft rock with σ_0 = 800 kPa. The whole bedrock is grade V secondary hard rock with σ_0 = 1000 kPa.
4. Gypsum bed (J^{ey})—Soft, fragile, and white gypsum bed, of type grade IV soft rock, is mainly spread in the Wenquan Valley area.

2.3.5 Triassic System

Comprised of schist, slate, sandstone, phyllite, and limestone, the Triassic system is mainly distributed in Budongquan, Wudaoliang, and Wuli Mountain area. Alternating beds of limestone and chert have developed in Tongtian River Small Bealock.

1. Schist (T^{Se})—Schist in this system is dark red, gray-red, and brown, with fractures, joints, and crushed rock stratum. Exhibiting weak weathering, this schist is grade IV soft rock and grade V secondary hard rock with grade IV surrounding rock and σ_0 = 400–600 kPa.
2. Slate (T^{Sl})—Slate in this system is dark red, gray-red, and brown, with fractures, joints, and crushed rock stratum. Exhibiting weak weathering, this slate is grade IV soft rock and grade V secondary hard rock with grade III and IV surrounding rock and σ_0 = 600 kPa.
3. Sandstone (T^{Ss})—Sandstone in this system is dark red and gray-red with fractures, joints, and crushed rock stratum. Exhibiting weak weathering, this sandstone is grade IV soft rock and grade V secondary hard rock with grade III and IV surrounding rock and σ_0 = 800 kPa.
4. Phyllite (T^{ph})—Phyllite in this system is dark red, with fractures, joints, and crushed rock stratum. Exhibiting weak weathering, this phyllite is grade IV soft rock with grade III and IV surrounding rock and σ_0 = 400–600 kPa.

5. Limestone (T^{lh})—In this system, limestone is interbedded with chert layer. These hard and off-white limestone sediments have 1–3-m-thick regolith and are grade IV soft rock with $\sigma_0 = 800$ kPa. The whole bedrock is grade V secondary hard rock with $\sigma_0 = 1000$ kPa.

2.3.6 Permian System

The Permian system outcrop area is mainly located in Luanshi Ditch, Wuli Mountain, and Kaixinling. The main lithology of Permian system is slate, schist, sandstone, shale, limestone, and phyllite.

1. Slate (P^{Sl})—Slate, presenting as gray-green or yellow-green, has joints, fractures, and weak regolith. This slate is grade IV soft rock and grade V secondary hard rock with grade III and IV surrounding rock and $\sigma_0 = 600$ kPa.
2. Schist (P^{Se})—Schist, presenting as gray-green or yellow-green, has joints, fractures, and weak regolith. This schist is grade IV soft rock and grade V secondary hard rock with grade III and IV surrounding rock and $\sigma_0 = 400$–600 kPa.
3. Sandstone (P^{Ss})—Sandstone, presenting as gray-green or steel-gray, exhibits joints and fractures and weak weathering. This sandstone is grade IV soft stone and grade V secondary hard rock with grade III and IV surrounding rock and $\sigma_0 = 800$ kPa.
4. Shale (P^{sh})—Shale, presenting as gray-green or gray, has fractures and joints and weak weathering. This sandstone is grade IV soft rock with grade IV surrounding rock and $\sigma_0 = 400$ kPa.
5. Limestone (P^{ls})—The gray and off-white limestone in this system is grade V secondary hard rock with and joints and fractures and $\sigma_0 = 1000$ kPa.
6. Phyllite (P^{ph})—Phyllite, presenting as gray-green or yellow-green, has fractures, joints, and crushed rock stratum. This weakly weathered phyllite is grade IV soft rock with grade IV surrounding rock and $\sigma_0 = 400$–600 kPa.

2.3.7 Paleozoic Group

With the main lithology of low metamorphic schist, sandstone, and limestone, the Paleozoic group mainly occurs as outcrops in the southern Dagangou Bedrock Mountain area and low mountain area. Granitic gneiss outcrops can be seen in Beisang Meander, the northern shore of Cuona Lake, and Yambajan.

1. Schist (P_z^{Se})—This group contains chlorite schist, sericite-quartz schist, tremolite, and actinolite schist, which are gray or gray-green with high schistosity and numerous joints. Squeezing movement has distorted the strata, and part of the

rock surface is severely weathered. These schist are grade IV soft rocks with grade III and IV surrounding rock and σ_0 = 600–800 kPa.

2. Sandstone (P_z^{Ss})—Sandstone, presenting as gray-green or gray, has joints and fractures and weak-to-strong weathering. These sandstones are grade IV soft rocks and grade V secondary hard rocks with grade III and IV surrounding rock with σ_0 = 800 kPa.
3. Limestone (P_z^{ls})—Limestone, coral, and other fossils can be found in gray or gray-green hard limestone. This limestone has joints and fractures and is mainly distributed near Naij Tal. It is a grade V secondary hard rock with grade V surrounding rock and σ_0 = 1000 kPa.
4. Granitic gneiss (Gn)—This system has off-white or gray-black granite sediments that exhibit a gneissic structure. The regolith is 1–5 m thick and is grade IV soft rock with σ_0 = 500 kPa. The whole bedrock is fifth-grade secondary hard rock with σ_0 = 1000 kPa.

2.3.8 Magmatic Rocks

1. Granite (γ_4, γ_5)—Granite γ_4, γ_5 in north Dagangou is formed in Variscan and that in the south in Yanshanian period. It is off-white or pale red with a medium-grained texture. These strong-to-weak weathered granites with joints and fractures are grade V secondary hard rock with σ_0 = 1000–1200 kPa.
2. Granite (γ_6)—Granite γ_6 is mainly distributing in Yambajan tunnels 1 and 2, in Liuwu tunnel, and from Gurong to Lhasa. This off-white or pale red granite has a massive and medium–coarse-grained structure with many joints. This 3–8-m-thick strongly weathered layer is grade IV soft rock with σ_0 = 500 kPa. The weakly weathered layer and the whole bedrock are grade V secondary hard rocks with σ_0 = 1000 kPa.
3. Gabbro (ν) and andesite (α)—Gabbro and andesite are mainly distributed in Kaixinling, Duopuer Meander, and Yangbaling. They are off-white or gray-black with an equigranular texture and a massive structure. This 2–4-m-thick strongly weathered layer is grade IV soft rock with σ_0 = 500 kPa. The weakly weathered layer and the whole bedrock are grade V secondary hard rocks with σ_0 = 1000 kPa.

2.4 Geological Structure

The Qinghai–Tibet Railway passes through the head of the Qinghai–Tibet Plateau eta-type tectonic system. Most of this area features intense uplift geosynclines, and linear fractures appear in bundles. The strata include many folds and faults from the Sinian Suberathem, and the new tectonics (mainly fractures) show obvious inheritance from and reconstruction of the old fractures with strike faults. In each generation, different tectonic movements generate different fracture systems.

The main folds along the Qinghai–Tibet Railway are the Naij Tal Anticlinorium, Hoh Xil Synclinorium, Fenghuo Mountain Syncline, Kaixinling Anticline, Tonglha Mountain Synclinorium, Amdo Arc Fault–Fold Belt, and Nagqu Compound Fold–Fault Belt.

The main faults along the line are the Golmud Concealed Fracture, Xidatan Concealed Compressive Fracture, Kunlun Mountains Bealock Compressive Shear Faults, the Northern Fenghuo Mountain Compressive Fracture, Wuli Fracture Zone, Yanshiping Regional Compressive Shear Faults, Tonghla Mountain Fault Zone, the Eastern Shore Fault at Cuona Lake, and the Nagqu–Damxung–Yambajan Fault Zone.

2.4.1 Folds

2.4.1.1 Naij Tal Anticlinorium

Located in the northern slope of Kunlun Mountains, Naij Tal Anticlinorium is a linear angular overturned and composite anticline comprised of Paleozoic metamorphic rocks. The nearly EW-trending axis of the anticline in the area from Naij Tal to Kunlun Bridge was formed in the Caledonian.

2.4.1.2 Hoh Xil Synclinorium

Belonging to Hoh Xil–Qumarlêb Synclinorium, the Hoh Xil Synclinorium has a NWW–SEE axis. The tightly distributed folds consist of the Late Triassic Bayanhar Group formed in the Indosinian.

2.4.1.3 Fenghuo Mountain Syncline

Fenghuo Mountain Syncline is comprised of lower Tertiary rocks formed in the Himalayan period. The axial direction of this gentle branchy syncline is nearly NW–SE, and the dip angle of axial part strata is less than 20°. The development of the north wing is incomplete because of fracture cutting, whereas the south wing has developed two secondary synclines. The axial direction of the first syncline is uncoordinated with the main syncline.

2.4.1.4 Kaixinling Anticline

With a nearly NS–SE axial direction, Kaixinling Anticline comprises the Permian system, with sandstone composing the axial section and limestone composing the two wings, which are almost identical. This anticline was formed in the Late Hercynian and was reformed by the late tectonic activity.

2.4.1.5 Tonglha Mountain Synclinorium

Tonglha Mountain Synclinorium is around 200 km long in the south–north direction. With a NW–SE axial direction, this synclinorium is composed of the Yanshiping group and was formed in the Early Yanshanian period. Five secondary synclines—Yanshiping, 102 Road, 104 Road to 106 Road, 111 Road, and 114 Road to 115 Road—and four secondary anticlines—Wenquan Military Depot, 103 Road, Tonglha Main Ridge, and Touerjiu Mountain—constitute the Tonglha Mountain Synclinorium.

2.4.1.6 Amdo Arc Fault–Fold Belt

Amdo Arc Fault–Fold Belt is composed of north Amdo Mountain and an arc structural valley, which has complex structural features: The arc roof is a convex northward arc structural belt, with the outer arc uplifted forming a high mountain; the dropped inner arc forms the valley.

2.4.1.7 Nagqu Compound Fold–Fault Belt

With similar anticlines and synclines and two symmetrical wings, Nagqu Compound Fold–Fault belt is distributed south of the Qinghai–Tibet Highway 126 Road, a nearly EW-trending undulate fold. The formation of these folds was accompanied by the formation of a few small nearly EW–NEE-trending fractures.

2.4.2 Faults

2.4.2.1 Golmud Concealed Fracture (F1)

The 400-km-long Golmud Concealed Fracture, a segment of Urt Moron–Nomuhong Concealed Fracture, lies to the south of Golmud. The Quaternary system in the southern and northern walls of the fault is 621 and >1000 m thick, respectively. Because of this fault, the Qinghai–Tibet Railway line from Golmud to Nanshankou is built as an embankment.

2.4.2.2 Xidatan Concealed Compressive Fracture (F2)

Located in the Xidatan Structural Valley, Xidatan Concealed Compressive Fracture is a part of Kunlun Lake–Xiugou Deep Fault, and it controls the southern boundary of the Paleozoic group. Magmatic intrusions are present along the fault, and the fracture surface is buried 300 m beneath the Quaternary. In this region, the Qinghai–Tibet Railway line has mainly used embankments.

2.4.2.3 Kunlun Mountains Bealock Compressive Shear Faults (F3)

1. D6-F1 and D6-F2 Compressive Fracture—This fracture is located in the valley area near the entrance of the Kunlun Mountains tunnel and intersects the Qinghai–Tibet Railway at DK976+058 and DK976+158. The D6-F1 fault strikes N72°W and dips southward; Quaternary deposits cover its surface, and the fault fracture zone is approximately 45 m wide. The D6-F2 fault strikes almost EW and dips 50°–70° northward; Quaternary deposits cover its surface. The fault fracture zone is about 50 m wide and consists of broken phyllite and slate, which was also found in the D6Z-247 drilling core. The railway line uses culverts to pass through this fault area, and the fractures slightly affect railway engineering in this area.
2. D6-F3 Compressive Fracture—This fracture, located north of Kunlun Mountains Bealock, intersects the line around DK979+520 at 45°. The fault strikes N75°W and dips northward at 70°–80°; Quaternary deposits cover its surface, and fault fracture zone is 20–27 m wide. Stratum lithology in the northern wall of the fault consists of Triassic slate with schist, and denudation in the southern wall of the fault has exposed Lower Pleistocene lacustrine silty clay and silt. The fault was activated again by the Kunlun Mountain Pass Earthquake on November 14, 2001. The Qinghai–Tibet Railway used embankments to pass through this fault area, and the fractures slightly affect railway engineering in this area.
3. D6-F4 Compressive Fracture—This fracture is distributed north of Kunlun Mountains Bealock, in the southern Kunlun Ditch. This normal fault intersects with the Qinghai–Tibet Railway line at 45° near DK981+850. The fault strikes N62°W and dips northward at 50°. The fault fracture zone is approximately 30 m wide and consists of broken phyllite and slate. Stratum lithology in the northern wall of the fault is Triassic slate with schist, and denudation in the southern wall has exposed Lower Pleistocene lacustrine silty clay and silt. The Qinghai–Tibet Railway line uses bridges to pass through this fault area.

2.4.2.4 Northern Fenghuo Mountain Compressive Fracture (F4)

Distributed in the north piedmont of Fenghuo Mountain and intersecting with the Qinghai–Tibet Railway in DK1147+420, the Northern Fenghuo Mountain

Compressive Fracture strikes almost NE and dips southward at 70°–85°; the fault zone is nearly 600 m wide. This fracture, produced in the Tertiary strata, is a part of Xijir Ulan Lake–Dam Qu Fracture, and the fault controls the upper Triassic Jieza group and the northern boundary of the Jurassic strata. The Qinghai–Tibet Railway line uses low embankments to pass through this fault zone, and the fracture has little influence on local railway engineering.

2.4.2.5 Wuli Fracture Zone (F5)

Wuli fracture zone is distributed in the piedmont area of Wuli Mountain and intersects the Qinghai–Tibet Railway line in DK1202+620 and DK1202+716. The fault strikes almost N78°W, and its surface is covered by Quaternary deposits. The fault fracture zone is approximately 96 m wide. Unfavorable geological phenomena, such as island taliks, hot springs, and frost mounds, are present along the fracture belt. The railway line uses bridges to pass through this fault area, and the fractures slightly affect railway engineering in this area.

2.4.2.6 Yanshiping Regional Compressive Shear Fault (F6)

Located in Yanshiping region, the Yanshiping Regional Compressive Shear Fault zone comprises a series of parallel faults and is hundreds of kilometers long. It controls the distribution of the Jurassic strata. In this area, the Qinghai–Tibet Railway line passes through seven buried faults, all of which are nearly orthogonal to the line (see Table 2.3). The D3-F1 fault, located in the Chaqingqu Bridge area, strikes almost N30°W and dips southward at 72°. The fracture affects bridge engineering here to some extent: in particular, the D3-F5 fault is located in the area with the middle section of the bridge, which contains gypsum; therefore, engineering treatment of the foundation is essential.

Table 2.3 Relationship between the Yanshiping Fault Zone and the Qinghai–Tibet Railway line

Fault	Railway boundary marker	Fault properties	Fault–railway intersection angle
D3-F1	DK1336+730	Reverse fault	70°, dips to N
D3-F2	DK1337+920	Reverse fault	70°, dips to N
D3-F3	DK1340+490	Reverse fault	50°, dips to N
D3-F4	DK1342+290	Parallel displacement fault	Orthogonal
D3-F5	DK1343+920	Reverse fault	Orthogonal, dips to S
D3-F6	DK1348+050	Reverse fault	Orthogonal, dips to N
D3-F7	DK1350+230	Reverse fault	Orthogonal, dips to S

2.4.2.7 Tonghla Mountain Fault Zone (F7)

The main strike of the Tonghla Mountain Compressive Fracture lies NWW and NE. Presenting as reverse faults and normal faults, the fault zone mainly occurs in the Mesozoic (Jurassic) strata, generally at an angle of 40°–60°. The main faults are described herein

1. D12-F1—A reverse fault, D12-F1 fault is distributed from DK1358+900 to DK1359+020; the fault fracture zone is 120 m wide, occurring at N69°W/79°S in the Jurassic strata. A gray–white gypsum layer constitutes the hanging wall of the fault, and a purplish-red conglomerate constitutes the footwall. A 10–20-m-thick Quaternary accumulation layer is present near the Qinghai–Tibet Railway; 200 m to the left of this line, the fault is exposed in the mountain slope, and the lithology is clear. No major projects have been implemented in the fault area. The railway line uses fills to pass through the fault zone, which slightly affects the railway engineering in this area.
2. D12-F12—Distributed around DK1360+000, the D12-F12 fault is a buried fault that strikes N42°E at a steep dip angle. Occurring in the Jurassic strata, the hanging wall of the fault consists of limestone and conglomerate, and the footwall consists of a gray–white gypsum layer. A 5–20-m-thick Quaternary accumulation layer is present near the Qinghai–Tibet Railway line. No major projects have been implemented in the fault area, which barely influences local railway engineering.
3. D12-F3—A buried normal fault, the D12-F3 fault (N20°E/60°S) lies to the right of DK1394+000. Occurring in the Jurassic strata, the lithology of this fault is mainly comprised of limestone, sandstone, and shale in this area. This fault is far from the Qinghai–Tibet Railway line, which uses subgrades to pass through the fault area; the fault exerts no effect on local railway engineering.
4. D12-F4—Distributed near DK1394+415–DK1394+430, the D12-F4 fault (N35°E/87°S) is a nearly 15 m wide buried fault. The fault fracture zone here contains fault breccia.
5. D12-F5—Occurring in the Jurassic strata, the D12-F5 fault is a buried reverse fault distributed near DK1396+300–DK1396+400. It strikes N38°E at a steep dip angle. The Buqu River Bridge is built in this fault region, which exerts a slight effect on local railway engineering.
6. D12-F5—Distributed in DK1398+400, the D12-F4 fault is a buried reverse fault present in the strata of Jurassic sandstone, shale, and limestone formations. It occurs at N39°E and has steep dip angle. Quaternary deposits are present near the Qinghai–Tibet Railway line in this area. No major projects have been implemented in the fault area, which barely influences local railway engineering.

2.4.2.8 Cuona Lake Eastern Coast Fault (F8)

To the east of Cuona Lake Eastern Coast Fault, which trends NEE, is a mountain ridge composed of Yanshanian Granite and Paleozoic metamorphic rock, and to the west is Cuona Lake. The fault fracture zone is 30–40 m wide. The fault extends to the north and intersects the North Sangkarigang Fault on the east coast of Sangqu. Along DK1553+000–DK1557+900, the Qinghai–Tibet Railway line passes 100–200 m away from the fault on average, with the closest approach at DK1579+100–DK1580+000 (25–50 m) and DK1580+000–DK1581+000 (<20 m).

2.4.2.9 Nagqu–Damxung–Yambajan Fault Zone (F9)

1. Nyainqentanglha Mountain Eastern Marginal Fault (D9-F1)—This fault (SN45°–60°E) is a Holocene active fault that strikes almost SN from Sangshung to Jiuzina and NE–NEE from Jiuzina to Gariqiao. The Qinghai–Tibet Railway passes close to the fault between Sangshung Station and Jiuzina Flyover. The fault exerts some effect on local railway engineering. Chun Lang Bridge (DK1797+449) and Jiuzina No. 1 Bridge (DK1802+916) as well as subgrade engineering are present in this section.
2. Damxung Basin Northern Marginal Fault (D9-F2)—A part of the eastern Nyainqentanglha Mountain Marginal fault, this fault strikes nearly EW and dips southward. Left-lateral strike slip has been occurring since the Holocene, and the normal dip-slip component is evident. In DK1801+000, the Qinghai–Tibet Railway uses a roadbed to pass through the fault zone.
3. Langguo–Sangqu Fault (D9-F3)—Starting from the toe of Baguran Gully, this fault cuts through the toe of Ribalang Gully in the east before entering Sangqu Ditch. In west Baguran, this normal fault (N70°E/70°N) is buried in the piedmont alluvial plain. The Qinghai–Tibet Railway line passes through a belt influenced by the faults. Apart from the sections with the Ribalang Bridge (DK804+970) and Baguran Bridge (DK1813+116), the line uses roadbeds to pass through this area.
4. Eastern Boundary Fault of Damxung–Yambajan Graben—This 20 m wide fault (N65°E/70°N) is spread along the Qinghai–Tibet Highway, and most of the fault along the Qinghai–Tibet Railway line is buried. The railway line passes through the influence belt of this normal fault in the DK1831–DK1846 section, where Benla Bridge (DK1840+374) and Jiagen Bridge (DK1841+960) are built. The line uses roadbeds to pass through the rest of the fault zone.

2.5 Hydrogeology

2.5.1 Regional Hydrogeological Conditions

The hydrogeological conditions along the Qinghai–Tibet Railway is influenced by the local climatic characteristics, topography, lithology, geological structure, ancient and recent tectonic activities, and other natural factors. The permafrost zone along the line (Xidatan to Amdo) can be classified into various hydrogeological units with differing hydrogeological conditions

1. Permafrost of a certain thickness can form a relatively complete and uniform aquifuge, and different types of groundwater, such as suprapermafrost water, infrapermafrost water, and meltwater, differ in their distribution characteristics and enrichment regularity.
2. The groundwater in this area is mainly recharged by atmospheric precipitation, snowmelt water, and glacier-melt water. The surface water accumulates and develops into streams, and in the runoff process, this surface water replenishes the suprapermafrost water and groundwater in the nonpermafrost regions. The frozen layer in the permafrost region prevents the infrapermafrost water from being directly recharged by the surface water. In addition, infrapermafrost water is present in low quantities because it is replenished only by the suprapermafrost water or surface water from rivers and lakes or the bottom of melting glaciers. Moreover, only 10% of the atmospheric precipitation infiltrates the ground to replenish the groundwater, with the rest either evaporating or being discharged as surface runoff.
3. Suprapermafrost water is excreted in the thaw area, and the runoff is slow. The outcrop of natural infrapermafrost water forms ice cones and frost mounds in the cold season.
4. The region along the Qinghai–Tibet Railway line can be categorized as the four large groundwater catchment areas of Kunlun Mountains, Tonglha Mountains, and Nyainqentanglha Mountains, the hydrogeological conditions of which differ widely because of the differences in the formation lithology, structural features, and topography.

   a. The region to the north of Kunlun Mountains is the groundwater area of Golmud Valley. Its hydrogeological environment is controlled by the latitudinal tectonic system. The bedrock here has fractures, and the valley is filled with loose gravel. The groundwater is of high quality, with a low mineralization degree and fast runoff.
   b. The region from Kunlun Mountains to Tonglha Mountains is the high plains groundwater basin area of the source areas of the Yangtze River, where a range of alternating low mountains and faulted basins constitute an undulating plain. This region contains Permian coal bearing strata and Mesozoic salts. The groundwater is of poor quality, with high mineralization degree, poor supply, and slow runoff.

c. The region to the south of Tonglha Mountains is the Za'gya Zangbo and Jiebu Qu groundwater area. Presenting as a series of stepped mountains, the hydrogeological environment here is controlled by an eta-type tectonic system. Clastic rocks and carbonate rocks of Middle Jurassic constitutes the main water-bearing strata. Groundwater is of high quality, with fast runoff.
d. Between Amdo and Samli, Quaternary pore phreatic water is mainly distributed in the valleys, plains, and interdune depressions. With a shallow depth of burial, the pore phreatic water is abundant and of high quality and is additionally supplied by surface water and atmospheric precipitation.
e. Between Samli and Lhasa, phreatic water is mainly distributed in the valley plain terrace and piedmont alluvial and diluvial plains; in bedrock fissures, water is distributed in the constituent sandstone and granite. Both the quantity and quality of this water is high.

5. Strong tectonic activities have formed structural valleys, mountains, and basins. These structural systems strongly influence the hydrogeological conditions in this area.

   a. The bedrock mountain area with folds and fractures has effective water storage structures, and the water quality is high.
   b. In the Early Cenozoic fault basin, the outwash sand gravel covering the layer is thin, and the groundwater is mainly hosted in the early Cenozoic lacustrine salt formation. The water quality is poor.
   c. In the structural valley, the groundwater is mainly hosted in the very thick outwash gravel deposits that formed in the Late Cenozoic, and the water is abundant and of high quality.

2.5.2 Groundwater Type

2.5.2.1 Groundwater in Nonpermafrost Regions

Table 2.4 summarizes the types and distribution of groundwater in nonpermafrost regions.

2.5.2.2 Groundwater in Permafrost Regions

The permafrost layer in permafrost regions increases the complexity of the distribution and burial of groundwater. The groundwater in such regions can be classified into three types: suprapermafrost water, infrapermafrost water, and meltwater.

1. Suprapermafrost water

Widely distributed in the permafrost regions of the Qinghai–Tibet Plateau, suprapermafrost water has an unstable water level and a changing phase state, with

Table 2.4 Characteristics of groundwater in nonpermafrost regions

Groundwater type		Distribution and burial	Features
Phreatic water	Pore water	Distributed in the valleys of Amdo, Xidatan, and Golmud River; alluvial and flood fans of Golmud River; and the pores of diluvial and ice-water sand gravel	High quantity and quality; most are related to river water
	Fissure water	Distributed in the weathering and structural cracks of igneous rock, metamorphic rock, sandstone, and shale in Kunlun Mountains and Tonglha Mountain area	Relatively high quality water, with uneven abundance
Piestic water	Karst water	Distributed in the fractures of marble and magmatic rocks in Kunlun River Valley	High quality and quantity
	Pore water	Interlayer water distributed in the toe of alluvial and flood fans in Golmud River	High quality and average quantity
	Fissure water	Distributed in the hanging wall of compressive fault zones in rocky mountains and the intersection of fracture zones with different mechanical properties	High quality and quantity

the burial conditions and water volume varying with the season. The thickness of the aquifers is controlled by the upper limit of the permafrost.

In this area, the upper limit depth of the permafrost is generally 1–3 m, and the aquifers have low thickness (~ 1 m). The distribution of the suprapermafrost water is affected by the runoff and vertical evaporation and is thus controlled by the topography; unified aquifers cannot form in higher grounds. In early April, the ground begins to thaw. The consequent melting of the ice body in the soil results in the gradual formation of a water-bearing layer in the active layer. By late September–early October, the thaw depth reaches the limit value, and the aquifer attains its peak thickness. By this time, the ground layer begins to freeze, and the frozen depth gradually increases until January; the soil layer remains completely frozen and connected to the permafrost at the end. Over the freezing period, the active zone of the groundwater becomes increasingly smaller, and the burial conditions change over time.

Suprapermafrost water can change from phreatic water in warm season to confined water in the cold season. When the hydrodynamic pressure becomes adequate to break through the upper soil layer, the confined water may run out of the ground at fragile sections and freeze into ice cones; alternatively, the water may move toward a weak underground soil layer, transforming the raised upper soil layer into frost mounds.

Suprapermafrost water is a type of phreatic water that is stored in a thin and unstable aquifer. Its burial and distribution is mainly determined by the distribution of the seasonal thawing layer and the baseplate shape of the permafrost meltwater.

2. Infrapermafrost water

A major form of groundwater in permafrost regions, infrapermafrost water has a stable phase state and is in a liquid state in all four seasons.

In addition to the edge of the Xidatan permafrost region, infrapermafrost water in this area is confined except in some places, where it may flow on to the surface. As the bearing roof, the thickness of the permafrost layer is directly influenced by the runoff conditions and the regularity of infrapermafrost water formation. This water has poor supply and runoff condition, mutable water quality, and highly nonuniform quantity.

The distribution of the infrapermafrost water is restricted by the permafrost layers. Depending on the characteristics of the water medium, the frozen layer can be classified as pore water, fissure water, pore–fissure water, and karst water (Table 2.5).

Table 2.5 Classification of infrapermafrost water

Groundwater type		Distribution and burial	Features
Pore water		(1) Distributed in glaciofluvial deposits and sandstone pores of the structural valleys of Xidatan, Wenquan, and Za'gya Zangbo, among others (2) Distributed in the half-diagenesis packsands and siltstone pores of fault basins of the Chumar River, Tuotuo River, and Tongtian River, among others	Water distributed in the ice-water sand gravel layer is abundant and of high quality whereas that distributed in the half-diagenesis layer is not
Pore–fissure water		Distributed in the mudstone and glutenite pores of fault basins in Hoh Xil Mountains, Fenghuo Mountain Piedmont, and Tongtian River, and Beilu River	Low quantity and quality, except in some fracture zones in the piedmont
Fissure water		Distributed in the structure cracks of igneous rocks, metamorphic rocks, sandstones, and shale in Kunlun Mountains and Tonglha Mountain Piedmont	Low quantity, except in the fault zone; quality varies widely
Karst water	Pore karst water	Distributed in the marl karst pores in the basins of Chumar River and Tuotuo River	Low quantity and quality
	Fissure karst water	Distributed in the karst fissures and pores of very thick carbonatite in Tonglha Mountain Piedmont	Relatively high quantity and quality

Table 2.6 Meltwater classification

Groundwater type	Talik type	Distribution and burial	Features
Piece melting zone water	Snow-covered talik	At the bottom of modern glaciers	High in quantity and quality; supplied by meltwater
	Lake talik	Around large lakes and at the bottoms of other lakes	High in quantity but of low quality; mostly from groundwater collection area
Zonal melting zone water	River talik	At bottom river beds and narrow strips of Tuotuo River, Tongtian River, and Buqu, among others	Structural fissure water in rocky mountains; pore phreatic water in basins and valleys; water quantity and quality is relatively high
Dotted melting zone water	Shallow circulating talik	At the intersection of shallow fractures and karst development zones	Mostly confined water of high quantity and quality and low temperature
	Deep circulating talik	At the intersection of deep fractures	Mostly confined water of high quantity and temperature but of low quality

3. Meltwater

The genesis of taliks is complex, but it can form in any terrain or structural unit. Some taliks (e.g., river taliks) can cover numerous topographical units and structural units, such as mountains, basins, and plains; pass through the distribution area of various rock strata; and contain various groundwater types. Therefore, the groundwater types in a talik cannot be classified with the usual method. Table 2.6 classifies meltwater on the basis of the distribution characteristics, burial conditions, and hydraulic properties of groundwater.

2.6 Earthquakes and Active Faults

Characterized by the wide distribution of large active faults, high-magnitude and frequent seismic activity, long surface rupture zones, and large co-seismic displacement, the Qinghai–Tibet Plateau is currently the most intense area of tectonic movement in the mainland of China.

Seismic problems were important in the investigation and design of the Qinghai–Tibet Railway, as earthquake faults cross the line numerous times. The high seismic intensity and the high risk of surface rupture in the plateau strongly affect the selection of local schemes and the design of engineering structures.

2.6.1 Regional Neotectonic Movement

Neotectonic movement refers to the tectonic movement that has occurred since the Late Tertiary; the tectonic movement in the Qinghai–Tibet Plateau in this period has been highly intensive and frequent. Earthquakes are closely associated with such movement.

The neotectonic movement in the Nanshankou to Kunlun Pass section is characterized by intracontinental compressional orogeny and a strong uplift. The main characteristics of this movement are fold uplift, thrust and strong uplift, large horizontal shrinkage and thickening of the crust, intermontane basin expansion, and obvious vertical movement. The plateau gradually formed after the large horizontal shrinkage and thickening of the crust and expansion of the intermontane basin. Strong earthquakes, hydrothermal activity, and Cenozoic magmatism have been common in the key section from Sangshung to Yambajan. The plateau is the most intense area of modern crustal deformation. Along the Qinghai–Tibet Railway, the highest earthquake intensity regions are between Tuoru and Yambajan in the Nyainqentanglha Mountains area.

2.6.2 Basic Seismic Intensity Zoning

According to the Earthquake Engineering Research Center of China Seismological Bureau, the Qinghai–Tibet Railway line passes through areas with high seismic intensity (intensity > 7). Of the total 1142 km of the Qinghai–Tibet Railway line from Golmud to Lhasa, approximately 620 km (54%), 307 km (27%), and 215 km (19%) passes through seismic zones of intensity 7, 8, and 9, respectively (see Table 2.7).

Table 2.7 Basic seismic intensity zoning of the Qinghai–Tibet Railway

No.	Mileage of the section beginning and end	Section	Basic seismic intensity	Length (km)
1	K814+150–DK866+500	Golmud–Ganlong	7	52.35
2	DK866+500–DK1001+750	Ganlong–Budongquan	8	123.00
3	DK1001+750–DK1585+850	Budongquan–Cuona Lake	7	567.65
4	DK1585+850–DK1712+500	Cuona Lake–Tuoru	8	115.00
5	DK1712+500–DK1933+500	Tuoru–Yambajan	9	215.00
6	DK1933+500–DK2005+918	Yambajan–Lhasa	8	69.00

Table 2.8 Major active faults affecting the engineering of the Qinghai–Tibet Railway

No	Fracture	Details of the line—fault intersection
1	Xidatan Concealed Compressive Fracture	
2	Kunlun Mountains Bealock Compressive Shear Faults	Line passes through three compressive fractures
3	Northern Fenghuo Mountain Compressive Fracture	
4	Wuli Fracture Zone	
5	Yanshiping Regional Compressive Shear Faults	Line passes through seven faults
6	Tonghla Mountain Fault Zone	Line passes through six faults
7	Za'gya Zangbo Active Fault Zone	
8	Touerjiu Mountains Fault Zone	Three faults affect the line
9	Amdo Beishan Fault	
10	South Boundary Active Fault in Amdo Basin	
11	Cuona Lake Eastern Coast Fault	
12	Northern Shengeligong Mountain Fracture	
13	Southern Shengeligong Mountain Fracture	
14	Liantong River–Dimaer Fracture	
15	Tuoru Basin Active Fault	
16	Bengcuo Active Fault	
17	Nyainqentanglha Mountain Eastern Marginal Fault	
18	Damxung Basin Northern Marginal Fault	
19	Langguo–Sangqu Fault	
20	Eastern Boundary fault of Damxung–Yambajan Graben	

2.6.3 Distribution Characteristics of Active Faults

The main structural fault line affecting the Qinghai–Tibet Railway trends nearly EW. The railway line runs NS and passes through the following mountain ranges: (north to south) Kunlun Mountains, Hoh Xil Mountains, Fenghuo Mountain, Kaixinling, Tonglha Mountains, Touerjiu Mountain, Sangxiongling, Jiuzina, and Yangbaling. The line intersects the main structural line at a large angle. Table 2.8 lists the main active faults affecting the railway line.

2.7 Distribution of Permafrost Along the Line

Permafrost is one of the major problems that was encountered in the construction of the Qinghai–Tibet Railway. Permafrost is widely distributed in the Qinghai–Tibet Plateau and is highly sensitive to the environmental conditions, especially temperature. The main railway engineering concerns in the permafrost region are thawing settlement of the permafrost subgrade, cutting slope collapse and mudflow, tunnel heaving and thawing cracking, icing inside the tunnel, frost heaving and thawing settlement of bridge and culvert foundation, erosion of the concrete surface. Usually, Railways passing through permafrost have a high disease rate after construction, both in China and worldwide. Therefore, solving these problems is key to the success of the Qinghai–Tibet Railway.

2.7.1 *Basic Distribution Characteristics*

The Qinghai–Tibet Plateau is located in a low latitude area. In contrast to permafrost in high-latitude areas, permafrost here has poor thermal stability, wide distribution of high temperature and high ice content permafrost, and high sensitivity to climate change.

The development and distribution of permafrost along the Qinghai–Tibet Railway is influenced by three directional zonalities: vertical zonality, caused by the altitudinal variation of heat and moisture; latitudinal zonality, caused by latitudinal (north–south) heat difference; and precipitation zonality, which varies with distance from the ocean and with the characteristics of atmospheric circulation. For every 100 m increase in altitude, the annual average temperature decreases by 0.5–0.6 °C and the thickness of the frozen soil increases by approximately 20 m. Similarly, for every 1° southward change in latitude, the altitude increases by 100–130 m, and the lower boundary elevation of the permafrost changes as well.

Permafrost in the plateau is a product of geological development; its occurrence, development, distribution, and evolution is influenced by many geological and geographical factors, such as geological structure, geomorphology, lithology, surface water, groundwater, vegetation, and snowfall, as well as climatic fluctuation over time. The climatic conditions of the plateau—for example, high altitude (approximately 960 km of the plateau is at >4000 m elevation), low pressure (544–600 mb), cold climate (annual average temperature = − 8.5 to −2 °C), long freezing period (permafrost is fully frozen for >7 months/year and frozen during nighttime in the warm season), low annual average temperature of frozen soil (−5 to 0 °C), thin snow, and short preservation time—are the factor that most strongly influences the development and persistence of permafrost. Because of the low latitude, high altitude, and thin snow covering, solar radiation on the plateau is strong.

The annual average temperature in the southern boundary of the Qinghai–Tibet Railway is −3 to −2 °C, which is 1–2 °C lower than that of the southern boundary of the permafrost regions in high-latitude regions. A 120–160-m-long area in the plateau has an annual average temperature of −5 °C. In some regions, the heat flow is high, and thaw areas often form along the structural geotherm. The thick ice layer is mostly distributed in the high moisture-content clay area, and more than 80% of such layers are up to 5 m thick. Typically, atop the ice layer lies a natural upper limit of permafrost. Because of the shallow burial depth, the top surface is easily affected by variations in the surface conditions and climate, resulting in the frequent development of thaw slumping, thaw lakes and ponds, ice cones, and hummocks.

2.7.2 Distribution of Permafrost Along the Line

Exactly 546.43 km of the Qinghai–Tibet Railway crosses the permafrost zone, with the northern boundary in the Xidatan (DK957+640, altitude = 4350 m, and annual average temperature = − 3.6 °C) and the southern boundary in Amdo Valley (DK1513+770, altitude 4780 m, and annual average temperature = − 2.9 °C). Taliks are distributed within these boundaries.

The permafrost regions of the Qinghai–Tibet Railway can be divided into 15 units. A comprehensive evaluation of the engineering geological conditions of each unit is as follows:

1. Xidatan Structural Valley (DK957+640–DK973+700)

In the Xidatan Structural Valley, the Qinghai–Tibet Railway line enters the permafrost region at DK957+640 and passes through the border zone of permafrost and nonpermafrost regions. Four sections totaling 8.518 km, namely DK957+640–DK960+260, DK964+500–DK968+190, DK969+650–DK971+940, and DK972+550–DK973+700, are the permafrost regions and account for 53.5% of the railway mileage in this unit. The remaining sections are nonpermafrost regions or thaw areas.

The main lithology in this unit is alluvial pebble soil and thick pellet soil. The upper limit of permafrost is generally 2.8–3.5 m, the annual average ground temperature of frozen soil is more than −0.5 °C, and the permafrost layer is 5–20 m thick. Ice-rich and ice-saturated frozen soils are present in two segments, DK965+410–DK965+740 and DK972+640–DK973+700. Permafrost with high ice content is 1.39 km long (16.3%). Overall, the engineering geological conditions in this unit are favorable, with subgrade being the major engineering structure.

2. Kunlun medium-high mountain area (DK973+700–DK1005+500)

In this unit, the line passes from Luanshi Ditch in Kunlun medium-high mountain area (DK973+700) to Budongquan, crossing Kunlun Mountains Bealock at DK983+600 and Budongquan Valley south of the Kunlun Mountains. A 0.93-km long (2.9% of the line in this unit) talik is present between DK1004+570 and DK1005+500, and the rest of the line passes through 60–120-m-thick permafrost with a 1.5–2.5 m natural permafrost table.

Two sections, DK997+300–DK1002+600 and DK1004+000–DK1004+570, are high-temperature zones of length 5.87 km (12.4%); the remaining 25 km (78.6%) passes through low-temperature zones. Frozen soil with high ice content is distributed mainly along beach land and slopes along 1.39 km of the line (56.1%).

The main lithology in Kunlun Mountain area is weathered broken slate and schist; fine granular soil, such as silt and powder clay in Kunlun Mountain Bealock Basin, and spall soil in Budongquan Valley and beach land. Unfavorable geological phenomena such as thaw slumping, ice cones, and frost mounds are common in the permafrost regions. The railway line passes through seven thaw slumping areas. Overall, the engineering geological conditions in this unit are poor, and filling is the major engineering measure.

3. Chumar River high plain (DK1005+500–DK1072+500)

In this unit, the railway line crosses Qingshui River, Baladacai Meander, and Chumar River. The lithology along the line is fine granular soil such as silt soil, powder clay, and fine sand. A 2.5-km long (3.7% of the line length in this unit) talik is present between DK1072+000 and DK1072+500, accounting for 3.7%. The rest of the line passes through 15–40-m-thick permafrost with a 2.0–5.0 m natural permafrost table.

The line spans 34.9 km (80.5%) and 29.6 km (44.2%) in high- and low-temperature regions, respectively. Frozen soil with high and low ice content is present along 53.93 km (80.5%) and 13.07 km (19.5%) of the line, respectively. No clear pattern is evident in the distribution of high ice content permafrost.

Thaw lakes and ponds, most of which are seasonal, are spread over the plain, and the major engineering measures applied in this unit are embankments and culverts. The ground temperature of the frozen soil here is high, and ice-rich permafrost is widely distributed. With harmful frozen-soil phenomena prevalent, the local engineering geological conditions are adverse.

4. Hoh Xil Mountains (DK1072+500–DK1124+500)

In Hoh Xil Mountain area, the line is mainly distributed over low mountains and hills, with permafrost present throughout. Thaw areas totaling 0.107 km (0.2% of the line length in this unit) are distributed between DK1083+108 and DK1083+215. The rest of the line passes over 30–100-m-thick permafrost with a natural permafrost table of 2.0–3.0 m.

The line passes 42.876 km (81.8%) and 9.41 km (18%) over low- and high-temperature regions, respectively. Ice-rich permafrost (38.775 km; 74%) is the dominant type, with frozen soil with low ice content accounting for only 13.618 km of the line (26%).

The upper section of the ground on which the line passes is mainly fine sand, silt, and other fine granular soil, the a weathered layer of mudstone and marl buried beneath. Because of harmful cryogenic phenomena, thaw lakes or ponds and frozen-soil wetlands are present in this unit. The line uses subgrades to pass through these regions. Nevertheless, annual average ground temperature here is low and the engineering geological conditions average.

5. Beilu River Basin (DK1124+500–DK1145+500)

In Beilu River Basin, the line mainly passes through Xiushui River and the floodplains and terraces of Beilu River. Permafrost is present throughout. A 0.09-km-long talik area (0.4% of the line length in this unit) is present between DK1124+800 and DK1124+910. Accounting for of the whole segment, the length of the talik region is. The rest of the line passes over 10–50-m-thick permafrost with a natural permafrost table of 2.0–3.0 m.

The line passes 15.9 km (75.7%) and 5.1 km (24.3%) over high- and low-temperature regions, respectively. Frozen soil with low ice content is common (13.974 km; 66.5%), and the ice-rich permafrost is mostly located in the Beilu River test Sect. (6.936 km; 33.0%).

Lithology along the line is mainly fine sand, silt, and powder clay, and the overall engineering geological condition is poor.

6. Fenghuo Mountain area (DK1145+500–DK1165+500)

Permafrost is present throughout Fenghuo Mountain area, except for the talik area between DK1149+300 and DK1149+800 (0.5 km long; 2.5% of the line length in this unit); the rest of the regions are mostly permafrost of thickness 50–120 m with a 1.0–2.5 m natural permafrost table.

The line passes 15.3 km (76.5%) and 4.2 km (21%) over low- and high-temperature regions, respectively. Frozen soil with low ice content is the main type (11.49 km; 57.5%), but ice-rich permafrost is mainly distributed in the low hills and depressions (8.01 km; 42%).

The main lithology in the northern and southern Fenghuo Mountain area is powder clay and alluvial spall soil, and interbeds of sandstone and mudstone in the mountain area; the weathered layer is 15–20 m thick. Frozen-soil wetlands are distributed in low hills, over which the line passes through using fills. The slope area over which the line passes is long and is influenced by many unpredictable factors. Hence, ensuring safety was of prime importance during railway construction.

7. Chiqu Valley (DK1165+500–DK1193+200)

In Chiqu Valley, except for the 0.79-km-long (2.9% of the line length in this unit) talik between DK1165+380 and DK1187+170, the railway lines passes through mostly permafrost of thickness 10–50 m with a 2.0–4.0 m natural permafrost table.

The line passes 20 km (72.2%) and 6.91 km (24.9%) over high- and low-temperature regions, respectively. Frozen soil with low ice content is the main type (15.803 km; 57.1%), and ice-rich permafrost present along 11.07 km (40.1%) of the line.

Lithology along the line is mainly brecciated soil and gravelly soil, along with powder clay and fine sandstone in some areas. Few harmful frozen-soil phenomena are present here. The majority of this unit is unstable and has high ground temperatures; the overall engineering geological conditions are average.

8. Wuli Basin (DK1193+200–DK1202+500)

All of Wuli Basin is covered in 50–120-m-thick permafrost with a natural permafrost table of 1.0–2.5 m; no taliks are present in this unstable, high-temperature region, whose annual average ground temperature exceeds −0.5 °C. Ice-rich permafrost is the main type of frozen soil here (5.395 km long; 58% of the line length in this unit), and 3.905 km (42%) of the line in this section is over frozen soil with low ice content.

In this basin, the line passes over fine granular soil such as powder clay and fine sandstone. Thaw lakes and ponds and frozen-soil wetlands are major harmful cryogenic phenomena here, and the engineering geological conditions are therefore adverse.

9. Wuli Mountain (DK1202+500–DK1217+700)

Taliks dominate the Wuli Mountain area, accounting for 13.513 km of the line (89% of the line length in this unit). Permafrost of thickness 5–20 m with a 2.0–3.0 m natural permafrost table accounts for 1.687 km (11%). The main type of permafrost in this unit is ice-rich frozen soil (1.186 km), as permafrost with low ice content is present for only 0.501 km. The permafrost here is largely unstable and of high temperature.

In this unit, thaw areas are more widely distributed than are permafrost areas, and the overall engineering geological conditions are favorable, except in the permafrost area.

10. Tuotuo River Basin (DK1217+700–DK1245+000)

Islet permafrost and thaw areas (12.223 km, 44.8% of the line length in this unit) are alternately distributed in this Tuotuo River Basin, where permafrost of thickness 5–30 m with a 2.0–4.0 m natural permafrost Table (15.077 km; 55.2%) is mainly distributed in the low-lying valleys of the northern and southern banks of Tuotuo River; taliks are mainly distributed in the foreland pluvial fan and the first and second terraces of the river.

Apart from the segment between DK1241+850 and DK1245+000, the permafrost here is generally unstable and of high temperature. Permafrost with low ice content is dominant, accounting for one-fifth of the line length, whereas ice-rich frozen soil accounts for only a third of the line in this unit (5.411 km).

Cailloutis soil, thick pellet soil, and other coarse-grained soils are widely distributed here. The line passes through sections of freezing and thawing transition several times, which combined with the high annual average ground temperature makes the local engineering geological conditions adverse.

11. Kaixinling Mountain area (DK1245+000–DK1252+800)

Kaixinling Mountain area has an alternating distribution of thaw areas and permafrost of thickness 20–40 m with a natural permafrost table of 1.5–2.5 m; the thaw areas are dominant. Taliks (7.3 km; 93% of the line length in this unit) are mainly distributed on gentile hills in Kaixinling, whereas permafrost is mainly distributed in the gullys and intermountain depressions.

Permafrost in this region is unstable and of high temperature and is narrowly distributed. The main type of permafrost in this section is frozen soil with low ice content, and the railway line mainly passes over brecciated soil and gravelly soil. The major harmful frozen-soil phenomenon is frozen-soil wetlands, over which the line uses fills and bridges. Overall, the engineering geological conditions here are not too unfavorable.

12. Tongtian River Basin (DK1252+800–DK1282+800)

In Tongtian River Basin, taliks are distributed between DK1280+500 and DK1282+800 for 2.389 km (8.0% of the line length in this unit); permafrost of thickness 20–40 m with a natural permafrost table of 1.5–3.0 m covers the other areas along the line.

In this unit, the line spans 15.8 km (52.7%) and 11.8 km (39.3%) in the high- and low-temperature regions, respectively. The main type of permafrost in this section is ice-rich frozen soil (20.725 km; 69.0%), as frozen soil with low ice content accounts for only 6.886 km (23%) of the line. The line mostly passes through regions with fine granular soil such as powder clay, silt, and fine sand, and thaw lakes and ponds are prevalent. Overall, the engineering geological conditions here are adverse.

13. Buqu Valley (DK1282+800–DK1360+800)

Apart from some taliks distributed between DK1332+600 and DK1360+800 for 36.045 km (46.2% of the line length in this unit), Buqu Valley is covered in permafrost of thickness 5–40 m with a natural permafrost table of depth 2.0–5.0 m between DK1282+800 and DK 1332+600 for 41.955 km (53.8%).

The permafrost area has poor stability and is of high temperature between DK1352+560 and DK1356+530 (3.97 km; 5%). The main type of permafrost in this unit is frozen soil with low ice content (22.093 km; 52.7%), as ice-rich frozen soil accounts for only (19.862 km; 47.3%).

Ice-rich and low ice frozen soils are distributed rather equally in the permafrost region. The line mainly passes through regions with brecciated soil and gravel soil. Overall, the engineering geological conditions here are of average favorability. However, as the line passes sections with freezing and thawing transition several times, the engineering geological conditions can be adverse in such Sects.

14. Wenquan Fault Basin (DK1360+800–DK1394+800)

An alternating distribution of permafrost and thaw areas can be observed in Wenquan Fault Basin. The thaw area contains river taliks and structural taliks for (9.31 km, 27.4% of the line length in this unit), and permafrost of thickness 5–20 m with a natural permafrost table of 2.0–3.0 m accounts for 24.69 km of the line.

This fault basin has poor stability, and the annual average ground temperature of permafrost is high (more than −0.5 °C). Ice-rich permafrost is rare (2.26 km;

6.6%), and low ice frozen soil accounts for the majority of the line length (22.43 km long; 66%).

Lithology along the line is mainly thick pellet soil and brecciated soil, and permafrost wetlands are the main harmful cryogenic phenomena. The line uses roadbeds and bridges to pass through these regions. This fault basin area is a high-temperature region, and the freezing and thawing transition sections make the local engineering setting adverse. Nevertheless, the overall engineering geological conditions are of average favorability.

15. Tonglha Mountains and Intermontane Basin (DK1394+800–DK1485+200)

In Tonglha Mountain area (DK1394+800–DK1513+770), the lines pass over regions with alternating permafrost and taliks (3.831 km; 36.4% of the line length in this unit).

Mostly, permafrost is present between DK1405+320 and DK1456+000 in Tonglha Intermontane Basin area, and a few thaw areas are present in the DK1440+800–DK1443+200 and DK1446+500–DK1449+500 sections for 5.4 km (10.7%).

The talik from Za'gya Zangbo to Amdo (DK1456+000–DK1513+770) spans 12.817 km (22.2%), with permafrost of the thickness 10–120 m with a natural permafrost table of 1.5–3.5 m comprising the rest of the line length.

Most regions (~74.7 km; 62.8%) are high-temperature zones (I, II), and the remaining sections (~31.45 km; 25.3%) are low-temperature zones (III, IV).

Overall, compared with permafrost areas in high-latitude regions, permafrost areas of the Qinghai–Tibet Railway can be characterized as long, ice-rich, and of high temperature. The thermal stability of the frozen soil in the Qinghai–Tibet plateau is dependent on the climatic conditions. Overall, the engineering geological and hydrogeology conditions along the line are complex, the frozen-soil environment complicated, and the ecological environment fragile, all of which rendered difficult the construction of the Qinghai–Tibet Railway.

2.7.3 Types and Distribution of Thaw Areas in Permafrost Regions

2.7.3.1 Types of Thaw Areas

A thaw area (talik) is a geological body containing either normal-temperature water, no water, or liquid water at subzero temperatures. The occurrence, development, and characteristics of thaw areas in permafrost regions are dependent on such factors as climate, geological structure, hydrological factors, and land cover. Depending on the objective, thaw areas can be classified according to various principles and methods. Table 2.9 presents a classification based on thaw-area genesis.

Table 2.9 Talik classification on the basis of genesis

Genetic type	Talik type	Subclassification
Structural	Structural talik	Structural–geothermal talik
		Structural–surface-water talik
Surface-water	Surface-water talik	River talik
		Lake talik
Infiltration and radiation effects	Infiltration–radiation talik	
Artificial function	Artificial talik	

1. Structural talik

Structural talik is of two types: structural–geothermal and structural–groundwater.

Structural–geothermal taliks are created by underground heat sources. Generally, underground hot water (water temperature > 45°) rises to the surface through the fault structure to form a hot spring, presenting as a geothermal anomaly. The talik area from DK1282+800 to DK1394+800—which is at the intersection of NW- and NNW-trending compressive shear faults and SN-trending tension–torsion fracture in Buqu Valley—is representative of a structural–geothermal talik. Obviously, the temperature around hot springs is high: around the original 103 Road, which is 100 m away from a modern spring, the geothermal temperature is 7.2 °C at a depth of 0.5 m, 12.8 °C at 2.5 m, and >50° at 40–50 m; 900 m from the spring, the geothermal temperature is 1.5 °C at a depth of 1 m, 3.5 °C at 3 m, and 12 °C at 30 m. The talik is 1.5–2.0 km wide in this area. Hot-spring outcrops can be observed along the intersection of these two groups of faults, such as at the original 90 Road, on both sides of the main ridge on Tonglha Mountains, and in the area between the original 115 and 116 Roads; structural–geothermal taliks can be seen in this outcrop area, the size of which is dependent on the spring temperature and the flow volume.

The formation of structural—groundwater taliks is dependent on the thermal influence of groundwater in the active fault zone. In contrast to structural–geothermal taliks, structural—groundwater taliks are characterized by a low groundwater temperature (generally below 10 °C), small size, and poor stability. Taliks of this type are mainly distributed in areas such as Budongquan, Erdaogou, and the Original 85 Road.

2. Surface-water thaw area

Surface-water thaw areas can be further categorized as river taliks and lake taliks.

River taliks exhibit a zonal distribution along the river and are confined to regions within the thermal influence of rivers. The width of these taliks depends on the river volume, water temperature, and the relationship between the river and the geological structure. Some penetrating taliks are distributed in Tuotuo River, Tongtian River, Buqu, and other rivers along the railway line.

The presence of ponding or a supply source for lakes leads to the melting of permafrost and the consequent formation of a penetrating or nonpenetrating talik. Lakes formed by the structure usually form large penetrating taliks. Many lakes and ponds of various sizes are present along rivers such as the Qingshui, Qumar, and Beilu. Taliks are often confined within the water area of these lakes and ponds. A penetrating talik may form at the bottom of lakes because of long-term hydrops, whereas a nonpenetrating talik can only develop in lakes with seasonal water accumulation.

3. Infiltration–radiation thaw area

An infiltration–radiation thaw area can form through two mechanisms: through large-scale surface absorption of solar radiation or through infiltration of atmospheric precipitation. In piece permafrost—islet permafrost transition zones, when all other conditions are almost the same, unequal solar radiation is incident on different slope surfaces, which may cause such phenomena as permafrost and talik in the north and south slopes, respectively. Solar radiation incident on the surface is also affected by the lithology and vegetation. Moreover, infiltration–radiation thaw areas can be distributed within the permafrost regions. Many such areas appear in open terraces or gentle watershed areas with relatively high temperature (more than −1 °C). These taliks differ from structural–geothermal taliks as the latter are formed because of recent strong tectonic movement and are geothermal anomalies; they also differ from river taliks, which are often subject to the thermal effect of dynamic water. Therefore, these taliks are often ignored by scholars, but they are nevertheless extensively distributed. For example, these infiltration–radiation thaw areas are spread across thousands of kilometers in Buqu Valley and along tens of kilometers of the northern Tuotuo River. These thaw areas, which are usually exposed, are covered by a layer of thick loose sand and gravel and thus have good drainage.

4. Artificial talik

Human economic activities, such as engineering construction, deforestation and reclamation, drainage and dewatering, water storage projects, and brick stacking, have led to the formation of artificial taliks. These activities have shifted the equilibrium state of the frozen-soil layer, causing heat accumulation at the upper limit of the frozen layer, which in turn results in the melting of the frozen soil or the descending of the upper limit. The range of these activities depends on the generated heat and is usually not much different from the place and scope of the human activities.

The forms and consequences of artificial taliks differ by region, and a lack of awareness of these consequences has caused many problems in engineering construction. Under conditions of low temperature and large thickness, large areas of continuous permafrost always develop nonpenetrating artificial taliks, a phenomenon called thaw bowl. In island permafrost regions, the formation of artificial thawing areas (including other types of financial areas) can reduce the area of permafrost islands or even eliminate them completely, extending the range of the taliks.

Table 2.10 Thaw areas of permafrost regions along the Qinghai–Tibet Railway line

No.	Mileage of the section beginning and end	Length (km)	Talik type	Location
1	DK969+540–DK969+580	0.040	Radiation talik	Frozen-soil boundary in Xidatan
2	DK1004+570–DK1005+500	0.930	Structural talik	Budongquan
3	DK1070+000–DK1072+650	2.650	River talik	Chumar River
4	DK1083+108–DK1083+215	0.107		
5	DK1124+820–DK1124+910	0.090	River talik	Beilu River
6	DK1149+300–DK1149+800	0.500		
7	DK1186+380–DK1187+170	0.790	River talik	Chiqu
8	DK1202+583–DK1202+700	0.117	Structural talik	Wuli Mountain area
9	DK1202+745–DK1206+150	3.405	Structural talik	Wuli Mountain area
10	DK1206+220–DK1209+870	3.650		
11	DK1210+495–DK1211+300	0.805		
12	DK1211+840–DK1214+797	2.957		
13	DK1214+868–DK1217+498	2.630		
14	DK1218+080–DK1221+920	3.840	River talik	Togton River
15	DK1229+476–DK1236+380	6.904	Infiltration–radiation talik	
16	DK1238+390–DK1238+450	0.060		
17	DK1245+210–DK1245+320	0.110	Infiltration–radiation talik	Kaixinling Mountain area
18	DK1245+730–DK1245+815	0.085		
19	DK1245+860–DK1245+930	0.070		
20	DK1245+980–DK1253+990	8.010		
21	DK1280+500–DK1282+889	2.389	River talik	Tongtian River
22	DK1309+000–DK1310+580	1.580	River talik	Buqu
23	DK1312+376–DK1313+136	0.760		
24	DK1313+666–DK1317+948	4.282		
25	DK1318+618–DK1319+340	0.722		
26	DK1319+500–DK1322+736	3.236		
27	DK1322+800–DK1322+903	0.103		
28	DK1322+942–DK1322+995	0.053		
29	DK1324+181–DK1324+596	0.415		
30	DK1325+570–DK1326+180	0.610		
31	DK1326+848–DK1327+374	0.526		
32	DK1327+474–DK1337+100	9.626		
33	DK1338+830–DK1341+740	2.910		
34	DK1341+860–DK1343+130	1.270		
35	DK1344+150–DK1345+200	1.050		
36	DK1346+152–DK1348+560	2.408		
37	DK1348+815–DK1349+390	0.575		
38	DK1349+530–DK1350+220	0.690		
39	DK1350+350–DK1351+090	0.740		
40	DK1351+390–DK1352+560	1.170		
41	DK1356+530–DK1356+900	0.370		
42	DK1357+350–DK1359+300	1.950		
43	DK1360+580–DK1360+700	0.120		

(continued)

Table 2.10 (continued)

No.	Mileage of the section beginning and end	Length (km)	Talik type	Location
44	DK1374+530–DK1375+940	1.410	Structural–groundwater talik	Wenquan Fault Basin
45	DK1378+850–DK1379+140	0.290		
46	DK1379+550–DK1379+650	0.100		
47	DK1380+400–DK1380+960	0.560		
48	DK1381+650–DK1382+400	0.750		
49	DK1382+950–DK1383+300	0.350	Structural–groundwater talik	Wenquan Fault Basin
50	DK1384+050–DK1384+200	0.510		
51	DK1384+550–DK1386+180	1.630		
52	DK1390+430–DK1394+500	4.070		
53	DK1394+710–DK1394+960	0.250	River talik Infiltration–radiation talik	Tonglha Mountain Area and Intermontane Basin
54	DK1395+800–DK1395+846	0.046		
55	DK1396+100–DK1396+450	0.350		
56	DK1396+900–DK1396+990	0.090		
57	DK1397+910–DK1397+950	0.040		
58	DK1398+940–DK1400+200	0.260		
59	DK1400+900–DK1401+460	0.560		
60	DK1401+750–DK1402+300	0.550		
61	DK1403+515–DK1403+700	0.185		
62	DK1404+610–DK1404+650	0.040		
63	DK1404+810–DK1405+320	0.510		
64	DK1441+080–DK1441+430	0.350		
65	DK1442+220–DK1442+300	0.080		
66	DK1442+850–DK1442+950	0.100		
67	DK1446+330–DK1448+700	2.370		
68	DK1458+446–DK1458+760	0.314		
69	DK1468+350–DK1469+290	0.940		
70	DK1469+780–DK1472+680	2.900		
71	DK1474+400–DK1477+100	2.700		
72	DK1480+360–DK1482+800	2.440		
73	DK1501+740–DK1502+030	0.290		
74	DK1509+980–DK1510+870	0.890		
75	DK1511+500–DK1511+950	0.450		
	Total	101.68		

In summary, the formation of artificial taliks affects the natural ecological environment and directly brings about grassland degradation, soil erosion, and other types of environmental deterioration. Therefore, while pursuing economic activities and immediate financial interests, we must not ignore the effects of these activities on the environment.

2.7.3.2 Distribution of Thaw Areas

Exactly 546.43 km of the Qinghai–Tibet Railway line is over the permafrost zone, whereas talik areas account for 101.68 km (18.6%). Table 2.10 summarizes the distribution of thaw areas along the line.

2.8 Unfavorable and Unique Geology Along the Line

The regional geological environment along the Qinghai–Tibet Railway is complex and is home to various unfavorable and unique geological phenomena, such as permafrost, geological hazards in active fault zones and slope areas, geothermal, dust storms, and snowdrift.

2.8.1 Permafrost

The major engineering geological problems in permafrost areas are frost heave, thaw settlement, and harmful frozen-soil phenomena. The geological conditions of permafrost along the Qinghai–Tibet Railway are complex and lead to various harmful frozen-soil phenomena.

Harmful frozen-soil phenomena, including unfavorable cryogenic phenomena, refer to the formation of medium–small terrain caused by the freezing and thawing of soil. These phenomena, examples of which include ice cones, ice mantle, frost mounds, gelifluction, thaw slumping, thaw lakes and ponds, frost soil swamps, and permafrost wetlands, adversely affects engineering project.

2.8.2 Geological Hazards in Active Fault Zones

The quality of fault rocks along both sides of the railway line is poor, with fault cliffs and fault triangular facets often occurring and forming a volley surface. Thus, geological disasters such as landslides and debris flow can be triggered easily along the line. North of the Qinghai–Tibet Plateau permafrost area, fault activities not only lead to roadbed deformation, pavement cracking, and engineering failure but also induce uneven frosting, structural fracture, migrating ice mounds, and other geological disasters.

A strong earthquake can produce remarkable surface ruptures and cause severe road deformation, bridge damage, and ground-surface subsidence as well as induce unfavorable geological phenomena such as sand liquefaction, collapse, and landslides, further aggravating the disastrous effects of the earthquake.

2.8.3 *Geological Hazards in Slope Areas*

Types of slopes in the Qinghai–Tibet Plateau include natural hillsides (e.g., along rivers, ditches, and reservoir banks) and general artificial slopes, and these slopes are prone to various geological diseases. Major geological hazards along the Qinghai–Tibet Railway line include dangerous rocks, rockfall, landslides, collapse, landslides, and debris flow; in addition, solifluction, thaw slumping, and slope wetlands occur frequently in permafrost slope areas. Some of these phenomena are explained herein

1. A dangerous rock refers to a rock mass or an individual block of rock on steep slopes and having a potential to collapse or fall, and rockfall refers to the phenomenon wherein an individual rock block suddenly falls or rolls down a steep slope under the effect of gravity. Collapse refers to the phenomenon wherein a rock mass suddenly falls down a steep slope under the effect of gravity or other forces (e.g., toppling, caving, and rolling). Slope deformation caused by these movements is also a type of collapse.
2. Because of either geological processes or human activities, a rock or soil mass may suddenly or quickly move entirely or partly down the slope of a weak surface; the resulting topography is referred to as a landslide.
3. Rainfall- or snowmelt-induced Debris flow is common in mountain areas. This phenomenon is a type of flood that carries a large amount of silt, stones, and other loose solid material.
4. Solifluction is the phenomenon wherein the structure of a turf and epipedon soil is damaged because of repeated freezing and thawing, following which the saturated soil slowly creeps down the hillside.
5. Slope wetlands are of two types: frozen-soil and nonfrozen-soil slope wetlands. Wetlands in permafrost slope areas are discussed comprehensively in the Sect. 5.2 of Chap. 5. Wetlands in nonpermafrost slope areas mainly contain cohesive soil. Strong infiltration and poor groundwater runoff in cohesive soil with or without rich humus gradually result in the formation of swamping wetlands.

2.8.4 *Geothermal Activity, Sandstorms, and Snowdrifts*

The area along the Qinghai–Tibet Railway has abundant geothermal and wind resources; however, it also has some unique and unfavorable geological phenomena such as geothermal activity, aeolian drift, and snowdrift, all of which adversely have affected the construction and maintenance of the railway line

1. Geothermal activities and human exploitation of geothermal resources may change the geological environment and trigger geological hazards, such as collapse and landslide due to the alteration of the structural characteristics of the

geological body, ground subsidence due to water drainage, and soil salinization due to geothermal fluid leaching. Furthermore, geothermal activity can alter the geological environment, either facilitating or hindering landslides, hydrothermal cementations, and soil salinization.

2. Aeolian drift in the area along the Qinghai–Tibet Railway, mainly presenting as severe sand burial and wind erosion of the railway line is a major effect of the characteristics of the plateau, namely strong wind, abundant sand availability, dry and warm climate, rapid population and livestock growth, and growth in the intensity of human and livestock activities.
3. Because of the wide distribution of snow on the plateau, snow damage such as avalanche and snowdrift (snow-bearing wind) is severe; snowdrift in particular adversely affects the Qinghai–Tibet Railway line.

2.9 Summary

Characterized by high elevation, low temperature, hypobaric hypoxia, and continuous permafrost, the Qinghai–Tibet Railway runs through a harsh environment with a fragile ecosystem.

This chapter introduced the geographical conditions along the Qinghai–Tibet Railway, including the local geography, climatic characteristics, soil and vegetation, wildlife, river system, and ecological environment. In particular, the geological environment along the railway line—including the topography, formation lithology, geological structure, hydrogeology, seismic and active faults, permafrost distribution, and geology—was described in detail.

The engineering geological conditions along the line—especially permafrost, earthquake intensity zoning, and geological hazards in slope areas—are complex, which along with the necessity of environment protection strongly influence line selection. In sum, a comprehensive analysis of the various engineering geological conditions along the line is essential for studying the line scheme.

References

1. Agricultural Resources Division Office, Qinghai Province. (1997). *Soil of Qinghai.* China Agriculture Press (in Chinese), Beijing.
2. Cao, D. Y., Lei, Y. D., Deng, X. F., et al. (2007). General engineering geological conditions along the Naij Tal-Tuotuo River section of the Qinghai–Tibet Railway. *Qinghai Environment, 17*(2), 78–80. (in Chinese).
3. Chen, R. H., & Wang, Z. T. (2003). Engineering-geological route selection of the gypsum section from Buqu to Wenquan of the Qinghai–Tibet Railway. *Journal of Glaciology and Geocryology, 25*(1), 14–16. (in Chinese).
4. Chen, M. X., Wang, J. Y., & Deng, X. (1994). *Geothermal resource of China: Forming characteristics and the potential assessment.* Beijing: Science Press. (in Chinese).

5. China Railway First Survey & Design Institute Group Co., Ltd. (1994). *The manual of routes* (2nd ed.). Beijing: China Railway Press. (in Chinese).
6. China Railway First Survey & Design Institute Group Co., Ltd. (2002). *Preliminary design of the Tonglha–Lhasa section of the Qinghai–Tibet Railway (Geology).* Internal information. (in Chinese).
7. China Stratigraphic Lexicon Editorial Board. (2000). *China stratigraphic lexicon: Quaternary.* Beijing: Geological Publishing House. (in Chinese).
8. Dai, J. X. (1990). *The climate of the Tibetan Plateau.* Beijing: Meteorological Press. (in Chinese).
9. Lanzhou Glacier Frozen Soil Research Institute, Chinese Academy of Sciences. (1983). *Collected papers of Tibet frozen soil research.* Beijing: Science Press. (in Chinese).
10. Li, W. Y. (1988). *Quatemary plants and environment in China.* Beijing: Science Press. (in Chinese).
11. Liang, S. H., Chen, J., Jin, X. M., et al. (2007). Regularity of vegetation coverage changes in the Tibetan Plateau over the last 21 Years. *Advances in Earth Science, 22*(1), 33–40. (in Chinese).
12. Liu, Z. P. (2008). Distribution characteristics of permafrost development along the Qinghai–Tibet Railway. *Railway Investigation and Surveying, 2,* 78–82. (in Chinese).
13. Liu, Z. Q., Xu, X., Pan, G. T., et al. (1990). *Geotecture and evolution of the Tibetan Plateau.* Beijing: Geological Press. (in Chinese).
14. Ma, T., Zhou, J. X., Zhang, X. D., et al. (2007). Preliminary studies on characteristics of vegetation distribution along the Qinghai–Tibet Railway line. *Research of Soil and Water Conservation, 14*(3), 150–154. (in Chinese).
15. Meng, X. L. (2006). Geological prospecting for permafrost engineering in the Qinghai–Tibet Railway. *Railway Investigation and Surveying, 3,* 28–31. (in Chinese).
16. Papers and reports on permafrost in the Qinghai–Tibet Plateau. *Annual Report of State Key Laboratory of Frozen Soil Engineering.* (in Chinese).
17. State Seismological Bureau of China. (1991). *Earthquake intensity zoning map of China (1990).* Beijing: Earthquake Press. (in Chinese).
18. Sun, Y. F. (2005). Permafrost engineering in the Qinghai–Tibet Railway: Research and practice. *Journal of Glaciology and Geocryology, 27*(2), 153–162. (in Chinese).
19. The China Society on Tibet Plateau. (1995). *Collected papers of seminar on the Tibetan Plateau and global change.* Beijing: Meteorological Press. (in Chinese).
20. The Expert Panels of the Qinghai–Tibet Projects. (1995). *Research on the formation and evolution of environmental changes and ecosystem of the Qinghai–Tibet Plateau.* Beijing: Science Press. (in Chinese).
21. The Glacier Permafrost Research Room, Institute of Geography, Chinese Academy of Sciences. (1965). *Investigation of frozen soil along the Qinghai–Tibet Highway.* Beijing: Science Press. (in Chinese).
22. Tong, C. J. (1996). *Geological evaluation and treatment offrozen soil in permafrost along the Qinghai–Tibet Highway.* Beijing: Science Press. (in Chinese).
23. Wang, Y. D. (2002). Preliminary exploration of geologic route selection in multiyear tundra of Qing–Zang Railway line. *Journal of Railway Engineering Society, 2,* 57–59. (in Chinese).
24. Wang, Z. H. (2006). Active tectonics and its secondary disasters along the Qinghai–Tibet Railway. *Journal of Railway Engineering Society, Supplements, 1,* 264–269. (in Chinese).
25. Wang, Z. H. (2007). Geological environment and disasters along the railway line in the Qinghai–Tibet Plateau. *Earth Science Frontiers, 14*(6), 031–037. (in Chinese).
26. Wang, Q. Q., Qin, N. S., Tang, H. Y., et al. (2007). Study on climate change facts and their characteristics in the Qinghai Plateau in the recent 44 years. *Arid Zone Research, 24*(2), 234–239. (in Chinese).
27. Wei, G. J. (2003). Geological line selection of the Xiaonanchuan–Wangkun unfavorable geological section of the Qinghai–Tibet Railway. *Geotechinical Engineering World, 6*(7), 41–45. (in Chinese).

28. Wu, Z. W., Chen, G. D., et al. (1988). *Roadbed engineering in frozen earth area*. Lanzhou: Lanzhou University Press. (in Chinese).
29. Wu, Z. H., Ye, P. S., Wu, Z. H., et al. (2003). Hazard effects of active faulting along the Golmud–Lhasa Railway across the Tibetan Plateau. *Modern Geological, 17*(1), 1–7. (in Chinese).
30. Wu, Z. H., Hu, D. G., Wu, Z. H., et al. (2006). Migrating pingos and their hazard effects in the vicinity of the railway across the northern Tibetan Plateau. *Geological Bulletin of China, 25* (1–2), 233–243. (in Chinese).
31. Yang, Y. F., Jiang, H., Niu, F. J., et al. (2007). Space-time variation analyses of air temperature over the Qinghai-Xizang Plateau in warm and cold seasons. *Plateau Meteorology, 26*(3), 496–502. (in Chinese).
32. Ye, D. Z., Gao, Y. X., Shen, Z. B., et al. (1979). *The meteorology of the Qinghai–Tibet Plateau*. Beijing: Science Press. (in Chinese).
33. Yi, M. C., Wu, Z. H., Hu, D. G., et al. (2003). N–S-trending active structures along the Qinghai–Tibet Railway and their influences on railway bed engineering. *Journal of Geomechanics, 10*(3), 343–354. (in Chinese).
34. Yi, Z. M., Zhou, J. X., & Zhang, X. D. (2006). Analysis of hydrologic conditions along the Qinghai–Tibet Railway. *Technology of Soil and Water Conservation, 4,* 14–16. (in Chinese).
35. Yin, A. (2001). Geologic evolution of the Himalayan–Tibetan orogen. *Acta Geophysica Sinci, 22*(3), 193–238. (in Chinese).
36. Zhang, G. X., Wang, S. J., & Zhang, Z. Y. (2000). *China engineering geology*. Beijing: Science Press. (in Chinese).
37. Zhao, J. C., Liu, S. Z., & Ji, S. W. (2001). Engineering-geological problems in railway engineering construction. *The Chinese Journal of Geological Hazard and Control, 12*(1), 7–9. (in Chinese).
38. Zhou, Y. W., et al. (2000). *China frozen soil*. Science Press. (in Chinese).
39. Zhou, J. X., Yi, Z. M., Li, D. X., et al. (2007). Distribution patterns of species diversity of natural vegetation along the Qinghai–Tibet Railway. *Journal of Soil and Water Conservation, 21*(3), 173–187. (in Chinese).

Chapter 3
Geological Line Selection in Permafrost Regions

Abstract Permafrost can adversely affect railway projects and can induce major disasters, such as roadbed frost heaving and thawing settlement. This chapter discusses the principles of geological line selection in permafrost regions, the classification of frozen soil and its engineering-geological properties, and the major engineering-geological structures in the permafrost regions along the Qinghai–Tibet Railway.

Keywords Geological line selection · Permafrost classification · Engineering-geological characteristics in permafrost regions · Engineering-geological problems in permafrost regions

Of the total 1142 km length of the Qinghai–Tibet Railway, approximately 550 km passes through permafrost districts. Engineering properties of permafrost in the Qinghai–Tibet plateau are complex and unique. Located in the hinterland of the Qinghai–Xizang Plateau, the Qinghai–Tibet Railway permafrost is a mid–low-latitude, high-altitude permafrost, which is characterized by poorer thermal stability, higher ice content, and more complex geological and climatic factors compared with high-latitude, low-altitude permafrost. In addition to permafrost, the line passes over island tundra and taliks and various frozen-soil phenomena, such as thaw slumping, thaw lakes and ponds, frost mounds, ice cones, permafrost swamps, deep iced grounds, and freezing-thawing (uplift) of the active layer. Both in China and across the world, railway lines over permafrost regions tend to have a high disease rate; moreover, new permafrost diseases tend to appear after the railway has been in operation for decades. Hence, the construction of the Qinghai–Tibet railway, especially over permafrost, entailed a series of unique engineering-geological problems.

J. Li et al., *Geological Line Selection for the Qinghai-Tibet Railway Engineering*,
https://doi.org/10.1007/978-3-662-55572-9_3

3.1 Classification of Permafrost

3.1.1 Fundamentals of Permafrost

Rock soil at subzero temperature and with ice is called frozen earth, and frozen earth that remains frozen for more than 2 years under natural conditions is called permafrost. Seasonal frozen layer refers to the veneer of crust that freezes in the cold season and thaws in the warm season in regions with above-freezing annual average temperatures. The substratum of this seasonally frozen layer is either a melted soil layer or a detachment of frozen ground. Similarly, a layer with a perennial frozen-earth substratum that freezes in the cold season and thaws in the warm season in regions with subzero annual average temperatures is called a seasonal thaw layer.

The lower limit of the permafrost's largest seasonal thaw layer is called the permafrost table and is referred to as the upper bound of the permafrost. The bottom of the permafrost is called the permafrost base and forms the lower bound of the permafrost. The upper bound formed under natural conditions is called the natural permafrost table, whereas the upper bound due to the influence of human activities is called the article permafrost table. Permafrost with partial melting zones between the lower and upper bounds is called continuous permafrost, and that without is called discontinuous permafrost. The area with and without a perennial melting layer between the seasonal thaw layer and the permafrost is called connected frozen ground and detachment of frozen ground, respectively.

On the basis of stability, permafrost regions can be classified as heartland and edge zone: Heartland permafrost is the one where the frozen depth is larger than the thaw depth within a year, which imparts the region with relatively high stability. The thaw depth exceeds the frozen depth in the edge zones, and such regions are consequently unstable. The bounds of a permafrost region vary with dynamic changes in the permafrost. These bounds are influenced by various zonal factors and are thus usually not straight.

The commonly studied ground-temperature characteristics include annual average ground temperature, ground-temperature changes with depth, annual average ground temperature below the active layer, and annual maximum and minimum ground temperatures.

3.1.2 Classification of Permafrost

3.1.2.1 Based on Planar Distribution

According to planar distribution, permafrost can be classified as continuous, island tundra and talik, and patchy permafrost. Continuous permafrost is patchy in distribution, and island tundra and taliks are insular. Patchy permafrost is similar to islands in nonglacial regions.

3.1.2.2 Based on Vertical Structures

Permafrost can be classified as connected frozen ground and detachment of frozen ground on the basis of vertical structures; in the former, the permafrost table is connected to the lower limit of the seasonal frozen layer, whereas in the latter, it is not.

3.1.2.3 Based on Water Content

On the basis of the total water content, permafrost can be classified into five types: low-ice frozen soil, ice-rich frozen soil, ice-rich soil, ice-saturated soil, and soil-containing ice layer (Table 3.1).

Frozen soil with less ice is referred to as low-ice frozen soil, and ice-rich frozen soil refers to ice-saturated soil and soil-containing layer. Ice-rich frozen soil has high stability for engineering activities and thus these ice-rich regions are excellent for geological railway line selection.

3.1.2.4 Based on Annual Average Ground Temperature

The annual average ground temperature of permafrost is the ground temperature of the region where the annual range of the ground temperature is zero. Table 3.2 presents the classification of permafrost on the basis of the annual average ground temperature.

Extremely unstable high-temperature permafrost and unstable high-temperature permafrost are both referred to as high-temperature permafrost, whereas cryogenic and stable permafrost is known as low-temperature permafrost. High-temperature permafrost affects engineering stability as well as permafrost distribution and is the most significant problem in railway geological line selection.

3.1.2.5 Based on Other Factors

1. Salinity—Salinity (%) is a measure of the concentration of a soluble salt per unit volume of permafrost. Permafrost whose salinity exceeds the boundary value listed in Table 3.3 is called saline frozen ground.

Table 3.1 Classification of permafrost on the basis of water content

Soil type	Water content ω_a (%)	Moisture level after melting	Permafrost type	Remarks
Debris soil, gravel sand, coarse sand, medium sand (silty soil and clay <15%)	<10	Wet	Permafrost with less ice	Low-ice permafrost
Debris soil, gravel sand, coarse sand, medium sand (silty soil and clay <15%)	<12	Slightly wet		
Fine sand, silty sand	<14			
Silty soil	<17			
Clay soil	$<\omega_p$	Hard		
Debris soil, gravel sand, coarse sand, medium sand (silty soil and clay <15%)	$10 \leq \omega_a < 15$	Saturated	Permafrost with much ice	
Debris soil, gravel sand, coarse sand, medium sand (silty soil and clay <15%)	$12 \leq \omega_a < 15$	Wet		
Fine sand, silty sand	$14 \leq \omega_a < 18$			
Silty soil	$17 \leq \omega_a < 21$			
Clay soil	$\omega_p \leq \omega_a < \omega_p + 4$	Hard plastic		
Debris soil, gravel sand, coarse sand, medium sand (silty soil and clay <15%)	$15 \leq \omega_a < 25$	Saturated, water exit (effluent <10%)	Permafrost with rich ice	Ice-rich permafrost
Debris soil, gravel sand, coarse sand,	$15 \leq \omega_a < 25$	Saturation		

(continued)

Table 3.1 (continued)

Soil type	Water content ω_a (%)	Moisture level after melting	Permafrost type	Remarks
medium sand (silty soil and clay < 15%)				
Fine sand, silty sand	$18 \le \omega_a < 28$			
Silty soil	$21 \le \omega_a < 32$			
Clay soil	$\omega_p + 4 \le \omega_a < \omega_p + 15$	Soft plastic		
Debris soil, gravel sand, coarse sand, medium sand (silty soil and clay <15%)	$25 \le \omega_a < 44$	Saturation, heavy water exit (10–20% effluent)	Ice permafrost	
Debris soil, gravel sand, coarse sand, medium sand (silty soil and clay <15%)	$25 \le \omega_a < 44$	Saturation, water exit (effluent <10%)		
Fine sand, silty sand	$28 \le \omega_a < 44$			
Silty soil	$32 \le \omega_a < 44$			
Clay soil	$\omega_p + 15 \le \omega_a < \omega_p + 35$	Flow plastic		
Debris soil, sand soil, silty soil	$\omega_a \ge 44$	Saturation, lots of water exit (effluent within 10–20%)	Ice layer containing soil	
Clay soil	$\omega_a \ge \omega_p + 35$	Flow plastic		

Notes 1 Total water content includes ice and (unfrozen) water; *2* Saline frozen ground, the formation of frozen soil, humus, and high-plasticity clay are not listed in the table; *3* ω_p = water content at plastic limit

Table 3.2 Classification of the permafrost based on annual average ground temperature (TCP)

TCP	TCP ≥ −0.5 °C	−1.0 °C ≤ TCP < −0.5 °C	−2.0 °C ≤ TCP < −1.0 °C	TCP < −2.0 °C
Permafrost type	Extremely unstable high-temperature permafrost	Unstable high-temperature permafrost	Stable low-temperature permafrost	Stable low-temperature permafrost
Remarks	High-temperature permafrost		Low-temperature permafrost	

Table 3.3 Salinity thresholds for different types of permafrost

Soil type	Coarse grain soil	Silty soil	Silty clay	Clay
Salinity	≥ 0.10	≥ 0.15	≥ 0.20	≥ 0.25

2. Degree of peatification—The degree of peatification (%) is the specific value of the mass of plant residue and peat and the dry density of permafrost per unit volume of permafrost. Coarse-grained permafrost and cohesive soil permafrost whose degree of peatification exceeds 3 and 5%, respectively, are called peat permafrost.
3. Degree of ice cementation and rheological property—On the basis of the degree of ice cementation and the rheological property, permafrost can be classified as hard permafrost, plastic frozen soil, and noncompact frozen soil. Permafrost whose coefficient of compressibility is less than 0.01 $(\text{MPa})^{-1}$ is called hard permafrost and can be considered incompressible soil, and when the coefficient of compressibility exceeds 0.01 $(\text{MPa})^{-1}$, the permafrost can be classified as plastic frozen soil and its compressibility varies with stress, which should be considered in engineering applications over such permafrost. Permafrost with <3% coarse-particle frozen soil is called noncompact frozen soil.

3.1.3 *Classification of Permafrost Thawing Settlement*

On the basis of the thaw-settlement coefficient, the thawing settlement of permafrost can be classified into five levels: no thawing settlement, low thawing settlement, thawing settlement, high thawing settlement, and thaw collapse (Table 3.4).

High ice-content permafrost has a large thawing settlement, a factor that must be considered during railway line selection.

Table 3.4 Classification of permafrost on the basis of thawing settlement

Coefficient of thawing settlement δ_0 (%)	$\delta_0 \leq 1$	$1 < \delta_0 \leq 3$	$1 < \delta_0 \leq 3$	$1 < \delta_0 \leq 3$	$1 < \delta_0 \leq 3$
Thawing settlement classification	I (no thawing settlement)	II (low-thawing settlement)	III (thawing settlement)	IV (high thawing settlement)	V (thaw collapse)
Permafrost type	Low-ice permafrost	Permafrost with much ice	Ice-rich permafrost	Ice permafrost	Ice layer containing soil

3.1.4 Frost-Heaving Classification of Seasonal Permafrost and Seasonal Melt Layer

Per the frost-heaving ratio, the seasonal permafrost and melt layer can be classified into five levels: no frost heave, weak frost heave, frost heave, strong frost heave, and very strong frost heave (Table 3.5).

3.1.5 Regionalization of Frozen Soil Along the Qinghai–Tibet Railway

3.1.5.1 Principle of Permafrost Regionalization

The permafrost engineering-geological zoning is performed considering such varied factors as the distribution and characteristics of permafrost along the railway line, annual average ground temperature, climate, underlying surface conditions, and human activities:

1. First level ofshyam regionalization—This level of regionalization is based on the permafrost distribution (i.e., sheet permafrost regions and patchy permafrost regions). From the engineering perspective, this level has a macroscopical guiding significance on engineering design and reflects the following characteristics of the permafrost region.

    a. Distribution range and thickness.
    b. Climatic characteristics, elevation, and latitude.
    c. Topography and landform characteristics.
    d. Formation lithology and geological tectonic characteristics.
    e. Harmful frozen-ground features.

2. Second level of regionalization—The main basis of this level of regionalization is the annual average ground temperature (TCP) of the permafrost, according to which, the permafrost region along the Qinghai–Tibet Railway can be regionalized as high-temperature extremely unstable permafrost (I), high-temperature unstable permafrost (II), cryogenic stable permafrost (III), and cryogenic unstable permafrost (IV), as follows:

    a. TCP ≥ -0.5 °C: I region.
    b. -0.1 °C $\leq$ TCP < -0.5 °C: II region.
    c. -0.2 °C $\leq$ TCP < -1 °C: III region.
    d. TCP < -2 °C: IV region.

    From the engineering perspective, the second level of regionalization has a stronger guiding significance in engineering design than does the first level. It highlights the following characteristics of permafrost with different temperatures:

Table 3.5 Frost-heaving classification of seasonal frozen soil and seasonal melting layer soil

Soil type	Natural water content before freezing ω (%)	Minimum distance between underground water level and frozen surface h (m)	Average frost-heave rate η (%)	Frost-heave grade	Frost-heave type
Gravel (pebble), coarse sand, medium sand (clay particles of size not more than 0.074 is <15%)	Not considered	Not considered	$\eta \leq 1$	I	No frost heaving
Gravel (pebble), coarse sand, medium sand (clay particles of size not more than 0.074 is <15%)	$\omega \leq 12$	>1.0	$\eta \leq 1$	I	No frost heaving
		≤1.0	$1 < \eta \leq 3.5$	II	Weak frost heaving
	$12 < \omega \leq 18$	>1.0			
		≤1.0	$3.5 < \eta \leq 6$	III	Frost heaving
	$\omega > 18$	>0.5			
		≤0.5	$6 < \eta \leq 12$	IV	Strong frost heaving
Silty sand	$\omega \leq 14$	>1.0	$\eta \leq 1$	I	No frost heaving
		≤1.0	$1 < \eta \leq 3.5$	II	Weak frost heaving
	$14 < \omega \leq 19$	>1.0			
		≤1.0	$3.5 < \eta \leq 6$	III	Frost heaving
	$19 < \omega \leq 23$	>1.0			
		≤1.0	$6 < \eta \leq 12$	IV	Strong frost heaving
	$\omega > 23$	Not considered	$\eta > 12$	V	Very strong frost heaving
Silty soil	$\omega \leq 19$	>1.5	$\eta \leq 1$	I	No frost heaving
		≤1.5	$1 < \eta \leq 3.5$	II	Weak frost heaving
	$19 < \omega \leq 22$	>1.5			
		≤1.5	$3.5 < \eta \leq 6$	III	Frost heaving
	$22 < \omega \leq 26$	>1.5			
		≤1.5	$6 < \eta \leq 12$	IV	Strong frost heaving
	$26 < \omega \leq 30$	>1.5			
		≤1.5	$\eta > 12$	V	Very strong frost heaving
	$\omega > 30$	Not considered			

(continued)

Table 3.5 (continued)

Soil type	Natural water content before freezing ω (%)	Minimum distance between underground water level and frozen surface h (m)	Average frost-heave rate η (%)	Frost-heave grade	Frost-heave type
Clay soil	$\omega \leq \omega_p + 2$	>2.0	$\eta \leq 1$	I	No frost heaving
		≤2.0	$1 < \eta \leq 3.5$	II	Weak frost heaving
	$\omega_p + 2 < \omega \leq \omega_p + 5$	>2.0			
		≤2.0	$3.5 < \eta \leq 6$	III	Frost heaving
	$\omega_p + 5 < \omega \leq \omega_p + 9$	>2.0			
		≤2.0	$6 < \eta \leq 12$	IV	Strong frost heaving
	$\omega_p + 9 < \omega \leq \omega_p + 15$	>2.0			
		≤2.0	$\eta > 12$	V	Very strong frost heaving
	$\omega > \omega_p + 15$	Not considered			

Notes 1 ω_p is the plastic limit of water content (%) and ω is the average natural water content in the frozen layer before freezing; *2* Salinization permafrost is not listed in the table; *3* Frost-heave grade decreases by 1° when the plastic index exceeds 22; *4* Soil containing more than 60% clay particles of size less than 0.005 mm is classified as no frost-heaving soil; *5* If the filler weight is more than 40% of the total weight, the type of frost heaving of the debris soil is determined on the basis of the filler type; *6* Frost-heave rate (η, %) is the rate of frost heaving in terms of surface area and the thickness of the frozen layer

a. Distribution range and thickness.
b. Topography and landform characteristics.
c. Formation lithology and geological tectonic characteristics.
d. Tabetisols type, distribution range, and characteristics.
e. Cohesiveness between the permafrost and the active layers.
f. Annual average precipitation and annual average temperature.
g. Major harmful features of the frozen ground.

3. Third level of regionalization—This level is mainly based on the ice-content characteristics (i.e., low-ice, more-ice, ice-rich, ice-saturated, and soil-containing ice-layer districts). From the engineering perspective, permafrost districts can be classified as low-ice districts and ice-rich districts. This level of regionalization has great guiding significance in engineering design, especially regarding the following factors:

   a. Types of distribution ranges, thickness, thaw settlement, and annual average ground temperature.
   b. Topography and landform characteristics, climate, and vegetation characteristics.
   c. Seasonal thaw depth and depth of frost penetration of the seasonal thaw region as well as frost heave of active layers.

Table 3.6 Length of the Qinghai–Tibet Railway in each permafrost zone (*Unit* km)

First zone		Second zone		Third zone		
Flake permafrost	Island permafrost (including melting zone)	TCP zoning	Length	Section of soil-containing ice layer	Section of ice-rich permafrost	Section of low-ice permafrost
403.15	143.28	TCP-I	199.75	11.0	75.63	113.12
		TCP-II	74.51	5.8	29.34	39.37
		TCP-III	110.75	24.2	36.64	49.91
		TCP-IV	59.75	16.2	24.35	19.19
		Melting zone	101.68			
Total amount			546.43	57.2	165.96	221.59
				223.16		

d. Hydrological and hydrogeological conditions.
e. Physical, thermophysical, and mechanical conditions.
f. Major harmful frozen-ground phenomena.

3.1.5.2 Result of Permafrost Engineering-Geological Zoning

On the basis of the development condition and distribution of permafrost along the Qinghai–Tibet Railway and through of analysis of the factors affecting permafrost engineering—in particular, annual average ground temperature distribution and ice-rich permafrost distribution—a three-degree zonation is proposed for the permafrost region along the railway (Table 3.6).

The warm and cryogenic permafrost regions account for 50.19 and 31.2% of the railway length along the permafrost region, respectively; ice-rich and low-ice permafrost account for 40.84 and 40.55%, respectively, with the remaining length being over thaw areas.

Figure 3.1 is a schematic of the distribution of permafrost ground temperature along the Qinghai–Tibet Railway, which we developed by reviewing the available partitioned measurements of permafrost annual average ground temperature as well as the ice-content characteristics.

3.2 Engineering-Geological Features of Permafrost Regions

The main engineering-geological features of permafrost regions include thaw collapse, frost heave, and harmful frozen-ground phenomena, all of which can severely damage buildings and the stability of their foundations. Intense activity in the

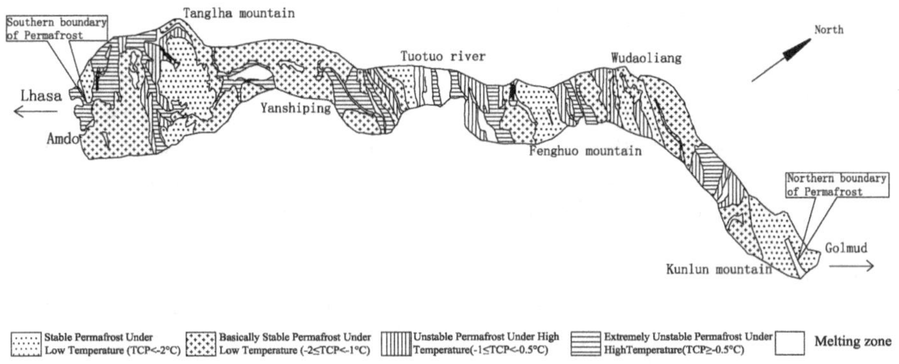

Fig. 3.1 Zoning schematic developed on the basis of the annual average ground temperature (TCP) of permafrost along the Qinghai–Tibet Railway

Qinghai–Tibet Plateau since the Late Geological period has contributed to the scattered development of strong hydrothermal activity and geothermal anomaly belts. In addition, thaw zones are present in the permafrost regions and induce deep frost penetration and strong freeze–thaw action, further contributing to the weakening of building stability.

3.2.1 Engineering-Geological Environment of Permafrost Regions

3.2.1.1 Topography

Collision between the Indian and Eurasian plates is causing continual crustal thickening of the Qinghai–Tibet Plateau. The crust of the plateau, hoiked in the ancient and the Quaternary and, is now called the "Roof of the World" and is still rising. It is only three to four million years old.

Mountain chains are spread vertically and horizontally across the Qinghai–Tibet Plateau. Hilly areas together with basins and valleys render the topographical feature in the plateau highly diverse. The northern Qinghai–Tibet Plateau area has an altitude of 4500 m and is of the high peneplain type. In the mountainous regions, such as the Kunlun, the Gangdise, and the Tonglha, except for >700 m relative height difference of the north face of the Kunlun, the mountains rise and fall similar to a quaquaversal; most of the relative height differences here are less than 300 m and is macroscopic open. Because of the exogenic geological and strong physical weathering processes, the mountains and hills are round with gentle slopes and are surrounded by numerous basins; this topography influences permafrost development.

3.2.1.2 Geological Structures

The fold belt of Kunlun Mountains, Hoh Xil–Bayan Har tectonic region, Qiangtang-southern Qinghai-Sanjiang tectonic region, northern Tibet tectonic region, and the belt of folded strata of the Himalayas are large geotectonic elements spread along the east–west direction. These tectonic regions and belts of folded strata are composed of secondary fault-fold zones or structures. The tectonic region links with secondary tectonic regions at deep fractures. Fractures in the north–east, north–west, and south–north directions intersect the main tectonic line running east–west. The aforementioned structural system is ordinal overlying. In addition, the more southern the structure, the newer it is.

In regions under the same climatic conditions, permafrost in mountainous regions has lower temperature and larger thickness because of its elevation is higher than that of permafrost in basins and valleys.

3.2.1.3 Formation Lithology

The northern Tibet area, which has drift beds and aqueoglacial deposits, has experienced five or six ice ages since the Quaternary period. Most of the drift beds are erratic blocks spread across high areas. The main constituent of aqueoglacial deposits is gravel pebble, which is distributed in relatively lower areas. The extensive glacial sedimentary facies evidence recent fluvial deposition. The strong physical weathering in the cold highland areas has contributed to the development of rock debris.

Soils of different lithology differ in their thermal properties, mechanical properties, surface-heat-transfer properties, water-holding capacity, and permeability; hence, the degree of permafrost development is nonuniform throughout the plateau. Under equal conditions, the upper limit of the permafrost over bedrock surpasses that over the Quaternary unconsolidated layer. In the unconsolidated layer, the upper limit of coarse-grained soil exceeds that of fine-grained soil. The Quaternary unconsolidated deposit in the Qinghai–Tibet Plateau is in the ground surface at depths of hundreds of meters, providing the material basis for the permafrost.

3.2.1.4 Climate and Vegetation

The region along the Qinghai–Tibet Railway is almost entirely arid and semiarid because of the surrounding mountains. Northern Tibet has a plateau continental climate, which is characterized by cold drying air, climatic variability but little seasonal change, thin air, and large day–night temperature difference. This region is frozen for more than half of the year, and the evaporation rate far exceeds the precipitation.

As the global climate has warmed since the 1970s, the annual average temperature of the Qinghai–Tibet Plateau has since increased by 0.2–0.4 °C. In particular, the winter temperature is increasing, with the annual temperature range consequently decreasing gradually. In the recent 20 years, the average ground temperature of patchy and continuous permafrost has increased by 0.3–0.5 and 0.1–0.3 °C, respectively. These changes substantially affect the permafrost formation and development.

Vegetation reflects direct sunshine and improves the water-holding capacity of the soil, thus decreasing the permafrost thaw depth. In addition, the plants that die in winter prevent soil heat loss. In general, vegetation has a cooling effect on the soil, the extent of which depends on species, coverage, and thickness of the plant and its root system. The Qinghai–Tibet Plateau has poor vegetation growth. Per the shape, vegetation on the plateau can be classified as moundy, maculosus, schistose, and squamous. Vegetation loss increases the seasonal thaw depth.

3.2.1.5 Hydrogeology

Hydrogeology is dependent on geographical factors such as regional climatic characteristics, geographical and geomorphic conditions, formation lithology, and geological tectonic characteristics. At a certain thickness, permafrost becomes completely impermeable, dividing groundwater into suprapermafrost water, infrapermafrost water, and melting-zone water. Groundwater is mainly recharged by precipitation and meltwater. Suprapermafrost water and the groundwater in frozen-soil regions are recharged by surface runoff. Because of the permafrost layer, subpermafrost water is recharged by lake meltwater. Springs can form ice cones and frost heaving in winter. Tonglha Mountains and Kunlun Mountains divide the permafrost region into three large groundwater collection basins. The hydrogeology in the permafrost is strongly dependent on the local geological structures.

The factor that most strongly influences the heat-transfer process in the plateau is the water migration caused by the moisture variation and temperature gradient in the permafrost. These factors affect the thawing settlement of permafrost and frost heave. In addition, the formation and development of deep iced ground, which affect the stability of the railway line, are strongly influenced by moisture. Apart from the several large rivers that are export-oriented, most rivers in the plateau are import oriented.

Penetrating tabetisols are present under perennial lakes and ponds. Especially in warm permafrost regions, no permafrost is present under the lakes because of low permafrost thickness. In some permafrost wetlands, the development of the bottom permafrost is rather strong because of the mulching function of stagnant water and hydrophilous plants. The resulting strong heat preservation function contributes to the development of the bottom permafrost. For example, permafrost distributed along the northern edge of the Tongtian River (DK 1254+580–DK 1255+470) has large thickness (>40 m) and is ice-rich.

Overall, the temperature of the Qinghai–Tibet Plateau is 20–25 °C lower than that of other regions at the same latitude, and the annual average temperature here is −7.0 to −1.0 °C. These temperature characteristics, together with the strong terrain incision, noncompact Quaternary debris, and appropriate natural conditions, afford the low-temperature environment that is essential for permafrost formation and preservation.

3.2.2 Factors Controlling the Engineering-Geological Conditions in the Permafrost Regions

The main factors that control the engineering-geological conditions of permafrost regions include the upper limit of permafrost, the ice content in soil and annual average ground temperature of permafrost.

3.2.2.1 Upper Limit of Permafrost

The upper limit of permafrost thawing refers to the maximum melting depth in the permafrost zone, namely, the top layer of the tundra, which is also the deepest position underneath the surface soil interface.

A thick layer of ground ice and ice-rich permafrost tends to be present in the vicinity of the permafrost. This layer is prone to collapse because of natural factors or human activities and is thus a major cause of ground deformation and failure in the permafrost region. For example, building a road necessitates slope excavation and laying of the foundation and the roadbed; however, these activities may change the permafrost table. To prevent the influence of such human activities, the construction must be carried out in the appropriate season, and appropriate insulation materials of adequate thickness must be used. Moreover, to determine the appropriate buried depth for structures such as bridges, culverts, and houses, the upper limit of the permafrost and the changes induced by the construction must be considered. Furthermore, tunnel trunks are buried deep underground; although they are far above the natural upper limit, they may induce a new melting circle after tunnel excavation. Overall, the most common undesirable geological phenomena on the Qinghai–Tibet Plateau are closely related to the upper limit of permafrost.

The formation of a thick underground ice layer is associated with the constant change in the permafrost table. The formation of such a layer results in heating thaw collapse, melting subsidence, and thaw lakes because of changes to the upper bound. The freeze–thaw condition and physics, thermal physics, and mechanical properties during the freeze–thaw process are all critical factors that must be considered in permafrost research, in particular, the layer near the upper bound. One of the key steps in studying permafrost engineering-geological problems is to locate the permafrost table and to predict possible changes to the upper bound after construction of any structure.

Factors affecting the upper bound can be classified into two categories: the nature of the medium and extraneous factors.

a. Nature of the medium—this includes factors such as soil moisture, particle size, and porosity. The thawing of the seasonal thaw layer in the permafrost area requires large amounts of latent heat. Thus, moisture strongly influences the upper bound, as demonstrated by the fact that under similar conditions, permafrost areas with higher soil moisture are thinner.
b. Extraneous factors—these mainly include factors such as temperature and ground surface condition. Absorption of solar radiant heat influences the ground temperature. In addition, the ambient geothermal temperature of the permafrost is commonly buried around 10 m below the surface, which is the bottom effect of the upper bound distribution.

3.2.2.2 Permafrost Ice Content

Permafrost engineering-geological properties are strongly influenced by the underground ice in the permafrost. The formation and thawing of thick underground ice and ice-rich frozen soil are the main causes of surface deformation in permafrost areas, heating thaw collapse, melting subsidence, thaw lakes, and damage to engineering structures. Hence, determining the ice distribution in the permafrost as well as the distribution of the ice-rich frozen soil layer is essential for engineering design. Because of the lack of a reliable and convenient test method for measuring ice content, moisture content in the total permafrost mass is often used as a proxy.

1. Distribution and characteristics

Underground water is generally most abundant in the section between the permafrost table and the annual change depth of the ground temperature along the railway, especially within areas below the upper bound. This section is close to the surface and is easily influenced by natural and human factors and is thus an important research object relevant to engineering design. The underground ice in this section is mostly composed of tectonic ice, which is created by for repeated epigenotype condensation actions. Although the distribution of the symbiotic in the ground ice is not as wide as that of the epigenotype, the symbiotic cannot be ignored as it is ice-rich. Underground ice (cold raw stereotype of frozen soil) near the upper bound is of five types: massive, banding, conglomerate, reticulated, and crevice-veinlet; the epigenetic structure of the thick underground ice layer is typically of the first three types.

Ice content in frozen soil is dependent on the local geology, hydrogeology, and thermophysical reactions. The major geological and hydrogeological factors include geotechnical composition (i.e., granularity, mineral composition, chemistry) and properties, genetic type, lithogenous level, burial condition, initial moisture content, and the aquiclude–aquifer relationship. Thermophysical reaction factors include the heat-transfer condition and geothermal gradient in the surface and ground and the presence of a frozen soil or tundra layer.

Engineering-geological explorations have revealed that thick underground ice exists in Kunlun Mountains, Chumar High Plains, Beiluhe, Erdaogou, Fenghuo Mountain, and Tonglha Mountains, usually along the gentle slopes and inter-mountain depressions and especially in areas with ice-rich cohesive soils. The distribution of thick underground ice in mountain areas differs from that in the high plains. In mountain regions (e.g., Wudaoliang, Hoh Xil, Fenghuo Mountain, and Tonglha Mountains), the underground ice is 1–4 m thick with 80% ice content (by volume), is mainly lenticular, is distributed underneath the upper bound, contains suspended bits of soil blocks, and exhibits visible or cryptic layering. By contrast, in the high plains (e.g., Chumar High Plains, Baladacaiqu, Qingshui River, and Chumar River), the underground ice is 10–20 m with 50% ice content (by volume), interbedded, and spread across valleys and depressions; irregular buried ice can also be seen as aqueoglacial deposits. Drills in the in the muddy sand and gravel layer of Qingshui District have revealed several layers of lenticular underground ice of thickness 0.2–0.5 m at depths of 2.1–5.27 m. Mudstone from Wudaoliang District contains several layers of lenticular underground ice of single-layer thickness 0.3 m at depths of 2.0–26.73 m. Cohesive soil and mudstone drilled at the exit of Fenghuo Mountain Tunnel contains several layers of vein ice of thickness 0.45–3.14 m at depths of 1.95–9.08 m. The hillsides on both sides of Beiluhe also contain underground ice of thickness up to 5.16 m (average 1.5 m; minimum 0.45 m) and ice content up to 70–80%. In sticky sand and sandy gravel from Erdaogou, banding underground ice of thickness 2.95 m appears at depths of 1.87–6.41 m. In cohesive soil and mudstone from Tonglha Mountains, banding underground ice of thickness up to 0.6 m is present at depths of 2.3–6.83 m.

2. Distribution of ice-rich frozen soil along the railway

Different types of underground ice are present in the permafrost layers along the Qinghai–Tibet Railway, which in addition with the varying soil result in differing frozen-soil tectonics. Banding ice formed through repeated condensation is the most common type; such ice layers have a shallow burial depth, forming ice-rich banding tectonic frozen soil and thick underground soil layers in the section between the upper bound and 20 m depth. The thawing of these layers is the main cause of damage to engineering structures in permafrost areas. The tectonic characteristics of frozen soil are influenced by various factors, such as ground temperature, terrain, lithology, and moisture, all of which differ widely along the railway line. Around the frozen soil and thawing area, where the soil grains are relatively coarser and the moisture content lower, permafrost is mainly massive, containing granular tectonic frozen soils with low-ice content.

In sheet permafrost regions, the condensation ice layer is spread widely, especially in lacustrine facies and diluvia plains; its burial depth is nearly identical to the permafrost Table (0.9–3.0 m, typically 1–2 m). Ice-rich frozen soil decreases with increase in depth. Generally, massive low-ice tectonic frozen soil occurs below the depth at which the ground temperature changes seasonally. The distribution and thickness of the ice layer differ by area and deposit environment along the railway.

In Kunlun Pass Basin and the Yangtze River Source High Plains (Baladacaiqu, Qingshui River, Chumar River, and Beiluhe), banding ice-rich reticulated tectonic permafrost is present in the third and fourth lacustrine fancy formation, where the ice layer is distributed fairly uniformly. These layers are usually 2 to >10 cm thick (maximum, 40–50 m), and the intervals between the ice layers is generally less than 10 cm. In 10–12-cm thick ice layers in the lacustrine fancy sand clay layer near Qingshui River at a depth of around 20.6 m, ice content in the soil layer is 30–50%. In the north shore terraces of Chumar River, ice content in the ice layer is 50% at a depth of 32.3 m.

Underground ice develops very well in the low mountain regions (i.e., Wudaoliang, Hoh Xil, Fenghuo Mountain, and Tonglha Mountains) of the Yangtze River Source High Plains. Here, its distribution characteristics are explained using Fenghuo Mountain area as an example. The burial depth of bedrock (sand and shale) here is less than 10 m in general, decreasing as one approaches closer to the upper hillsides. Thick underground ice only exists as unconsolidated formations, especially in colluvium, eluvial deposits, and mudflow deposits at the bottom of hillsides. In the middle or bottom of schattenseite and intermountain depressions, the thickness of the ice layer is in the range of dozens of centimeters to 6–7 m, and this layer is mainly concentrated between the upper bound and 8 m depth (i.e., between the bedrock and the depth at which the ground temperature changes seasonally). Ice content decreases suddenly with increase in depth. The ice layer contains sandy clay pieces and macadam suspended in ice, which makes up 50–90% of the layer volume. Compared with the ice layer in lacustrine facies, the ice layer here is thicker, the extension depth shallower, and the impurity content higher. In coarse deposits, such as water-saturated alluvial sand deposition pebble layer and residual clastic, along the railway, underground ice mainly appears as a type of cementation, often presenting as ice-rich conglomerate tectonic permafrost. In permafrost bedrock, the underground ice is occasionally distributed along crevices or as veinlets (e.g., in sand and shale), occasionally interbedding with mudstone and marlstone in a banding state.

Table 3.7 summarizes the distribution characteristics of thick underground ice in permafrost along the Qinghai–Tibet Railway.

3.2.2.3 Annual Average Ground Temperature of Permafrost

The annual average ground temperature of permafrost is the ground temperature at the depth where the annual range of permafrost is zero; this parameter is one of the major indicators used for classifying tundra regions and is an important index reflecting permafrost stability: the lower the annual average ground temperature is, the larger the cool storage capacity is. Permafrost is difficult to thaw and has high stability even when disturbed; by contrast, the higher the annual average ground temperature is, the smaller the cool storage capacity of permafrost is. Permafrost responds rapidly and is likely to thaw in responses to changes in various factors, including climate, vegetation, and human activities. Ground temperature changes

Table 3.7 Distribution of thick underground ice in permafrost along the Qinghai–Tibet Railway

No.	Mileage	Length (m)
1	DK986+030–DK986+170	140
2	DK989+000–DK989+370	370
3	DK992+550–DK992+740	190
4	DK992+820–DK992+960	140
5	DK1033+870–DK1035+700	1830
6	DK1035+880–DK1036+180	300
7	DK1036+670–DK1037+700	1030
8	DK1042+150–DK1042+610	460
9	DK1061+100–DK1061+850	750
10	DK1063+030–DK1064+150	1120
11	DK1064+740–DK1065+500	760
12	DK1079+560–DK1080+000	440
13	DK1080+200–DK1081+600	1400
14	DK1084+070–DK1085+250	1180
15	DK1091+900–DK1093+050	1150
16	DK1093+080–DK1097+800	4720
17	DK1097+880–DK1100+200	2320
18	DK1100+460–DK1105+970	5510
19	DK1110+460–DK1111+800	1340
20	DK1150+650–DK1152+435	1785
21	DK1169+820–DK1171+130	1310
22	DK1194+300–DK1194+600	300
23	DK1241+210–DK1244+600	3390
24	DK1257+000–DK1259+510	2510
25	DK1263+700–DK1267+260	3560
26	DK1272+300–DK1273+550	1250
27	DK1337+600–DK1337+900	300
28	DK1394+500–DK1394+650	150
29	DK1398+350–DK1398+920	570
30	DK1405+960–DK1406+400	440
31	DK1413+700–DK1414+950	1250
Total amount		41,965

with time and depth; therefore, it not only reflects the development and evolution of the frozen soil but also the current characteristics. Accordingly, the annual average ground temperature of permafrost is a factor that strongly influences engineering design principles.

In general, regions classified as an unstable or extremely unstable hypothermal region on the basis of their average ground temperature are more likely to geological defects in their ice-rich permafrost, which adversely affect railway engineering in such regions.

1. Factors influencing annual average ground temperature

The major factors that influence the annual average ground temperature in permafrost regions are altitude, latitude, climatic condition, geological tectonics, landform, surface water, vegetation, surface swamp, and formation lithology.

a. Altitude and latitude

Altitude and latitude determine permafrost formation and distribution. In general, permafrost stability decreases with increase in altitude and latitude.

b. Climatic condition

Climatic condition, especially air temperature (which in turn varies with the altitude and latitude), strongly and directly influence ground temperature. Usually, annual average ground temperature of permafrost is high in regions with high annual average air temperature.

c. Geological tectonics

Geological tectonics influences geothermal heat flow, which in turn is strongly related to ground temperature. In ancient geological tectonic regions with relatively high stability, the underground heat flow is low, whereas it is high in new geological tectonic regions with high tectonic activity. The Qinghai–Tibet Railway passes over different structural systems and through various tectonic zones, which are active in the Late Geological period. Groups of cold and hot springs are distributed in groups in regions with high fault density and intense hydrothermal activity, which increase the ground temperature of permafrost, even forming banding melting zones. This phenomenon can be seen in Xidatan, Budongquan, Fenghuo Mountain, and Wuli Mountain and around hot springs.

d. Landform

Land slope and gradient affect the ground temperature. Along the Qinghai–Tibet Railway, the annual average ground temperature of southern slopes is usually higher than that of northern slopes; similarly, the top regions of the slope are colder than are the bottom regions. For example, per drill DFZ-2, the annual average ground temperature of Fenghuo Mountain hillside is −3.00 °C, and drills GTZ-81 and CYZ-13 have revealed that the annual average ground temperature of the top region and southern slopes is −3.25 and −2.04 °C, respectively. Similarly, at Tonglha Mountains, drills CTZ-29 and CTZ-30 have shown that the annual average ground temperature of the northern and southern slopes is −2.15 and −0.06 °C, respectively. The trend holds at Kunlun Mountains as well, where the annual average ground temperature at the groove center (drill DZ-246) was 1.5–1.6 °C higher than that around the groove center (DZ-244 and D6Z-248).

e. Surface water

Surface water facilitates soil freezing through temperature storage and attemperation and thus affects the annual average ground temperature distribution of permafrost. Along the Qinghai–Tibet Railway, shallow and surface water protects permafrost by keeping the ground temperature low. Open talik have developed under perennial rivers with heavy flow, such as Ulan Moron, Tongtian River, and Buqu. Hyperthermic and extremely unstable permafrost regions have developed over the taliks, where the annual average ground temperature is higher than that of regions without surface water. The annual average ground temperature of permafrost decreases gradually with increasing distance to the talik.

f. Vegetation and surface swamp

In regions with vegetation, the annual average ground temperature of permafrost is lower than that in regions without, as evidenced in drill CTZ-40 at the southern boundary of the permafrost along the Qinghai–Tibet Railway. Permafrost islands develop in thickly forested regions. The annual average ground temperature of bare and rough surfaces is higher than that of turfed surfaces.

Surface swamps develop in regions with high water accumulation on the surface. Thicker peat and peat layers often exist in swamp districts, which can help maintain a low ground temperature. The annual average ground temperature below in such regions is usually lower than that in regions that have undergone paludification. For example, the large surface swamps near Kaixin Mountain pass keep the local annual average ground temperature low.

g. Formation lithology

Generally, under the same conditions, the annual average ground temperature of bedrock is lower than that of loose soil. Within loose soil, the annual average ground temperature of coarse-grained soil (gravel soil) is higher than that of soft sandy clay (cohesive soil and sandy soil). In the Tumengela formation south of Tonglha Mountains, the annual average ground temperature differs by 0.5–0.7 °C because of lithological variations.

2. Annual average ground temperature distribution along the railway

The annual average ground temperature of permafrost is closely related to such factors as the properties of the geomorphic units. On the basis of the landscape and the ground temperature characteristics, permafrost regions can be divided into three classes:

The first class is the middle and high mountain distribution region, where the annual average ground temperature of permafrost is relatively low. They are mainly spread across Kunlun Mountains (DK973+700–DK1005+500), Hoh Xil Range (DK1072+500–DK1124+500), Fenghuo Mountain (DK1145+500–DK1165+500), Wuli Mountain (DK1202+500–DK1217+700), and Tonglha Mountains (DK1394 +800–DK1485+200).

The second class is high plains, valley, and basin distribution region, where the annual average ground temperature of permafrost is relatively high. They are mainly spread across Chumar High Plains (DK1005+500–DK1072+500), Beiluhe Basin (DK1124+500–DK1145+500), (Tuotuohe Basin DK1217+700–DK1245

+000), Tongtianhe Basin (DK1252+800–DK1282+800), and Buqu Valley (DK1282+800–DK1360+800).

The third class includes permafrost islands and taliks in consistent permafrost. They are mainly distributed in Xidatan Island, Chumar High Plains (DK1005+500–DK1072+500), Tuotuohe Basin (DK1217+700–DK1245+000), Tongtianhe Basin (DK1252+800–DK1282+800), Buqu Valley (DK1282+800–DK1360+800), and hot spring fault basins (DK1360+800–DK1394+800). The annual average ground temperature is highest in this class.

3. Influence of annual average ground temperature on permafrost engineering-geological conditions

The variation of annual average ground temperature in permafrost regions, in particular, the increase of temperature in the recent years, strongly influences the engineering-geological conditions of permafrost regions. The increase in the annual average temperature may turn the continuous frozen soil discontinuous in the horizontal direction and degrade island permafrost into seasonal frozen soil. Similarly, in the vertical direction, convergent permafrost may be degraded to undovetail permafrost. As this degradation proceeds, the range of high-temperature permafrost and taliks widen. With increase in the thickness of the active layer and the expansion of the talik, the reservoir, storage, and movement conditions of groundwater change sharply, and the engineering and mechanical properties of frozen soil also change with increase in ground temperature. These changes adversely affect the construction and maintenance of railway lines.

3.2.3 *Major Engineering-Geological Problems in Permafrost Regions*

Frozen soil is a type of rock soil, and its composition and thermophysical and mechanical properties differ from that of normal soil. The main engineering-geological problems in permafrost regions are thaw collapse, frost heaving, and permafrost with poor geology.

3.2.3.1 Thaw Collapse

Thick underground ice or an ice-rich permafrost layer is always present near the permafrost table. Because of its shallow burial depth, this layer is easily affected by natural factors and human activities, resulting in thaw settlement. This phenomenon, also known as thaw collapse, is the primary reason for the settlement of foundation in permafrost regions. Thaw collapse in permafrost with low ice content is less obvious than that in ice-rich permafrost (Fig. 3.2).

Fig. 3.2 Thaw collapse in Beilu River

3.2.3.2 Frost Heave

Frost heave refers to the ground-heaving phenomenon due to water in the seasonally active layer freezing and expanding in the cold season. Frost heaving is highly destructive to engineering structures in permafrost regions. It occurs not only in the seasonal thawing layer but also in the seasonal freezing layer buried deep in the ground.

Typically, in stable high-temperature permafrost regions, the thickness of the seasonal thawing layer is relatively low; this region experiences two-way freezing at a low freezing rate with low moisture migration and relatively light frost heaving.

In unstable high-temperature permafrost regions, the thickness of the seasonal thawing layer is relatively high, and the freezing rate is low. Moreover, in the presence of fine soils and adequate moisture recharge, the high water migration may lead to severe frost heaving.

In permafrost thawing zones and seasonal frozen-soil areas, frost heaving may occur if the burial depth of the groundwater is shallow and if the soil is susceptible to freezing. In addition, uneven frost heaving may occur if the roadbed filling materials are uneven or if roadbed transition sections with different lithology and hydrogeological conditions are designed inappropriately. Furthermore, owing to moisture migration during freezing in roadbeds filled with clay or silt, frost heaving may occur due to ice accumulation.

3.2.3.3 Harmful Frozen-Soil Phenomena

Harmful frozen-soil phenomena, also known as unfavorable cryogenic phenomenon, is the formation of new small or medium-sized terrain because of the adverse effects of freezing and thawing of soil, such as thick underground ice, ice cones, ice mantles, frost-heave mounds, solifiuction, thaw slumping, thaw lakes, and permafrost bog. Several harmful frozen-soil phenomena occurring along the Qinghai–Tibet Railway are briefly described herein.

1. Thick underground ice layer

It forms because of the migration of vast amounts of water to a frozen peak surface due to the cold climate, abundant moisture recharge, and slow freezing rate during permafrost formation. Thick underground ice refers to underground ice whose thickness exceeds 0.25 m and to multiple ice layers of total thickness exceeding 0.25 m separated by intervals of 2–3 cm. Moreover, per the Provisional Rules of Permafrost Region Reconnaissance, underground ice layers of total thickness exceeding 0.25 m with 1 cm intervals can be considered thick underground ice, mainly because the melting of such layers would result in severe deformation and destruction of engineering structures. Of all harmful frozen-soil phenomena, thick underground ice is the one that is most sensitive to temperature variation and the one that affects railway roadbed construction the most; thus, regions with such layers should be avoided.

Thick underground ice of thickness <10 cm to several meters is distributed in the gentle slopes with good vegetation in Kunlun Mountains, Hoh Xil Mountains, Fenghuo Mountain, Kaixinling, Tonglha Mountains, and Touerjiu Mountain. The maximum thickness of the thick underground ice layer in Fenghuo Mountain is 4 m; its long-term strength and permissible load tend to zero. Because of its high sensitivity to temperature variation, artificial disturbances and roadbed excavation can result in the rapid thawing of such layers. Therefore, thick underground ice, which is difficult to avoid over large distances, creates great difficulties in the construction and maintenance of railway roadbeds.

The formation and melting of thick underground ice is the main reason for surface deformation, thaw slumping, heating subsidence, and the formation of thaw lakes and consequently for damage to engineering structures. In the permafrost region through which the Qinghai–Tibet Railway passes through, underground ice is abundant 0.5–1.0 m below the upper bound. Because this ice layer is close to the surface and is easy to be influenced by natural factors (e.g., temperature) and artificial factors (e.g., railway construction), it has been a key research object in the field of railway engineering and construction.

2. Ice cone and ice mantle

An ice cone, commonly known as salivary water, is the surface ice formed from repeated flow of underground water onto the surface. Ice cones differ in their shape and size: some are ice cones of diameter 2–3 m, whereas some present as ice slopes or ice mantles, which are ice cones spread across large areas on the order of several square kilometers, extending tens and even hundreds of meters with occasional overflows.

Pitons are widely distributed in permafrost regions, often appearing in wash-lands, terrace edges, alluvial front edges, and piedmont zones, because large volumes of groundwater flow on to the surface in these districts. When a frozen melt layer forms over a talik in the winter, the groundwater pressure increases, forcing the water onto the overlying land. This overflow freezes gradually in the form of a taper, extending along the original underground runoff and eventually forming an ice cone. Ice cones are not restricted to permafrost; they can form in any cold place with a nonfreezing water source (e.g., water channels and overflows). Depending on the water source, ice cones can be classified into three types: river, lake, and spring piton. In permafrost regions, ice cones are mainly distributed in the permafrost and seasonal frozen-soil regions, and the vast majority of ice cones are seasonal.

Accordingly, ice cones are widely distributed along the Qinghai–Tibet Railway, mainly in Budongquan Valley and the shoaly lands, low mountains, and hills north of Kaixinling and Tonglha Mountains. They are more prevalent in the southern regions than in the northern ones, in mountainous and hilly areas than in the high plains, in washlands more than in terraces, and in mountain passes more than on hillsides. In addition, if any structure blocks groundwater flow channels or if drainage channels are not appropriately designed, ice cones can form in the vicinity of such structures, endangering their structural integrity. Similarly, if the line passes over an aquifer, excavation can lead to the formation of ice cones, resulting in heavy structural damage due to rift-water accumulation.

3. Frost mound

Frost-heaving mounds (Fig. 3.3) are conical uplift hills with an icy core, formed by differential frost heaving. They are typical cold-season adverse geological phenomena in permafrost areas and can be used to estimate the quality of the permafrost and groundwater and underground ice conditions. Frost mounds inflict considerable damage on the railway roadbed and other engineering structures, as they generate a large heaving force, causing severe roadbed deformation; in addition, because frost-mound uplift can raise the embankment, frost mounds may tilt piers around them.

Fig. 3.3 Frost mound

The occurrence and development of frost mounds are strongly dependent on the environmental factors and human activities. For example, the type of groundwater strongly influences the frost-mound heaving, as frost mounds are always present in suprapermafrost water regions and regions around springs. The topography of and soil in the seasonal thawing layer also influence the formation of frost mounds and ice cone. Moreover, when engineering activities change the groundwater activity, frost mound often form around the built structures. For example, near frost-heaving mounds or in slope areas, new frost mounds are induced along the two sides of a road because the construction of the road changes the hydrogeological conditions. Therefore, adequate drainage measures and other necessary measures should be implemented when building engineering structures in such areas.

According to the material composition, frost-heaving mounds can be classified as clay frost mounds, coarse-grained frost mounds, mudrocks, and gravel frost mounds. According to the water source, they can be categorized as permafrost mounds recharged by infrapermafrost water and frost mounds recharged by suprapermafrost. Frost mounds recharged by infrapermafrost water are typically large, with diameters on the order of tens of meters and heights as high as 10 m. In the warm season, water overflowing from frost-mound cores and through the surface cracks forms springs. By contrast, frost mounds recharged by suprapermafrost water are typically small, with diameters on the order of a few meters (mostly) to up to 10 m, with heights less than 1 m. These mounds completely disappear in the warm season.

According to their generation and seasonality, frost-heaving mounds can be classified as seasonal frost mounds and permafrost mounds. Permafrost mounds are mainly distributed in Kunlun Mountain area, and they tend to be relatively large and have high damage potential. Therefore, the Qinghai–Tibet Railway line avoids passing over permafrost mounds, such as at of DK982+400, where the line lies 180 m to the right of such a mound. Permafrost mounds are also present along the slopes in Chiqu Valley. Annual frost-heaving mounds are mostly distributed in Budongquan Valley, Chumar High Plains, and Tuotuohe Basin, such as at DK1007 +096–DK1007+478, but these mounds are small are relatively harmless; the line passes over five such mounds and in the vicinity of ten mounds.

4. Thaw slump

On slopes with thick underground ice, the underground ice may be exposed because of human activities (e.g., engineering construction and soil excavation) or natural factors (e.g., river wetting the slope toe). In the melting season, the pit wall of the underground ice melts, resulting in the soil thawing; consequently, the soil collapses under the action of gravity. The collapsing materials cover the toe, exposing ice on both sides as well as new underground ice. Subsequently, the newly exposed ice melts, resulting in a new collapse. This repeated collapsing phenomenon is named thaw slumping (Fig. 3.4).

Fig. 3.4 Thaw slump

Depending on the development stage and the damage potential, thaw slumping can be classified as stable and active. Stable thaw slumping is caused by natural and artificial actions; such slumps eventually stabilize.

5. Thaw swale and thaw lakes

Natural or human activities (e.g., deturfing and deforestation) can extend the depth of the seasonal melting depth, resulting in the melting of underground ice and the partial melting of permafrost, which in turn causes surface subsidence and a negative terrain; this phenomenon is called thaw swale. When the underground ice melts and the melting (or surface) water gathers in the swale, the resulting lakes are called thaw lakes (Fig. 3.5).

Thaw lakes can induce damaging phenomena such as uneven frost heaving and subsidence of the subgrade and slope rifting. Moreover, waterlogging, which affects the heat balance of the frozen soil, can easily make the subgrade wet and soft, further aggravating the frost heaving and subsidence.

Thaw swales and thaw lakes develop mainly on transverse slopes in high plain regions when the temperature is less than 3 °C. Thaw swales can also develop in the presence of excess water on the upper frozen layer and when atmospheric precipitation exceeds evaporation. Thaw lakes can have diameter as large as tens or hundreds of meters but with a shallow depth of <2 m; they often cluster together across small areas as circles or ellipses.

Nearly 100 thaw lakes are distributed in Chumar High Plains, Hoh Xil Mountains, Tongtian River Basin, and Buqu Valley. Approximately 700 m of the Qinghai–Tibet Railway line passes over 13 thaw lakes, passes within 50 m of 31 thaw lakes (which affect the roadbed severely), passes within 50–100 m of 23 thaw lakes (which affect the roadbed moderately), and passes within 100+ m of 32 thaw lakes (which have little effect on the roadbed). The line uses bridges to pass over five large thaw lakes, and fills are used to pass over the rest.

Fig. 3.5 Thaw lakes

6. Permafrost swamp and permafrost swamp wetland

Permafrost swamps and permafrost wetlands are mainly distributed in high plains and mountain valleys with relatively flat landform and poor surface water drainage. Wetlands of diameter tens or hundreds of meters are formed by ice that melts in the warm season. The lithology is mainly fine soil in the Quaternary loose layer, with lush vegetation. The presence of a peat layer with high humus content and adequate moisture often leads to the formation of a permafrost swamp. Freezing and thawing grass mounds of various size and height (but typically 50–70 cm in diameter and 30–60 cm high) develop within permafrost swamp wetlands. Frost cracks are common between grass mounds. Such wetlands are mainly distributed in Fenghuo Mountain, Kaixingling, Tunggula Mountain Basin, Buqu River Valley, and Amdo Valley.

Permafrost swamp areas promote the development of a thick underground ice layer. Permafrost wetlands tend to cluster together in Buqu River Valley and the Tonglha Mountain region. Approximately 17 km of the Qinghai–Tibet Railway line passes through a total of 41 permafrost swamps. Fills are used along most of the line, but bridges are used at two instances.

3.3 Effects of Permafrost on Railway Engineering

Permafrost is extremely sensitive to moisture. External loads (e.g., embankments) can redistribute the moisture in the frozen soil, which together with the pressure changes the structure and density of the soil particles in the frost. When the frozen soil melts, the soil consolidates under its own weight as well as the external loads, leading to sedimentation. The active layer in the permafrost region undergoes thawing and freezing over the course of each year, leading to periodical frost heave and thaw settlement of the foundation and subsoil. This phenomenon induces a series of unique engineering-geological problems that strongly influence the railway construction and the project stability.

The construction of the roadbed changed the natural hydrothermal condition of the permafrost, consequently affecting the hydrothermal equilibrium and mechanical stability of the permafrost and causing severe problems such as frost heaving, thaw and hole collapse, freezing–thawing, and frost boiling, frost heaving and thaw settlement of abutments and culvert foundations, bridge jumping, frost cracking of the tunnel lining, and ice caves. This section discusses the effect of these problems on the design and construction of railway projects.

3.3.1 Frost Heave and Thaw Collapse and Frost Boiling of Roadbed

Frost heaving and thawing settlement is mainly due to ponding or inadequate roadbed height; this phenomenon leads to the heat thawing of the permafrost under the roadbed and uneven subsidence of the subgrade. During seasonal thawing, layer deposition is dependent on the cohesiveness of the soil, and large frost heaving occurs because of the subgrade thaw water. The polyethylene ice phenomenon, which occurs during refreezing, often produces large frost heaving, which during the melting season further deteriorates the thaw subsidence. This repeated freezing and thawing deformation, which is difficult to control, severely affects the roadbed stability and account for nearly 85% of the roadbed diseases along the line.

Subgrade construction alters the water heat exchange conditions of the surface layer, damages the surface soil, and changes the depth of the natural upper limit of the permafrost. Road embankments increase the thermal resistance, raising the natural upper limit of the permafrost and inducing the melting of some frozen soil, leading to the drainage and consolidation of the natural foundation soil. During the warm season, when the embankment height exceeds a threshold in the high-temperature permafrost zone, the thawing soil creates an embankment nucleus, resulting in the ice melting of the ice within the permafrost and consequently the sinking of the embankment. When the embankment is too high (e.g., east–west embankments), the effect of slope on the upper limit of the permafrost increases, which may lead to the uneven embankment settlement. Subgrade construction also alters the surface-water and groundwater runoff conditions, and inadequate drainage measures worsen subgrade ponding. The thermal effect of the water body reduces the upper limit of the frozen soil and melts the underground ice, resulting in roadbed subsidence. Furthermore, adverse permafrost phenomena that harm the roadbed formation may develop in some cases—such as when the height of the roadbed is lower than the minimum height required for permafrost stability, when no special treatments are done, when the necessary protective and drainage measures are not implemented, when the design is inappropriate, when soil and grass are removed, and when human activities further damage the natural upper limit of the permafrost.

Railway subgrades in permafrost regions are susceptible to the following frost-heaving diseases:

1. A rough subgrade surface together with thawing-induced ponding can induce frost heaving, typically less than 25 mm, generally 30–40 mm, and occasionally more than 50 mm; 30–50 mm frost heaving under the subgrade usually begins in mid-October and stabilizes by the end of November.
2. In the presence of ponding, an unclean (i.e., mixed with dirt) or severely polluted gravel or ballast bed can produce frost heaving. When the mudstone content is 20–50%, frost heaving can be as high as 10–20 mm. The frost heaving of ballast beds starts in mid-October and stabilizes by early November.
3. Surface water- or shallow groundwater-induced uneven wetting of the subgrade soil can induce frost heaving.
4. Uneven embankment filling an inhomogeneous cutting embankment soil can induce various frost-heaving diseases because of the differences in the soil properties and structure.
5. Dissimilar roadbed conditions can induce uneven frost heaving. For example, if the railway line runs east–west, the differences in the adret and ubac roadbeds, the moisture content of the embankment fills in the winter, and the freezing depths result in unilateral frost heaving.

In addition, subgrade frost heaving and thaw settlement often occurs in areas susceptible to thick underground ice, ice cones, and frost mounds. Thick underground ice is highly sensitive to temperature variations, and artificial disturbances (e.g., subgrade excavation) can cause it to melt rapidly, severely affecting the construction and maintenance of the subgrade.

If ice cones cover the surface during the winter, they can cause subgrade frost boiling and subsidence in the summer. Ice cones can damage the subgrade in two ways: they can intrude the subgrade basement, causing frost heaving and thaw collapse, or they may block the groundwater, especially around engineering structures. If the permafrost under the embankment blocks natural water flow, ice cones can develop on the upstream side. Similarly, if the water channel is blocked or altered by excavation, the water can flow on to the road surface and form ice cones on the cutting slope, affecting the functionality of the road. The tremendous heave force of frost-heaving mounds can severely deform roadbeds.

Freezing–thawing and frost boiling, caused by the freeze–thaw cycle, are a unique frost-boil phenomenon that affects subgrade beds in cold regions. Subgrade beds with high water content but low water-carrying capacity are susceptible to soft deformation under the effect of train vibration loads, because of which slurry may leak out along the road in the shape of a mushroom. In cold regions, because of the long-term effects of subzero temperatures in winter, moisture in the soil may migrate to the upper soil, leading to ice accumulation and an increase in the humidity of the upper soil. The bed bottom does not melt fully during the spring thawing period, whereas the water content in the upper layer reaches or exceeds the liquid limit. When this excessive moisture cannot be discharged in a timely manner, bed-frost boiling may occur; this roadbed disease is unique to permafrost regions and is also called freezing–thawing and frost boiling.

This phenomenon may last for a short while, but if can affect railway safety if a train passes over it directly.

3.3.2 Cutting Slope Collapse and Mudflow

Permafrost affects the cutting slope in two aspects. First, in the presence of shallow but thick underground ice on the cutting slope, permafrost partially melts because of various anthropogenic actions during the construction and operation of the railway, consequently leading to the subsidence of the overlaying soil. As subsidence intensifies, the cutting slope collapses, inducing mudflow. Solifluction and heating thaw collapse might not only bury roads and block bridges and culvert but also accelerate roadbed melting and collapse.

Second, when the cutting slope blocks the groundwater flow, groundwater flows out of the cutting slope and freezes, resulting in the formation of water balloons at the bottom, hanging ice on the slope, and even ice-curtain tracks, all of which severely hinder the operation of the railway. These phenomena, which mostly occur on slopes with thick layer underground ice, can be triggered by human activities (e.g., excavation soil) as well as natural factors. When the line passes through hilly and mountainous areas, the soil sampling sites must be located far from the roadbed. In addition, eliminate the seeper to avoid damaging the natural ground during sampling.

3.3.3 Frost Heaving and Thaw Collapse, Deformation and Cracking, and Bridge Jumping of Bridge Foundation

Owing to the cyclical characteristics of the freeze–thaw process, permafrost affects bridges differentially. For example, bridges in permafrost regions are affected by diseases almost always at their foundations, as frost-heaving uplifts foundations in the cold season and the melting in the warm season sinks it. Uneven settlement and frost heaving and thaw collapse of the foundation can induce various phenomena such as bridge pier abutment cracking, leaning, collapsing, and bridge jumping, all of which can severely endanger routine railway operations and bridge stability. Rivers, riverbeds, and the bedrock underneath frozen soil are most susceptible to frost heaving, especially in the presence of groundwater in the riverbed. Bridge construction may affect the underground runoff, which in turn may reduce the flow section areas and thus increase the probability of frost heaving mounds. Uplift of frost-heaving mounds can also heave the embankment, and frost-heaving mounds near bridge piers can result in pier tilting, abutment tilting, and collapse.

Large differential settlement at the interface of abutment structures and the roadbed behind the abutment together with steep slopes or dislocations near the back wall of the abutment alters the profile of the roadbed. Trains moving at high speed over such uneven profiles may experience a "jumpy" ride; this phenomenon, called bridge jumping, is severe in frozen-soil regions and may be triggered by multiple mechanisms: the uplift or settlement of the abutment and transition zone of a stable bridge foundation; the frozen uplift and thaw settlement of an abutment of a stable bridge foundation and transition zone; the frost heaving and thaw settlement of the bridge foundation when both ends of the abutment and transition zone are stable; and uneven frost heaving and thaw collapse of the bridge foundation and its abutment in the roadbed transition zone.

The effects of permafrost on bridge engineering in the permafrost regions along the Qinghai–Tibet Railway cannot be ignored because the lines use around 53 extra-large bridges, 260 large and medium bridges, and 138 small bridges.

3.3.4 Frost Thawing and Boiling of Culverts and Basements and Slump at the Inlets and Outlets

More than 660 culverts of different forms and sizes, with a total length of 13,266.03 m, are present in the permafrost regions along the Qinghai–Tibet Railway. Culverts can induce many diseases in permafrost regions because their construction has a large footprint. For example, both culverts and basements can be affected by strong and repeated freezing–thawing, and this deformation is bidirectional. Some sections of culverts, such as inlet and outlet sections and their wing-walls, are more susceptible to thaw settlement than to frost heaving, whereas the middle section is more susceptible to frost heaving uplift. The common culvert deformations in permafrost regions are as follows:

1. Cracking and subsidence of the culvert pier—Along the flow direction, culverts are usually more than 10 m in length, with integrated and subsectioned basements. Because of their shallow burial depths, culvert basements can be damaged by the strong effects of the freeze–thaw cycle (i.e., frost heaving in the cold season and thaw settlement in the warm season), which induce such deformations as sinking and inclining. If this deformation is not adequately addressed, water passing through the culvert may permeate through the bedding layer of the culvert bottom and the roadbeds surrounding the culvert pier. This phenomenon carries the latent heat of the water into the perennial tundra, which in turn affects the hydrothermal balance of the subsoil. Consequently, the settlement and cracking of the culvert pier intensifies and finally destroys the culvert body.
2. Damage to culvert inlets and outlets—Deformation of the culvert inlets and outlets, such as cracking of the headwall and wing-wall and the leaning or sinking of the wing-wall, is among the most severe diseases in culvert engineering in permafrost regions.

3. Damage to the culvert bottom pavement—Damage to culvert bottom pavement mainly manifests as paving fracture in the pavement, settlement of soil and foundation, and frost-heave cracking and crushing failure of pavement.

Culvert construction severely damages the perennial frozen soil and vastly alters the natural steady state, destabilizing the frozen-soil layer. In particular, 1 or 2 years after construction, the culvert basement exhibits large deformation because of the large temperature variations. In addition, some culverts undergo freezing damage early in the construction stage if the culvert axis is such that sunlight is nonuniformly incident on the culvert, which in turn leads to a large temperature difference around the axis. For example, the ratio of the annual maximum thaw depths on the east and west side of the culvert of National Highway 10 is 1:0.5–0.85, as the culvert section under the road has a relatively low temperature. The annual maximum thawed depth along the culvert axis exhibits a concave distribution: asymmetrically large at both ends and small in the middle.

Runoff in the culverts induces large temperature differences in the culvert basement and thus worsens freeze damage. Typically, if the runoff period is long or if the runoff is high, the damage is likely to be severer, but ventilative culverts, which carry runoff intermittently (e.g., after precipitation) are less prone to such damage. For example, the runoff period at the peak of Tonglha Mountains lasts nearly 180 days; this long runoff period has damaged 90% of the culvert body.

3.3.5 Cracking Due to Frost Heaving and Freezing of Tunnel Lining

Rail tunnels in perennial frozen-soil areas are subject to a series of problems different from and more challenging that those encountered in nonfrozen-soil areas. Tunnels alter the direct sunlight radiation conditions and increase convective heat transfer. After construction, tunnels develop a new frozen layer irrespective of whether permafrost rocks surround the tunnel. In regions with frozen soil, the freezing period of the tunnel lasts 100–210 days, with a wall-rock reiterative freezing and melting depth of at least 1.0 m, which includes the lining layer. This deteriorates the strength of both the lining layer and the wall rocks, particularly the lining layer of saturated concrete; this loss cannot be ignored, especially for weathering wall rocks such as schist, shale, and other terrene surrounding rocks. The maximum strength loss generally exceeds 20% but can reach as high as 50% after two or three freeze–thaw cycles.

Weathered rocks are common around tunnels in the Qinghai–Tibet Plateau. The wall rocks behind the lining generally freeze every year. Water-bearing wall rocks are susceptible to frost heaving, which imposes a freeze-heaving force if the heaving is restrained by the lining layer and the wall rocks. The water temperature in wall-rock aquifers in frozen-soil areas is generally low, 6–8 °C in the mid-section of seasonal frozen soil, 4–6 °C deep in seasonal frozen soil, and 0–4 °C in perennial frozen-soil areas. This low temperature facilitates freezing as the water seeps into the tunnel or amasses in ditches.

The major diseases of railway tunnel in permafrost regions are lining cracking and wall-rock water icing due to frost heaving, icing due to water seepage and icing of the spring snowmelt, frozen drains, snow blocking the exits, and hanging and lateral ice at the exits.

The frost heaving-induced cracking of the lining results in lateral, portrait, and heterotopic damage cracks as well as the dropping, loosening, and collapse of blocks of various sizes. Moreover, percolating water cause numerous ice-induced damages, such as silver thaw, ice cones, ice accretions, roadbed flooding, and icebergs, that last at least 7–8 months every year. These damages block or slow the traffic and may even destroy the tunnels. Furthermore, the blind shaft and blind drain sets on both sides of the tunnel lose their functionality when water freezes in them because of their burial shallow depth or proximity to the lining layer.

Precipitation in the Qinghai–Tibet Plateau is low, mostly less than 300 mm a^{-1}. However, because of snowdrift, thick snow may accumulate in the low-lying regions, especially in mountainous areas, where precipitation increases with altitude. For example, precipitation outside tunnel inlets and outlets in the Qinghai big board mountain is more than 800 mm a^{-1}, whereas the maximum thickness of snow exceeds 3 m. Similarly, tunnel inlet and outlet cutting can also create conditions that facilitate snow accumulation and can severely affect the train traffic if the accumulated snow is not cleared. In addition, during the spring melting period (March–April), meltwater flowing into the cutting and tunnel along the slope from the cutting slopes freezes at night, leading to ice-induced diseases.

Water-bearing surrounding rocks around the tunnels in permafrost regions may bring about sudden water outbursts in the cold season (especially November–December) as these rocks (as well as the concrete lining) are affected by the frost layer, which gradually increases in depth as the cold season progresses. Consequently, the water pressure increases, and the confined water can burst out at any time from the partial melting zone or from the weak spots in the lining. During an outburst, a large volume of water flows out for a few hours to up to 1–2 days, following which the flow shrinks rapidly. The outburst water usually freezes in situ. Water outbursts can also occur if the frozen layer is heated rapidly, which reduces the intensity of the frozen layer and thus creates conditions that facilitate the pumping out of confined water. This type of outburst typically occurs in March–April, when the tunnel temperature is almost below zero, meaning that the outflow can trigger other cold-weather diseases.

3.3.6 *Effects of Permafrost on Stations and Other Railway Engineering Structures*

Building railway stations in permafrost regions entails a series of unique engineering problems. All the problems discussed in this chapter severely or slightly station construction, as station construction can alter or damage the surface-water conditions, water–heat exchange conditions, topsoil, and natural upper limit depth of the permafrost; these damages may result in frost heaving and thaw collapse of foundations near the station, especially around stations built over unfavorable or unstable permafrost. In addition, if the station is built over a surface with poor drainage conditions, various ice-induced diseases might develop, threatening the structural safety of the station. Thus, strong frost-heaving areas, such as large seasonal melt layer regions, ice-rich frozen regions, and sections with thick underground ice, are unsuitable for station construction.

Unfavorable permafrost phenomena not only substantially increase the complexity of railway site selection and engineering design but also influence the design of railway tracks, power supply, site management, and environmental protection. Thus, these phenomena must be considered in railway site selection and construction to ensure safe and stable railway operation.

3.4 Principles of Geological Route Selection in Permafrost Regions

3.4.1 Overview of Geological Route Selection in Permafrost Regions

Geological line selection in permafrost regions differs from that in other regions and should be based on adequate meteorological, hydrological, and geological data; the permafrost characteristics; and the distribution, type, scale, and severity of and solution to unfavorable geological phenomena.

In addition to the general principles, railway line selection in permafrost regions should consider the following:

1. Avoid unstable permafrost regions and warm areas where the annual average ground temperature exceeds $-0.5°$.
2. To ensure foundation stability, prefer embankments over cutting so that the frozen soil remains undisturbed. Minimize zero fill and low-fill sections (height <1.0 m) to avoid permafrost damage, which may affect roadbed stability.
3. In hilly areas, lay the line in the upper part of gentle slopes. Excavating tunnels is preferred over going around the hill (e.g., in Fenghuo Mountain and Kunlun Mountains). When the formation conditions are similar, to minimize damage of the frozen soil around the tunnel body and to minimize the infrastructure investment (e.g., thermal insulation), a single long tunnel is safer than are multiple short tunnels, even if this necessitates merging of short tunnels.
4. In river valley areas, locate the line in river melt zones or permafrost areas with high terrace stability rather than in the transition section of river melt zones. When unavoidable, minimize the route length in such transition sections.

5. Avoid regions with thick underground ice, thaw slumping, lakes, and ponds caused by thawing, icing, frost mounds, and freezing swamps. When unavoidable, determine the direction and position of the line according to the following principles:

   a. In areas with thick underground ice, the line should pass from a region with narrow width to a region with low thickness.
   b. In area susceptible to heating thaw collapse, use embankments below the landslide area.
   c. In areas with thaw lakes, use embankments of height 0.5 m above the highest water level.
   d. In areas with groundwater, ice cones, and frost mounds, consider the probability of the development of new ice cones and frost mounds due to construction-induced changes to the local hydrogeological conditions, and avoid excavations where this probability is high. Instead, use embankments that are at least 2.0-m high, as adequate height can facilitate drainage, thus preventing damaging phenomena such as ice sheets. Locate the lines far from ice cones and frost mounds on the downstream side.
   e. To ensure a stable roadbed in thin and narrow permafrost swamp areas, use embankments of height at least 1.0 m above the largest ponding water level of the swamp.

6. Locate the line in the dry zones and areas with little permafrost, for example, in zones with rock, gravel soil, and coarse sand. If the line has to pass over the fine granular soil, such as clay and silt, locate the line over less frozen land with little water content to the extent possible, and avoid passing over ice-full or ice-rich zones, both of which are susceptible to large frost heaving and thaw collapse, and over humus soil, cohesive soil, and fine sands, all of which are difficult to stabilize.
7. Avoid building tunnels over regions with thick underground ice and groundwater development. Avoid placing the tunnel exits over permafrost regions with unfavorable geological phenomena; when unavoidable, minimize excavations and prefer open-cut tunnels. To protect the exit piers against asymmetrical pressure, open-cut tunnel must be designed such that asymmetrical pressure and single pressure are minimized.
8. River regions along the Qinghai–Tibet Railway permafrost have unique hydrological characteristics and unfavorable permafrost phenomena because of the effects of landform and climate. The design characteristics of bridges (e.g., bridge structure, span, and foundation) used to cross these rivers should account for the effect of the local climate and environment, and bridges should not be built in regions with unfavorable geological phenomena such as thick underground ice, ice cones, frost mounds, heating thaw collapse, thaw lakes, and solifluction. Large and medium bridges should be built on sections with good hydrological conditions where rivers are connected to taliks and on sections with less permafrost in unconnected taliks. Over taliks and permafrost, ensure that the bridge piles have the same type of foundations. In addition, avoid large

and medium bridges over sections with large ice cones. In sections with thick underground ice, prefer sections with thin and narrow ice. Located bridges at sections where the river sections have not been treated in any manner. For ease of construction and maintenance, prefer simply supported bridges. To prevent frost heaving and thaw collapse-induced damage, the bridge foundation should be located outside the influence area of the riverbed thermal regime.

9. Account for soil conditions; establish centralized borrow areas and specific earth-moving routes, and strictly avoid borrowing earth from the sides of the route. Ensure that earth-moving or borrowing way does not lead to the formation of thaw lakes.
10. Wherever the railway line is parallel to the Qinghai–Tibet Highway, locate the line close to the highway; this avoids additional adverse effects on the environment and the ecology. However, do consider a buffer zone of more than 100 m between the railway and the highway to account for future reconstruction and extension.
11. The basic criteria for site selection for stations are as follows:

    a. Locate station in zones with bedrock and coarse-grained soils and not in ice-rich permafrost or areas with thick underground ice.
    b. Avoid stations in areas with unfavorable permafrost phenomena and unstable permafrost.
    c. Large seasonal melting areas are prone to strong frost-heaving and are thus unsuitable for building stations.
    d. Build stations in area with good drainage conditions, and locate the station as close to the water source as possible.
    e. Considering the harsh conditions of life and work in the Qinghai–Tibet Plateau and to minimize damage to the natural environment, build as few as stations as feasible. Locate the stations near the inhabited areas.

3.4.2 Conditions Along the Qinghai–Tibet Railway Permafrost Regions

Using comprehensive engineering-geological data from the permafrost regions, we should optimize the railway route scheme considering the distribution of ice-rich and high-temperature permafrost as well as that of the various unfavorable permafrost phenomena, especially geothermal zones. Geothermal zones along the railway are as Zone I (total length = 199.75 km, 36.6% of the overall length), Zone II (74.51 km, 13.6%), Zone III (110.75 km, 20.3%), Zone IV (59.75 km, 10.9%), and taliks (101.68 km, 18.6%).

High-temperature permafrost is widely distributed along the line. We optimized the route by selecting the route with the shortest length over unstable and high-temperature regions, thus improve the line stability and safety and reducing the investment.

3.5 Comparison of Main Line Routes in Permafrost Region

3.5.1 Col Route Across Tangalle Mountain

3.5.1.1 Overview

Spread across the middle of Qinghai–Tibet Plateau, Tonglha Mountains is the highest mountain along the Qinghai–Tibet Railway and has thus been a topic of intense research, exploration, and design. Numerous theoretical and field-based engineering and technical studies have examined surveyed col route selection across Tonglha Mountains and have reported that the high col scheme running along the Qinghai–Tibet Highway (hereafter referred to as the highway col scheme, highway scheme, eastern col scheme, or ICK) and the low col scheme running far from the Qinghai–Tibet Highway (hereafter referred to as the railway col scheme, railway scheme, western col scheme, or CK) are superior to other schemes. Hence, one of these two schemes can be implemented. Accordingly, supported by the growing understanding of the engineering-geological properties of the plateau permafrost and more than 30 years of scientific advances in permafrost and railway engineering, we comprehensively and systematically investigated the route schemes. For approximately 40 km in the east–west direction, we perform a preliminary precision survey and a feasibility study on the highway and railway col schemes by using aerial remote sensing and global positioning system technology.

3.5.1.2 Topography and Landform Along the Route

The study route runs between Tonglha Mountains and other mountains. Tonglha Mountains, over which runs the Yangtze River, is the watershed area between Za'gya Zangbo Continental river and Nujiang River. Buss Kangen peaks (elevation, 6602 m), where snow persists throughout the year, is the primary front in the study area. The elevation of the eastern ridge is 5200–5800 m, that of ICK is 5237 m; while the western ridge elevation is 5050–5400 m, and that of CK is 5072 m. These two cols are nearly 28 km apart.

Both the eastern and western cols of Tonglha Mountains are relatively thick, and the topography is low and mild. Therefore, subgrade engineering must be implemented along most of the study route. Gullies on both sides of the ridge zone have developed well. The valley water system north of the ridge is shaped like a tree fork, with Buqu River Bend as the mainstream and Rongma Bend and Baiduo River as the tributaries. The valley water system south of the ridge is arborized parallel to the mountains and mainly consists of Za'gya Zangbo and its tributaries such as Buniu River Bend, Hughdong River Bend, Xianggaer River, Daoxiong River Bend, and Anazangbu. The source of Za'gya Zangbo is located south of Tonglha Mountain highway col and is parallel to Touerjiu Mountain. The elevation of the mainstream and the col of Touerjiu Mountain, which is one of watersheds between Za'gya Zangbo and Nujiang River, is 5346 and 5172–5185 m, respectively. Lari Bend and its branches, such as Lebu Bend, Jiebu Bend, and Duopuer Bend, lie in southern Touerjiu Mountain.

3.5.1.3 Overview of the Route Schemes

We performed a preliminary precision survey on two schemes (Fig. 3.6): CK (i.e., western col) scheme and ICK (i.e., eastern col) scheme.

CK—This route runs for 132.01 km as follows: Starts from Buqiangge station, turns southwest, heads upstream along Baiduo river, heads down along the left bank of the small bend after crossing the western col in Tonglha Mountains (elevation, 5072 m) by using roadbeds, crosses Xiudong Bend and the soil-door highway to enter Za'gya Zangbo Basin, turns southeast at the mouth of the canyon, heads up along Anazangbu, crosses the offset of Touerjiu Mountain and turns into the upper reach of Lebu Bend, heads down along the bend, enters the Duopuer Bend Canyon, and finally heads down the river to Amdo. The earthwork volume is 7,692,000 m^3 and masonry volume 95,800 m^3; this route requires one large bridge (0.56 km) and 47 medium bridges (6.82 km). The estimated total investment is 2034 million yuan.

Baiduo River, Za'gya Zangbo, and Duopuer River Bend groove are all 4 km long each. Apart from a few steep slopes, most of the slopes along this route are mild. This route mostly passes over aqueoglacial deposit layers with little unfavorable geological and permafrost phenomena. The route spans 123.5 km in permafrost regions (including 16.4 km in taliks), of which 28.5 km spans ice-rich permafrost. The terrain in this scheme is open, except for the canyon areas along Baiduo River and Za'gya Zangbo.

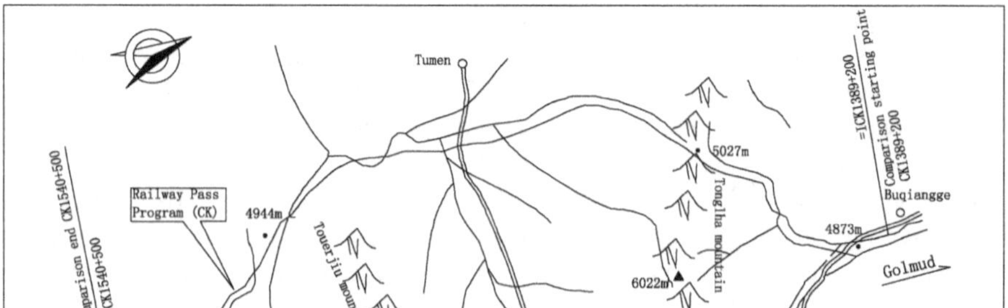

Fig. 3.6 Comparison of selection schemes for Tonglha Mountain Pass area

ICK—The route runs for 143.22 km as follows: turns southeast at the starting point, goes upstream along Rongma River Bend, crosses the east col on Tonglha Mountain (elevation, 5237 m) by using roadbeds because some sections of the south slope are steep, U-turns to Buniu River Bend, and arrives at primary track Sect. 109 after crossing Buniu River Bend twice. Subsequently, the route heads south along the west side of the Qinghai–Tibet Highway, crosses Tumen Highway and Za'gya Zangbo to Daoxiong River Bend, heads westward from Touerjiu Mountain extension (elevation, 5069 m) by using a Ω-shaped downlead to Jiebu River Bend south of Touerjiu Mountain, heads down along the river via Lari River Bend, and finally arrives at the destination, Amdo. The earthwork volume is 9,991,000 m^3 and the masonry volume 156,000 m^3; this route requires two large bridges (1.83 km) and 38 medium bridges (10.16 km). The estimated investment is 2751 million yuan.

The terrain in this scheme is unfavorable for 5 km of the route along Rongma River Bend north of the ridge and river valley of Lari River Bend, 51 km away from the south col. Slopes along the route are mild, except at some sections of Tonglha Mountains and Touerjiu Mountain. The route spans 116.9 km in permafrost regions (including 13.6 km in taliks), of which 37.0 km are over ice-rich permafrost. The route lies on slopes when crossing mountainous areas, which are susceptible to overland flow, permafrost swamps, thaw lakes, and heating thaw collapse. For example, large permafrost swamps are present in Buniu River Bend; nummuloidal thaw lakes are distributed south of Tumen Highway; and ice-rich permafrost is present along the banks of Buqu River.

Table 3.8 summarizes the technical and geological conditions of these two schemes.

3.5.1.4 Evaluation of the Railway Scheme

1. Topographicalal conditions

The study route passed through Tonglha Mountains and Touerjiu Mountain. The elevation of the railway scheme passes through the two cols is lower than that of the highway by 165 and 125 m. Both lines in this scheme drift off the airline, avoiding the high mountainous areas, and passing through low and even cols using subgrades; the advantages are that both sides of the railway col are valleys, and the longitudinal slope of the riverbed is small; in particular, the line avoids the Touerjiu Mountain, meaning that only one mountain needs to be crossed. However, the highway col has a high elevation, and its line transects the drainage of the southern slope. Because the line crosses Tonglha Mountains, the offset of Touerjiu Mountain, and the triangular watershed area constructed by the Yangtze River System, Za'gya Zangbo, and Nujiang River System, the topography is relatively harsh. Moreover, a long section of the line passes through valleys: as the northern longitudinal slope of Tonglha Mountains is steep and so is the northern longitudinal slope of Touerjiu Mountain, the line follows a horseshoe-curve and bypasses Touerjiu Mountain.

Thus, from the topographical perspective, the railway scheme, which uses the low col and passes alongside wide valleys, is more suitable for the study route.

2. Geological conditions

Table 3.8 clarifies that the geological condition of both schemes are relatively unfavorable, with the primary difference being the geological condition of the frozen soil.

The railway scheme mainly passes through the piedmont water deposition sloping plains, where the stratigraphy consists of coarse particles deposited by the ice water of the Quaternary system. The line crosses short sections of less-harmful frozen soil. The length of the railway scheme passing through the ice-rich frozen soil—the only adverse geological phenomenon along the line—is 8.5 km shorter compared with the highway scheme. By contrast, the highway scheme passes through more low and high mountain areas, and a substantial section of the stratum consists of fine particles (cohesive soils) of the Quaternary system. In addition, some sections of the highway scheme are prone to dangerous rocks and rockfall. Thus, the railway scheme is more favorable than is the highway scheme.

3. Line-design technology conditions

As explained in Table 3.8, compared with the highway scheme, the railway scheme passes through sections with lower elevation and requires a lower climb; the angle is small, the line flat, and the vertical section good. Specifically, the length of the railway scheme passing at elevations exceeding 5000 m is 44 km shorter compared with the highway scheme, with a maximum gradient of 16% (vs. 20%). Furthermore, the railway scheme is shorter by 11.2 km and straight and has one fewer station, which saves 7,230,000 yuan annually.

Thus, from the technology perspective as well, the railway scheme is more advantageous than is the highway scheme.

4. Engineering technology conditions

The railway scheme passes through wider valleys than does the highway scheme, which makes roadbed engineering easier in the former. Specifically, the railway scheme saves 2,299,000 m^3 of stone and 60,200 m^3 of masonry compared with the highway scheme. In addition, subgrade construction accounts for 35 and 56% of the line length in the railway and highway schemes, respectively. The length of the railway scheme passing through ice-rich frozen soil and wetlands is 8.5 and 31.88 km shorter compared with the highway scheme. In addition, in the railway scheme, the retaining wall along the Re-Rong Lake is shorter by 0.3 km.

Table 3.8 Technical route indicator and geological conditions

Project		Unit	Scheme of railway col (CK)	Scheme of highway col (ICK)	*D* (relative to CK)
Route length		km	132.01	143.22	11.21
Coefficient of the extension line		1.28	1.39	CK scheme is better	
Height of col	Tonglha Mountains	m	5072	5237	165
	Touerjiu Mountains	m	4944	5069	125
Height of lifting	Up	m	548	720	172
	Down	m	362	534	172
Altitude more than 5000 m	Length of route and sections	km/section	24/1	68/2	44/1
	Number of stations	Unit	2 (3)	3 (3)	1
Foot slope length		km	4	14.8	10.8
Curve length and specific weight		km/%	50.5/38%	61.6/43%	11.1/5%
Minimum radius and number of curves		m/unit	800/26	600/17	CK scheme is better
Unfavorable and unique geology	Permafrost	km	123.5	116.9	3.9
	Taele with high ice	km	28.5	37	8.5
	Taele with little ice	km	78.6	66.3	12.3
	Tabetisol	km	16.4	13.6	2.8
	Permafrost swamp	km	10.5	46.7	36.2
	Permafrost wetland	km	0.22	2.41	2.19
	Hot melt pond	km	0.3	1.8	1.5
	Crag and rockfall	km	–	1.6	1.6
Subgrade length		km	124.6	131.2	6.6
Excavation length		km	5.85	5.37	0.48
Subgrade earthwork/9 km conditions		10,000 square km	769.2	666.1/6.98	229.9/1.15
Masonry/9 km masonry		10,000 square km	9.58/0.07	15.60/0.11	6.02/0.04
Subgrade worksite	Slope protection and taele subgrade with high ice	km/section	2 802/21	26.28/84	1.74/3
	Foundation treatment wetlands	km/section	11.01/34	42.89/44	31.88/10
	Hot melt lake foundation treatment	km/section	0.12/1	0.17/1	0.05/0
	Cutting slope protection	km/section	2.22/11	1.57/6	0.65/5
	Retaining wall engineering	km/section	–	0.3/1	0.3/1
	Scour protection engineering	km/section	1.77/8	1.33/5	0.44/3
	Collapse regulation project	km/section	–	1.05/1	1.05/1
	Frozen-layer water-treatment engineering	km/section	0.79/1	–	0.79/1

(continued)

Table 3.8 (continued)

Project		Unit	Scheme of railway col (CK)	Scheme of highway col (ICK)	*D* (relative to CK)
	Total engineering	km/section	43.9/136	73.6/142	29.7/6
	Proportion of subgrade worksite	km/section	35%	56%	21%
Bridge opening worksite	Extra-large bridge, large bridge, and medium bridge	km/section	7.38/48	11.99/40	4.61/8
	Small bridge and culvert	km/section	5.39/324	6.07/260	0.68/64
	Proportion of bridge length	km/section	5.60%	8.40%	2.80%

Compared with the highway scheme, the railway scheme requires lower bridges, and the span of large and medium bridges is shorter by 4.61 km, thus easing the engineering workload and bridge-foundation management.

In sum, considering the engineering technology, the railway scheme is more favorable than is the highway scheme.

5. Construction and operation conditions

The lines in the railway scheme are mostly located approximately 30 km beside the Qinghai–Tibet Highway, and pioneer roads of length 132 km need to be constructed. By contrast, the highway scheme largely follows the Qinghai–Tibet Highway 9 km beside it, and about 42 km (in two sections) of pioneer roads need to be constructed. The construction conditions are relatively poor and inconvenient in the railway scheme and thus require an additional 90 km of pioneer roads compared with the highway scheme. The standard of the 132-km pioneer road can later be improved so that it can be used as the maintenance passageway.

6. Environmental protection

The railway scheme is shorter than the highway scheme and thus requires less land and exerts less damage to wetlands and the alpine meadow ecosystem. Moreover, the railway scheme has less unfavorable geological frozen-soil conditions and environmental diseases. Although the highway scheme entails more cutting than does the railway scheme, the scale of environmental damage in the former is smaller because it runs along the Qinghai–Tibet Highway.

7. Project investment

Table 3.9 details the major material volumes and project investment in the railway and highway schemes. Given the advantages discussed in the foregoing paragraphs, the railway scheme is simpler, and its static investment saves 7.17 million yuan compared with the highway scheme.

Table 3.9 Estimation of the main engineering material volumes and investment costs

Project		Unit	Railway scheme CK1389+200–CK1540+500	Highway scheme ICK1389+200–ICK1540+500	Difference
Length of path		km	132.01	143.22	11.21
Subgrade	Earthwork	$\times 10^4$ m^3	769.2	999.1	229.9
	Masonry	$\times 10^4$ m^3	9.58	15.6	6.02
	Fill-in permeability of soil and water	$\times 10^4$ m^3	18.72	32.02	13.3
	Rubble stone	$\times 10^4$ m^3	81.14	100.84	19.7
	PU	$\times 10^4$ m^3	11.4	13.2	1.8
	Composite geomembrane	$\times 10^4$ m^3	89.4	96.5	7.1
	Geogrid	$\times 10^4$ m^3	41.6	43.8	2.2
	Geocell	$\times 10^4$ m^3	15.9	16.5	0.6
	Permanent ground	Mu	7962	8914	952
	Temporary ground	Mu	5952	7667	1715
Bridge	Extra-large bridge	m/seat	560.54/1	1825.9/2	1265.36/1
	Large and medium bridge	m/seat	6818.78/47	10,165.27/38	3346.49/9
	Small bridge and culvert	m/seat	5388.86/324	6068.12/260	679.26/64
Static investment		10,000 yuan	203.409	275,066	274,862.591

3.5.1.5 Conclusion

Both the railway and highway schemes pass through regions with unfavorable engineering-geological conditions. Nevertheless, long-term research and the operating experience of the Qinghai–Tibet Highway have shown that all such conditions can be managed by taking appropriate project measures. Thus, from a technical perspective, both schemes are feasible. However, the poor operating conditions notwithstanding, the railway scheme has numerous advantages over the highway scheme, as the former has more stable frozen soil, shorter and straighter line, and savings of 717 million yuan. Thus, the railway scheme, which is far from the highway, is the preferred scheme.

3.5.2 Comparison of Roadbed and Bridge Schemes Across Qingshui River

Chumar River High Plains has unstable high-temperature or ice-rich permafrost, and the stratigraphy is fine sand prone to heavy subsidence. The upper limit of permafrost here is between 1.9 and 4.8 m, and the annual average ground temperature is higher than −0.5 °C. Various unfavorable permafrost phenomena and partially aeolian sand phenomena, such as heat thawing lakes and ponds, heating thaw collapse, permafrost swamps, ice cones, and ice mantles, are present here. Per the experience with the Qinghai–Tibet Highway, sections passing through these high plains are among the most dangerous in terms of structural diseases. Thus, to ensure safety, we decided to replace the roadbed in this section with a bridge. Figure 3.7 presents a comparison of the roadbed (DK) and bridge (IDK) schemes.

DK scheme—This roadbed scheme spans 18 km and complies well with the landform. In addition, the geological conditions of this scheme are more favorable than those of the IDK scheme. Although 3.54 km of this scheme passes over ice-containing soils, it is shorter, thinner, and less expensive than the IDK scheme. However, this scheme is 43.85 m longer than the IDK scheme.

IDK scheme—The bridge scheme is a cut-off scheme of length 17.956 km. Despite its shorter length, its major drawback is the poor geological conditions, such as at IDK1034+000–IDK1041+000, which is 7 km of thick (5–20 m) ice-containing soils, while the bridge foundation is 30 m thick. In addition, although an 8-m-wide light bridge is adopted here, it is economically infeasible because of the high unit price of the bridge. Using large-span bridges may influence the laying and erecting scheme, thus necessitating additional studies on the design of bridges over a very thick ice-containing soil layer, which further lengthens the

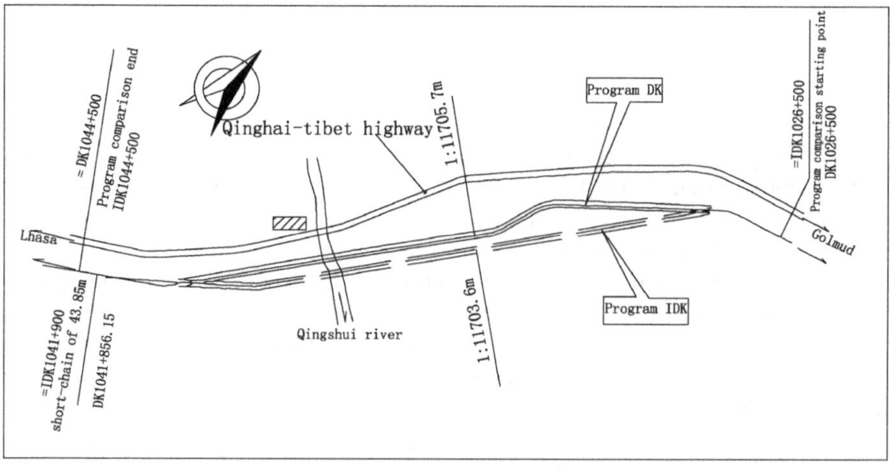

Fig. 3.7 Comparison of the roadbed and bridge schemes across Qingshui River

schedule of the Qinghai–Tibet Railway. Overall, engineering of the IDK scheme becomes more difficult as a larger bridge of span 8–16 m is required at IDK1041 +610 to cross thaw lakes; the static investment for the bridge scheme is 19.87 million yuan.

Thus, the DK scheme is more apt for the local geological conditions, is less expensive, and is thus recommended.

3.5.3 Comparison of Bridge Schemes Across Chumar River

The surface of Chumar River High Plains has silty clay, silty sand, fine sand, gravelly sand, breccia soil, and ground gravelly sand, the underlying layers are marls of the Tertiary, and the bedrock surface is stable. The upper limit of the permafrost is 2.5–4.0 m, with an annual average ground temperature exceeding −0.5 °C. Under the constraints of the local geological characteristics, we examined and compared the preliminary design of two schemes wherein the roadbed is replaced by a bridge (Fig. 3.8): the passing round scheme (DK) and the short-direct scheme (IDK). The DK and IDK schemes span 10.3 and 10.925 km, respectively.

Both these schemes pass over 3.84 km of ice-rich and ice-full permafrost and hence their geological conditions are the same. Nevertheless, the IDK scheme is more advantageous because it is shorter than the DK scheme by 41 m, and the Hoh Xil Station is only 100 m away from the highway. Thus, the IDK scheme was implemented.

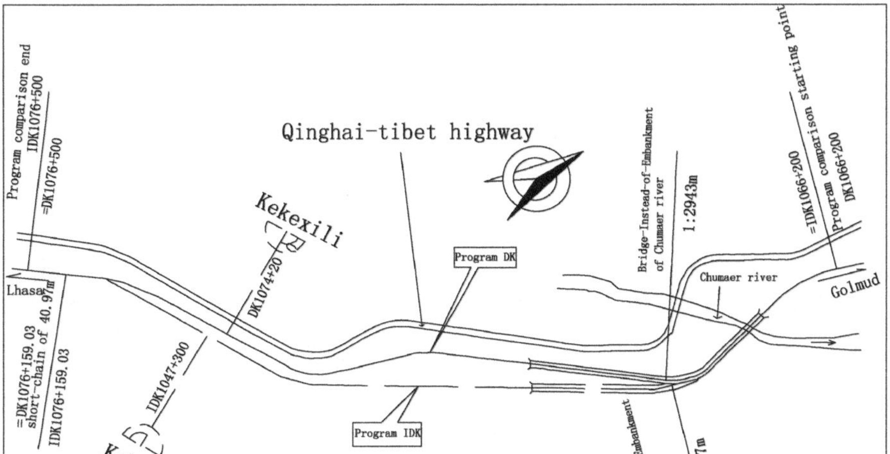

Fig. 3.8 Bridge and roadbed schemes across Chumar River

3.5.4 Comparison of the Schemes to Cross Thaw Lakes and Ponds Between Kaixinling and Tongtian River

Large and small thaw lakes and ponds are distributed widely between Kaixinling and Tongtian River. Unauthorized borrow pits on both sides of Qinghai–Tibet Highway, created during the highway construction, easily ponds and induces diseases. For example, a large thaw lake and pond of approximate length and width 350 m is present 60-m away from the west edge of the highway, at around the 3190 km milestone.

Considering that thaw lakes and ponds develop only in the warm season, we examined the effects of these features on the DK and DIK schemes (Fig. 3.9).

DK scheme—This 10.44-km route here first heads south parallel to the Qinghai–Tibet Highway and finally arrives at the destination after passing over large thaw lakes and the highway. Its major disadvantages are its proximity to the Qinghai–Tibet Railway and its inability to completely avoid heat thawing lakes and ponds.

DIK scheme: This scheme gradually deviates from the highway toward the right, runs through zones with high terrain, and finally arrives at the destination after bypassing the large heat thawing lake and pond. This scheme is advantageous because it completely avoids the large thaw lake, is more than 100 m away from the highway, and is relatively less expensive. However, it is 189-m longer than is the DK scheme.

Overall, adding some length not lead to large quantities for this section is subdued to pography. Thus, the DIK scheme is adopted.

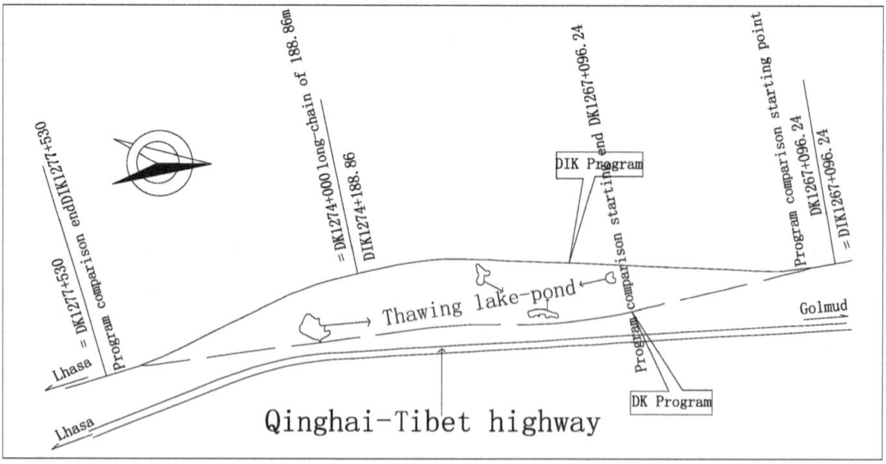

Fig. 3.9 Comparison of two schemes from Kaixinling to Tongtian River across the intervening thawing lakes

3.6 Summary

Approximately, 550 km of the Qinghai–Tibet Railway runs over long sections of high-temperature and ice-rich permafrost, which have poor heat stability and are thus susceptible to various unfavorable permafrost phenomena. Line selection in this region is based on the guiding ideology of reducing disturbance to the permafrost environment.

Ideally, the line must run along slopes exposed to sunlight for long periods as this would ensure a high water evaporation rate, relatively low surface, and low underground water content, all of which reduce the probability of ice and other diseases. Thus, the top section of mild slopes is an ideal choice to cross-mountain areas. If the line runs along a large river valley, cross the permafrost edge zone such that the line length over the permafrost is the shortest, and avoid alignment with the permafrost edge in large river melt zones. Moreover, the line should avoid to the extent possible zones with frozen soils, such as rocks, cabby soils, gravelly soils, coarse sand, medium sand, fine sand and clay, clayey sand, and sandy soils with low water content. In regions with rich permafrost, especially ice-rich or ice-full permafrost, avoid locating the line over humus soil, clayey sand, sandy clay, and silty sand. In addition, avoid unfavorable geological phenomena such as thick underground ice, heating thaw collapse, ice cones, frost mounds, and permafrost swamps, and where unavoidable, use bridges instead of roadbeds.

To ensure foundation stability, prefer embankments over cutting so that the frozen soil and the environment remain undisturbed. Determine the height of the earth fill on the roadbed on the basis of the principle of protecting frozen soil and on the basis of the carrier design and topographical conditions; in other words, avoid to the extent possible high fills and deep cuts, zero fills, and zero cuts, semifills, and semiexcavation engineering. Treatment requirements for the earth fill differ depending on the permafrost characteristics. Thus, the final scheme should be determined considering the treatments required, the optimal route longitudinal slope, and surveys conducted to compare various route schemes in permafrost regions.

References

1. Chen, L. T. (1998). Test and application of the relationship between anomalous snow cover in interspring over the Qinghai-Xizang Plateau and the first summer rainfall in southern China. *Quarterly Journal of Applied Meteorology, S1,* 1–8. (in Chinese).
2. Chen, R. H., & Wang, Z. T. (2003). Engineering-geological route selection of the gypsum section from Buqu to Wenquan of the Qinghai–Tibet Railway. *Journal of Glaciology and Geocryology, 25*(1), 14–16. (in Chinese).
3. China Railway First Survey, & Design Institute Group Co., Ltd. (1994). *The manual of routes* (2nd ed.). Beijing: China Railway Press. (in Chinese).
4. China Railway First Survey, & Design Institute Group Co., Ltd. (1999). *The railway engineering geology manual.* Beijing: China Railway Press. (in Chinese).

5. China Railway First Survey, & Design Institute Group Co., Ltd. (2000). *Comprehensive planning at nature reserve in Nacuo*. Beijing: Internal Information. (in Chinese).
6. China Railway First Survey, & Design Institute Group Co., Ltd. (2001). *Roadbed engineering construction drawing of ice-rich frozen soil of test section in Beiluhe*. Beijing: Internal Information. (in Chinese).
7. China Railway First Survey, & Design Institute Group Co., Ltd. (2001). *Feasibility study report of the Qinghai-Tibet Railway*. Beijing: Internal Information. (in Chinese).
8. China Railway First Survey, & Design Institute Group Co., Ltd. (2001). *Line survey and instruction of the Qinghai–Tibet Railway*. Beijing: Internal Information. (in Chinese).
9. China Railway First Survey, & Design Institute Group Co., Ltd. (2001). *Railway engineering geology reconnaissance specification*. Beijing: China Railway Press. (in Chinese).
10. China Railway First Survey, & Design Institute Group Co., Ltd. (2001). *Geotechnical investigation code for areas unfavorable for railway engineering (TB10027-2011)*. Beijing: China Railway Publishing House. (in Chinese).
11. China Railway First Survey, & Design Institute Group Co., Ltd. (2002). *Geotechnical investigation of Qinshui River experiment section*. Beijing: Internal Information. (in Chinese).
12. China Railway First Survey, & Design Institute Group Co., Ltd. (2002). *Preliminary design of the Tonglha–Lhasa section of the Qinghai–Tibet Railway (Geology)*. Beijing: Internal Information. (in Chinese).
13. Chinese Geological Atlas Editorial Board. (1996). *Geological atlas of Chinese*. Beijing: Geological Press. (in Chinese).
14. Dai, J. X. (1990). *The climate of the Tibetan Plateau*. Beijing: Meteorological Press. (in Chinese).
15. Duo, J. (2003). Basic characteristics of the Yangbajing geothermal field: A typical high-temperature geothermal system. *Engineer Science, 5*(1), 42–47. (in Chinese).
16. First Highway Survey and Design Institute, Ministry of Communications. (1996). *Compilation of research and design documents for the renovation project of the Qinghai-Tibet Highway*. Beijing: Internal Information. (in Chinese).
17. Jiang, Z. F. (1996). Geological hazards, forming conditions, and control along Sichuan-Xizang highway. *Chengdu Hydrogeological and Engineering-Geological Team, 3* (16), 243–249. (in Chinese).
18. Liu, J. K., Tong, C. J., & Fang, J. H. (2005). *Introduction to geotechnical engineering in cold regions*. Beijing: China Railway Publishing House. (in Chinese).
19. Meng, X. L. (2006). Geological prospecting for permafrost engineering in the Qinghai–Tibet Railway. *Railway Investigation and Surveying, 3,* 28–31. (in Chinese).
20. Meng, H., Zhang, Y. Q., & Yang, N. (2004). Analysis of the spatial distribution of geohazards along the middle segment of the eastern margin of the Qinghai–Tibet Plateau. *Journal of China Geology, 2*(31), 218–224. (in Chinese).
21. Ministry of Railways. (2006). *Code for design of railway line (GB50090-2006)*. Beijing: China Plan Press. (in Chinese).
22. Shu, L., Lou, W. H., & Wang, L. J. (2003). An analysis of the ground temperature of Yangbajing Tunnel. *Journal of Glaciology and Geocryology, 25*(1), 24–28. (in Chinese).
23. Sun, Y. F. (2005). Permafrost engineering in the Qinghai–Tibet Railway: Research and practice. *Journal of Glaciology and Geocryology, 27*(2), 153–162. (in Chinese).
24. The Glacier Permafrost Research Room, Institute of Geography, Chinese Academy of Sciences. (1965). *Investigation of frozen soil along the Qinghai–Tibet Highway*. Beijing: Science Press. (in Chinese).
25. Tibet Science and Technology Commission, Archives of Tibet Autonomous Region. (1982). *Tibet geological historical data compilation*. Lhasa: Tibet People's Publishing House. (in Chinese).
26. Tong, C. J. (1996). *Geological evaluation and treatment offrozen soil in permafrost along the Qinghai–Tibet Highway*. Beijing: Science Press. (in Chinese).
27. Wang, Y. D. (2002). Preliminary exploration of geologic route selection in multiyear tundra of Qing–Zang Railway line. *Journal of Railway Engineering Society, 2,* 57–59. (in Chinese).

28. Wang, Z. J. (2002). Permafrost engineering in the Qinghai–Tibet Railway construction. *Chinese railways, 12,* 31–37. (in Chinese).
29. Wang, Z. H. (2003). Landslides along the Qinghai–Tibet Railway and Highway. *Modern Geological, 17*(4), 355–362. (in Chinese).
30. Wang, S. L., & Mi, H. Z. (1993). Change in permafrost under roadbed with asphalt pavement along the Qinghai–Tibet Highway. *Glaciology and Geocryology, 15*(4), 566–573. (in Chinese).
31. Wei, G. J. (2003). Geological line selection of the Xiaonanchuan–Wangkun unfavorable geological section of the Qinghai–Tibet Railway. *Geotechinical Engineering World, 6*(7), 41–45. (in Chinese).
32. Wu, Z. W., Chen, G. D., et al. (1988). *Roadbed engineering in frozen earth area.* Lanzhou: Lanzhou University Press. (in Chinese).
33. Wu, Z. H., Ye, P. S., Wu, Z. H., et al. (2003). Hazard effects of active faulting along the Golmud–Lhasa Railway across the Tibetan Plateau. *Modern Geological, 17*(1), 1–7. (in Chinese).
34. Wu, Z. H., Hu, D. G., Wu, Z. H., et al. (2006). Pressure ridges and their ages of the Xidatan strike-slip fault in south Kunlun Mountains. *Geological Review, 52*(1), 15–25. (in Chinese).
35. Ye, D. Z., Gao, Y. X., Shen, Z. B., et al. (1979). *The meteorology of the Qinghai–Tibet Plateau.* Beijing: Science Press. (in Chinese).
36. Yi, Z. M., Zhou, J. X., & Zhang, X. D. (2006). Analysis of hydrologic conditions along the Qinghai–Tibet Railway. *Technology of Soil and Water Conservation, 4,* 14–16. (in Chinese).
37. Yuan, D. Y., Zhang, P. Z., Liu, X. L., et al. (2004). Tectonic activity and deformation features during the Late Quaternary of Elashan Mountain active fault zone in Qinghai Province and its implication for the deformation of the northeastern margins of the Qinghai–Tibet Plateau. *Earth Science Frontiers, 11*(4), 393–402. (in Chinese).
38. Zang, E. M., Wu, Z. W., et al. (1999). *Permafrost degradation and road engineering.* Lanzhou: Lanzhou University Press. (in Chinese).
39. Zhang, G. X., Wang, S. J., & Zhang, Z. Y. (2000). *China engineering geology.* Beijing: Science Press. (in Chinese).
40. Zhao, J. C., Liu, S. Z., & Ji, S. W. (2001). Engineering-geological problems in railway engineering construction. *The Chinese Journal of Geological Hazard and Control, 12*(1), 7–9. (in Chinese).
41. Zhou, Y. W., et al. (2000). *China frozen soil.* Beijing: Science Press. (in Chinese).

Chapter 4
Line Selection in Active Fault Zones and Meizoseismal Areas

Abstract This chapter discusses the spatial distribution of the Kunlun Mountain, Bengcuo, and Yambajan–Damxung–Gulu seismotectonic zones and explains the relationship between active faults and earthquakes, which can severely damage railway structures. The line selection and engineering properties of the Qinghai–Tibet Railway in high-intensity earthquake zones and active fault zones are implemented not only as per the geological-engineering line-selection principles for high-earthquake-intensity regions but also considering the railway safety, reliability, and cost.

Keywords Active fault zone · Seismic belt · Geological line-selection principle Railway engineering · Scheme comparison · Engineering settings

Active fault zones and meizoseismal areas are highly active, often inducing geological disasters that strongly affect railway engineering projects. In the Qinghai–Tibet Plateau, the new tectonic movement is strong, and seismic events occur frequently, influencing the Qinghai–Tibet Railway. The chapter discusses the spatial distribution of active fault zones, temporal, and spatial distribution of seismic belts, effects of active fault zones, and meizoseismal areas on the railway, and geological line selection.

4.1 Spatial Distribution of Active Fault Zones

The Qinghai–Tibet Plateau has rock formations from different periods, namely Proterozoic, Paleozoic, Mesozoic, and Cenozoic, along multiregional tectonic movements, all of which contribute to the complex lithostratigraphy and geological evolutionary history. These formations have gradually sculpted the crustal structure and the tectonic landscape, strongly influencing the environmental and global climate changes of the Late Cenozoic. Currently, intensive tectonic movements and

J. Li et al., *Geological Line Selection for the Qinghai-Tibet Railway Engineering*,
https://doi.org/10.1007/978-3-662-55572-9_4

seismic activities are the strongest tectonic units of tectonic movements on Earth's surface (i.e., the continental dynamic field), and these movements are popular geological research topics globally. The eastern Qinghai–Tibet Plateau is composed of six terranes, specifically, five suture zones and one deep fault (Fig. 4.1): (north to south)—Qilian–Arkin suture zones, Qilian Mountain terranes, Qinghainanshan, and northern Huaiyang Faults, eastern Kunlun–Qaidam terranes, eastern Kunlun–Animaqin suture zones, Songpan County–Ganzi County–Hoh Xil terranes, Jinsha River suture zones, Qiangtang terranes, Bangong Lake–Nujiang river suture zones, Lhasa terranes, Yarlung Zangbo river suture zones, and Himalayan terranes. Each period of ocean-basin curtailment and terrane collision has resulted in the five long and narrow suture zones along the Qinghai–Tibet Plateau. Newer (older) suture zones are found in the north (south), evidencing that the Asian continental plate is gradually moving southward. The Indian subcontinental plate collided with the Asian continental plate, closing the Tethys residual sea, which was the outlet of the Yarlung Zangbo River, in a slow and phased process due to the tectonic activity and the consequent ground uplift.

The Qinghai–Tibet Railway line crosses the following major active faults.

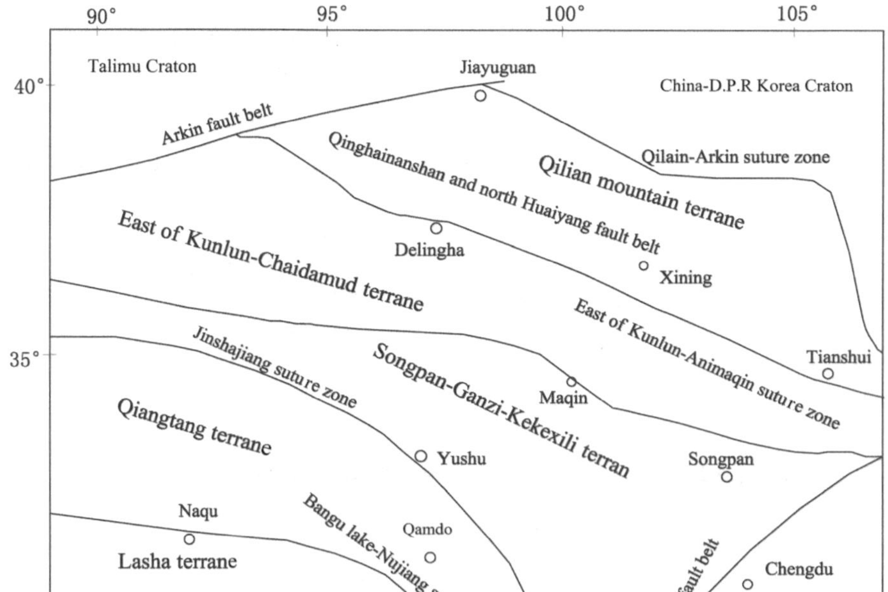

Fig. 4.1 Major tectonic units of the Qinghai–Tibet Plateau

4.1.1 *The Active Faults of Eastern Kunlun–Qaidam Terranes*

Eastern Kunlun–Qaidam terranes are composed of the northern Kunlun and Qaidam Basins, which lie in the mountains south of Qinghai and between the northern Huaiyang ruptures and eastern Kunlun–Animaqin suture zones; they mainly contain the Wahong Mountain–Ela Mountain Faults, guide basin faults, the active faults of eastern Kunlun. Among them, the Qinghai–Tibet Railway is mainly affected by the active faults of eastern Kunlun along the Kunlun Mountain areas.

The active faults of the eastern Kunlun lie at the boundary of Hoh Xil and the active blocks of Kunlun, which belong to the left-lateral giant strike-slip active fault belt system that extends more than 2000 km along the northern Qinghai–Tibet Plateau. East of eastern Nagqu in Gansu province, the fractures run through Maqin Xian County, Tuosuo Lake, Alake Lake, Dongxidatan, and Hoh Sai Lake and extend west until the western Cetacean Lake at the boundary of Qinghai and Xinjiang Uygur Autonomous Region (Fig. 4.2).

The active faults of Kunlun are strongly seismically active in northern Qinghai–Tibet Plateau. According to the faulted structures and their types and occurrence, the faults can be classified as six secondary seismic rupture structures: (west to east) Hoh Sai Lake Faults, Xidatan Faults, Alake Lake Faults, Tuosuo Lake Faults, Dongqing Ditch Faults, and Nagqu Faults.

West of the active faults of eastern Kunlun, Hoh Sai Lake Fracture, and Xidatan Fracture present a complex juxtaposition relationship that is prone to earthquakes, with obvious strike-slip and tectonic extrusion motion. Figure 4.3a, b depict the 24-m left-laterally leaping secondary terrace of Dongdatan Faults and the southern margin of the larger extrusion ridge tectonics of Xidatan Luanshi Ditch, respectively. From Holocene onward, the average slip rate of each section of the active faults of eastern Kunlun has been high. Specifically, from mid-Late Pleistocene onward, the slip rate of Kunlun Mountain Pass has been 13–14 mm a^{-1} (Table 4.1).

In the past hundred years, many strong earthquakes have occurred along the active faults of Kunlun, such as Ms8.1 Kunlun Mountain Pass earthquake

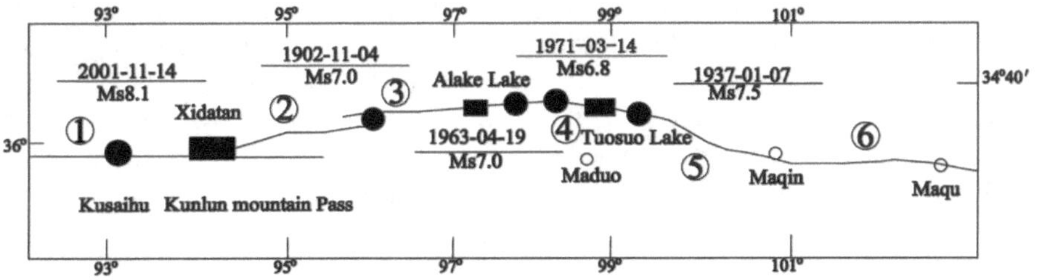

Fig. 4.2 Earthquakes along eastern Kunlun active faults

A. Dongdatan Faults' left-laterally leaping secondary terraces of 24 m (lens facing the north)

B. Southern margin faults of larger Xidatanjingxiangu extrusion ridge tectonics (lens facing the west)

Fig. 4.3 Landscapes of the active faults of eastern Kunlun

Table 4.1 Average slip rate of the active faults of eastern Kunlun

Fault segment	Period(s) of activity	Average slip rate ($mm\ a^{-1}$)
Kusai Lake–Kunlunshan pass	Mid-term of Q_3 to initial stage of Q_4	13–14
Dongxidatan–Tibet Dagou	Q_4	7.5–9.0
Alake Lake–Tuosuo Lake	Final phase of Q_3 to Mid-term of Q_4	5.0–6.0, 8.0
Tuosuo Lake–Dongqinggou	Final phase of Q_3 to Mid-term of Q_4, final phase of Q_4	4.0–6.0, 6.0–8.0
Maqin–Maqu	Final phase of Q_3 to modern times, final phase of Q_4	9.0, 12.6

(November 14, 2001), Ms6.8 Tuosuo Lake earthquake (March 24, 1971), Ms7.0 eastern Alake Lake earthquake (April 19, 1963), Ms7.5 Huashixia earthquake (January 7, 1973), and Ms7.0 eastern Alake Lake earthquake (November 4). Many strong ancient earthquake events have occurred in the Holocene along Xidatan strike-slip fractures. For example, three or four series of strong earthquakes of magnitude Ms7–8 have occurred in the modern Holocene, forming huge earthquake sites that show seismic ruptures, earthquake scarps, extrusion ridges, sag ponds, multiperiod extrusion ridges, earthquake wedges, colluvial wedges, and multiperiod pearl-shaped earthquake bulges with oblique distribution. Along the strike-slip faults of Hoh Sai Lake in the southern margin of Kunlun Mountains, a strong earthquake of magnitude Ms8.1 and rupture length of 426 km occurred in 2001.

4.1.2 The Active Faults of Songpan County–Ganzi County–Hoh Xil Terrain

The Songpan County–Ganzi County–Hoh Xil area is spread across eastern Kunlun–Animaqin suture zones, Jinsha river suture zones, and Longman Mountains Fault zones and includes the strike-slip active faults along the eastern Hoh Xil areas, Xianshui River, Ganzi County–Yushu City, and Longmen Mountain thrust belts. In this terrain, the Qinghai–Tibet Railway mainly runs across eastern strike-slip active faults along Hoh Xil.

The eastern strike-slip active faults include the active faults of Wudaoliang and Hoh Xil, northern Hoh Xil Mountains, and southern Ekexili Mountains (Fig. 4.4).

The active faults of Wudaoliang, whose overall strike emerges in the EW direction, extends more than 50 km and has clear left-lateral strike characteristics; specifically, the ruptures are active left-lateral tension–torsional fractures of varying (five) sizes. These active faults have a tendency relatively steep inclination of 10°. Structural fractures and frost heave groups that run nearly parallel to the fractures form almost vertical secondary tension cracks. Springs, fault scarps, and beaded depression ponds are present along the fractures. All other active faults strikes are in the EW direction and are relatively steep, with linear remote sensing image characteristics. In addition, fault facets and fault scarps are present, as are linearly distributed melting lakes, linear spring groups, oblique spring layers, structural fractures, and significantly abnormal radon levels.

The active faults of eastern Hoh Xil Mountains are composed of four faults, most of which trend nearly north and are relatively steep. Located in the northern Wudaoliang Mountain, the main body is distributed on the northern side of the basins and at the boundary of Hoh Xil Mountains. The active faults of northern Hoh Xil Mountains extend more than 50 km in the latitudinal direction and are characterized by a large-scale, deep-cutting, and long-term movement, with left-lateral and tension–shear fractures. The faults left-laterally cut gullies and mountain ridges of the Holocene and partly cut the lacustrine sedimentary strata of the Late Pleistocenetrata, with a left-lateral displacement of 50–150 m of the Late Pleistocene–Holocene. The fracture is due to a regionally strong fault-structure movement of the Holocene. Fault scarps, fault facets, ground fissures, thaw lakes and ponds, linear spring groups, and oblique ice cumulus are present along the fractures. The active faults on the northern boundary of Hoh Xil Mountains belong to the main faults of Hoh Xil Mountains and trend 345°–355°. The linear depression and oblique ice expansion cumulus here develop into secondary tensile cracks along the fractures. The faults have left-lateral slide valleys, terraces, and mountain ridges; 25-m-wide terraces of the Holocene and 3–4 m-elevation faulted scarps displace the 50-m-wide terraces of the Late Pleistocene and 50-m-elevation fault scarps.

The active faults of southern Hoh Xil Mountains are characterized by an obvious left-lateral strike-slip movement, a part of which forms the left-lateral shift gullies and the river terraces of the Holocene. From the Holocene movement onward, the active fractures have developed into obvious fault scarps, structural fractures, frost

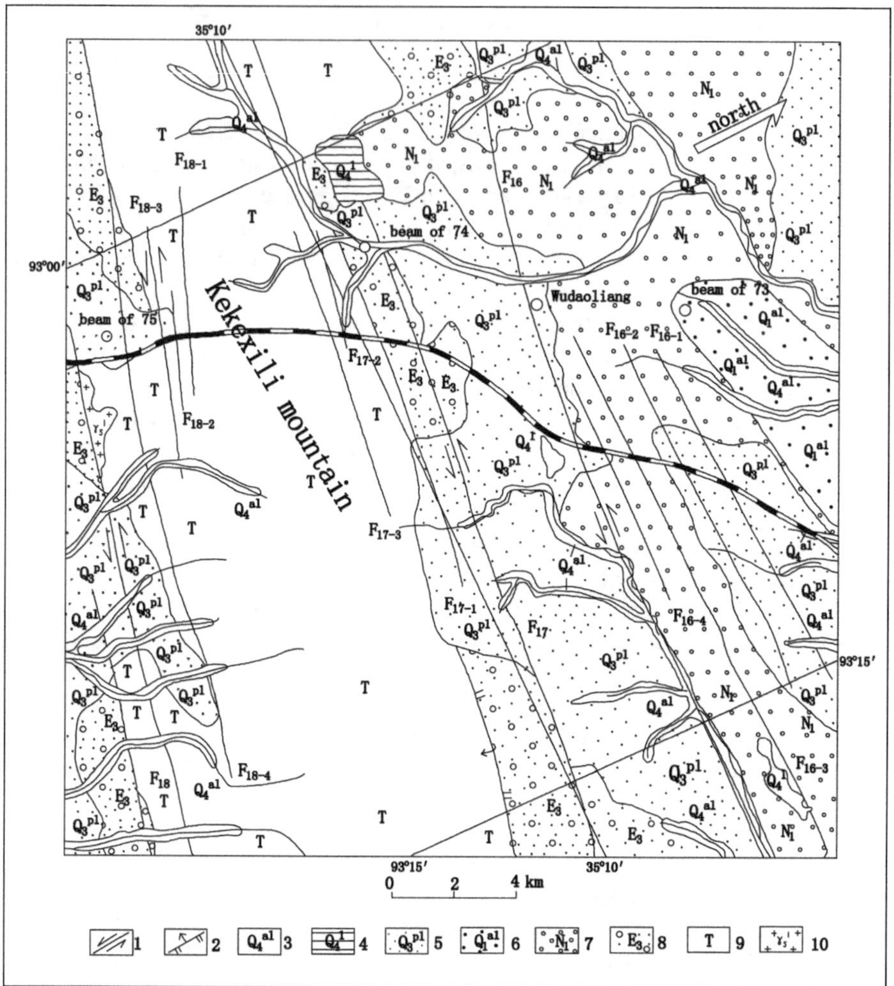

1: Active fault, 2: Ancient thrust fault, 3: Holocene river deposit , 4: Holocene lacustrine deposit, 5: Late Pleistocene moraine, 6: Early Pleistocene river deposit, 7: Miocene sandy conglomerate, 8: Oligocene sandy conglomerate, 9: Triassic slate and epimetamorphic sandstone, and 10: Indo-Chinese epoch granite

Fig. 4.4 Distribution of active faults in eastern Hoh Xil Mountains

heaving groups, thaw lakes and ponds, fault facets, and fault valley terraces, and the ruptures have developed into friction mirrors, horizontal scratches, and terraces along the faulted planes. The active faults of southern Hoh Xil Mountains are mainly composed of five scales of active fractures, with faults of southern Hoh Xil, which trend 110° and lie at the boundary of the southern margin of Hoh Xil Mountains, being the main boundary of the southern basins and mountain. Here, bedrock fracture zones, springs, frost heaves, fault scarps, and bedrock fault facets are distributed in the fault layers.

4.1.3 Active Faults of Qingtang Terrains

Qingtang terrains lie between the suture zones of Chin-sha River and that of Bangong Lake–Nujiang, and the major active faults are those of Fenghuo Mountain, Wuli, Tongtian River, and western boundary of Wenquan Basins. Per the left-lateral strike-slip movements, the main active faults of the terranes are mostly EW–NWW the strike-slip fault in Late Cenozoic in the Qinghai–Tibet Plateau.

4.1.3.1 Active Faults of Fenghuo Mountain

The active faults of Fenghuo Mountain lie in northern Fenghuo Mountain area in the Qinghai–Tibet Plateau, which are an important regional active structural system composed of pull-apart basins and a series of active faults trending NWW. The main body of the active faults of Fenghuo Mountain lies in Fenghuo Mountain zones and include the active faults of Fenghuo Mountain on the northern side, of Fenghuo Mountain, of the basin margin of the Erdao Ditch, and of the Erdao Ditch (Fig. 4.5). The majority of active faults trend NW–NWW and are strike-slip fractures. Some active faults trend NE, forming the boundary of the pull-apart basins. Fault scarps, linear spring groups, beaded ice cumulus, left-laterally sliding gullies, mountain ridges, alluvial and flood fans, displaced gravel rocks of the Late Pleistocene–Holocene are present along the main active fault.

The active faults of Fenghuo Mountain on the northern slope mainly trend NWW. Regionally extending more than 200 km, these left-lateral and compresso-shear active faults are composed of three left-lateral strike-slip active faults trending NWW. The active faults of Fenghuo Mountain present on the northern side of the mountain, along the early unconformity surface, and thrust faults of the anterior Quaternary; the faults of the Late Quaternary are reactive and have developed into fault facets, fracture zones, and fault gouges. From the Late Pleistocene onward, the fault activity is obvious and has developed into zonal fault scarps, ground fissures, and bedrock glide planes. Figure 4.6a depicts left-laterally leaping gullies of length 33 m, whose newest sediments have an optically stimulated luminescence age of 7.65 $\pm$ 0.65 ka, and left-lateral strike-slip faults with an average rate of 24 $\pm$ 4.3 mm a^{-1}. In winter, the active fractures develop into ice cumulus, as shown in Fig. 4.6b, which depicts an ice cumulus with the largest hummock diameter of 62 m and height of 1.8 m along the active ruptures in the western railway.

The active faults of Fenghuo Mountain are composed of a series of active fractures trending NW–NWW, most of which, according to the thrust ruptures in the Paleogene, belongs to regionally active faults. The linear remote sensing image characteristics of the active faults of Fenghuo Mountain is obvious and is characterized by large-scale regional elongation, further developing into linear fault facets, fault valleys, and oblique ice cumulus. Fault zones and the top cleavage belts are

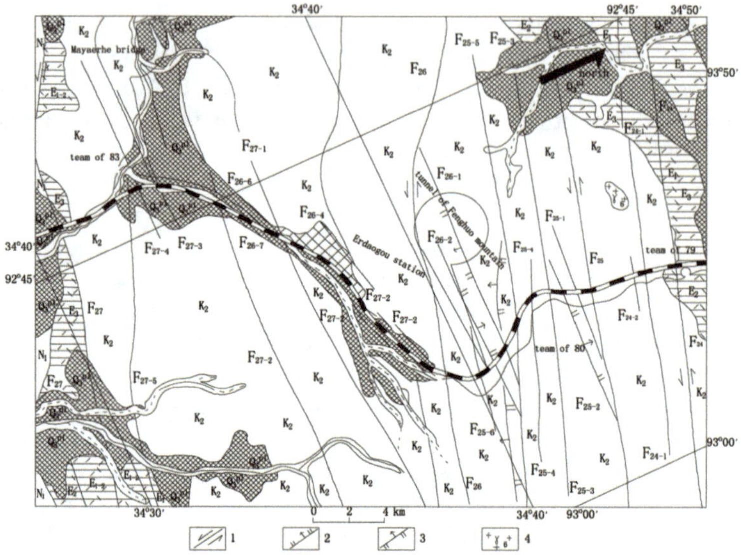

1: Strike-slip active fault, 2: Normal active fault, 3: Ancient thrust fault, 4: Himalayas granite,
Q_3^{pl}: Late Pleistocene river deposit, N_1: Miocene Sandy conglomerate,
E_3: Oligocene sandy conglomerate, E_{1+2}: Paleocene–Eocene sandy conglomerate, K_2: Late Cretaceous red beds

Fig. 4.5 Geological map of the active faults of Fenghuo Mountain

A. Active faults of northern Fenghuo Mountain scarps (lens facing the east)

B. Ice cumulus along the active faults of northern Fenghuo Mountain in winter (lens facing the east)

Fig. 4.6 Disaster effects of the active faults of Fenghuo Mountain covered by modern humus soil, forming obvious landscape scarps, evidencing that the fractures of the Late Quaternary undergo certain structural movements.

The active faults of the margin of Erdaogou Basin lie in the periphery of Erdao Ditch Basin and consist of the various boundaries of the active faults basins. The majority of the faults are active fractures of the Quaternary that cut off early; old faults in different directions have developed into obvious fault scarps and fault facets that have since displaced the alluvial and flooding fans of the Late Pleistocene and the gullies. The faults on the northern side of Erdaogou Basin turn off the red beds in the upper Cretaceous and shift the modern gullies, forming fault scarps of height 5–6 m.

The active faults of Erdaogou are mainly distributed on the southern side of Erdaogou Basin and are mostly NW–NWW-trending active faults, including the main fault on the southern side of Erdaogou and its two branch faults. The main fault is a product of the reactive, steep, and NW-trending thrusting faults of the Paleogene. The Paleogene faults are due to large-scale thrusting, and the faults in the Quaternary have developed into left-lateral tension–torsion faults that belong to the active fractures of Ulan Ul Lake–Yushu, which contain obvious linear remote sensing image characteristics, fault facets, fault scarps, beaded faulted lake basins, and en echelon hummocks with a 160° long axis. The fault movement in the Holocene is obvious and has developed into structural tension–torsion fractures and linear spring groups. The average rate of the left-lateral strike-slip movements and vertical motion of the faults is 2.61–3.52 and 0.12 mm a^{-1}, respectively.

4.1.3.2 Active Faults of Wuli

The active faults of Wuli are the regionally important active structural systems on the northern side of the Qinghai–Tibet Plateau, in Wuli Mountain. The active fractures on the northern side of Wuli Mountain (84th team), of the margin of Wuli Basin (85th team), and of southern Wuli Mountains (86th team) are composed of 17 strips of active faults with different scales, directions, properties, and sequences. The active faults on the northern side of Wuli Mountain consist of the northern terrains of the Wuli Basin, which belong to the active faults of the Late Quaternary and deposited layers of the Middle Pleistocene with early cutting. The ruptures develop into linear scarps, fault valleys, linear spring groups, deflection ice-cumulus structure, and structural fractions. The active faults of the margin of Wuli Basin, where the faults are distributed more densely, consist of eight strips of different sequences of active ruptures. Most faults that strike 90°–120° are distributed in the faulted basin of Wuli and peripheral mountains. Most active faults of Wuli Mountains margin have linear remote sensing image characteristics, which develops into obvious fault scarps, fault valleys, structural sags, structural cracks, linear spring groups, and deflection ice-cumulus structures. The active faults on the southern side of Wuli Mountain are distributed in southern Wuli Mountain Basin, which consists of eight strips of active faults with different sequences and scales. Most fractures whose strike direction are 80°–120° have left-lateral oblique-slip or left-lateral strike-slip motion characteristics, developing into fault scarps, structural fractures, linear spring groups, and beaded ice cumulus, with linear remote sensing image characteristics. Some sections of the faults clearly cut off the strata of the Late Pleistocene and modern water systems. Wide fault valleys in the 86th team are formed by the faults on the eastern and western sides along the Qinghai–Tibet Highway, regions which are favorable for migration and enrichment of groundwater and development of structural taliks, low-temperature hot springs in summer, and migrating ice cumulus or ice cumulus of different scales.

4.1.3.3 Active Faults of Tongtian River

The active fractures of Tongtian River are left-lateral strike-slip active structural fault systems in the northern mountains of the Qinghai–Tibet Plateau, distributed across the Tuotuo River Basins (high plains), Kaixinling Mountains, and Tongtian River Basins (high plains); they are mainly composed of the active faults of the northern side of Tuotuo River Basins, of southern Tuotuo River Basins, of Kaixinling, of Bumalangna, of Tongtian river, of southern Tongtian River, and of the southern boundary of Tongtian River Basins, all of which control the evolution of the pull-apart basins in the various directions, the majority of which develop into ice cumulus with beaded deflection-structure distributions.

The active fractures on the northern side of Tuotuo River Basins are distributed in the margin of the northern fault basins; these fractures strike 100°–120° and consist of one main active fault and a few branch faults. They have obvious linear remote sensing image characteristics, with the large fractures controlling the evolution of the Tuotuo River Basin. The faults have developed into fault scarps, expanding ice-mound groups, sinter platforms, tension–torsion structural fractures, linear scarps, en echelon hummocks, and ancient sinter mounds, cutting off and leaping lake sedimentary soil of the Late Pleistocene and gravel rocks of the Holocene along the active faults of the northern basin of Tuotuo River.

Active ruptures of the southern basin of Tuotuo River lie in the southern margin of Tuotuo River Basins and belong to the active faults of the southern boundary of these basins. The left-lateral strike-strip fault has obvious linear remote sensing image characteristics and manifests in the landscape as a visible linear fault scarp of strike N68°W and length >40 km. The majority of branch faults trend EW–NEE and are present in the main fault periphery, along with the derivation faults, which have developed into obvious fault scarps, ice cumulus with deflection structures, linear spring groups, and structural fracture zones along each secondary active fault. A part of the secondary faults has developed into EW-trending Wenquan groups because of sandblasting and water erosion.

The active fractures of Kaixinling are composed of six strips of faults with different directions, scales, properties, and sequences. The main fault has obvious linear remote sensing image characteristics in the NW direction. NW-trending left-laterally leaping gullies and mountain bodies in the 91th team cut off and displace the alluvial and flood fans in the Holocene and thus have developed into linear spring groups along the fault scarps.

The active faults of Bumalangna are composed of one NW-trending main fault and some EW-trending branch faults. The main fault of Bumalangna, whose length and strike are >33 km and N65°W, respectively, has obvious linear remote sensing image characteristics. The left-laterally displacing Tongtian River branches have developed into beaded faulted lakes and large-scale ice ridges in the EW direction along the fractures. EW-trending branch faults of the active faults of Bumalangna form the boundary of the miniature pull-apart basin in the EW direction, which have developed into linear fault scarps, beaded ice cumulus, and numerous EW-trending structural fractures.

The active faults of Tongtian River are distributed in the Tongtian River Basin, which consist of the EW-trending main fault and the NW-trending branch faults. The main fault is distributed in the main channel and the southern bank, has an obvious linear remote sensing image characteristics, and presents as linear scarps in the landscape, developing into linear spring groups and Wenquan groups along the fault scarps, which themselves have developed into linear ice cumulus along the main fault. The main fault cuts off the permafrost and underground ice layers and is characterized by a left-lateral strike-slip motion; these faults have developed into beaded deflection-structure ice cumulus along the faults at faulted compound positions and in the faulted and shattered water-rich zones.

The active fractures of southern Tongtian River Basin, which strike N60°W and are more than 57-km long, engender miniature pull-apart basins in the EW direction, linear fault scarps, expanding ice mounds with the beaded arrangement and ice expansion ponds, and hummocks along the southern active faults of Tongtian River Basins.

The active fractures of the southern boundary of Tongtian River are distributed on the northern side of Yanshiping Mountains and belong to NWW-trending left-lateral strike-slip faults, which have developed into the linear spring groups and deflection-structure ice cumulus along the active faults of Yanshiping Mountain, manifesting as linear scarps and fault valleys in the landscape.

4.1.3.4 Active Faults of the Western Boundary of Wenquan Basin

The active faults of the western boundary of Wenquan Basin lie on the northern side of Tonglha Mountains, whose main body trends SN. The large-scale extensional tectonics of northern Qinghai–Tibet Plateau are composed of a series of boundaries of SN-trending active faults with deflection combination; they control the spatial distribution, and formation and evolution of Wenquan Basin. Wenquan Basin has a width of 8–12 km, length of >40 km, and elevation of and 4800–4900 m; its strike is mainly SN; it lies in the south-central region of Qingtang terrain and is close to the suture zones of Bangong Lake–Nujiang River on the southern side. Most peaks have an elevation of 5400–5600 m, and a few peaks in the periphery mountain basin have elevations exceeding 5800 m. The mesozoic and

Late Cretaceous–Paleogene strata has undergone compression deformation in the SN direction in Wenquan Periphery Basin, forming SW-trending folded structures and thrusting faults. However, in the SN direction, Wenquan Basin crosscuts the early deformation bands with EW-trending compression structures. In inner Wenquan Basin, active faults in different directions compose the extensional activity tectonic system, which controls extensional tectonic evolution, underwater thermal cycling, and linear Wenquan formation during the Quaternary; multiphase events here have formed multifault facets of piedmont, fault scarps, multilevel fault-scarp landscapes (Fig. 4.7a), and linear sinter platforms (Fig. 4.7b). Many events have occurred in the paleoseismic and tectonic settings of the earthquake of Ms6.0–7.0 in the future. The structures and landscapes here indicate that the vertical activity of the faults occurred in the Late Quaternary. Fault scarps, fault cliffs, and multistage sinter platforms of varying heights and from various periods manifest as fractures of at least seven grades along the faulted zones of the piedmont. The fault scarps exhibit seven levels of vertical displacements: 0.3–0.6, 2.2–3.9, 4.6–6.5, 6.2–10.5, 14–24, 29–37, and 50–60 m. Of these, the recent vertical activity caused by surface-crushing ancient earthquakes is the least displayed, and the other six levels of fault scarps and fault cliffs have displaced and developed alluvial terraces or contemporaneous alluvial fans and sinter platforms (T2–T7) in the foothill belt. Through the U-series, OSL, and ^{14}C testing of calcareous cements, pebble calcic coatings, carbon-containing accumulations, silt layers associated with sinters, flood fans, and terrace gravel layers of the Late Quaternary sediments (Table 4.2), we find that fault cliffs of height 35–40, 18–19, 8–10, 5–6, and 2.5–3.2 m represent the vertical activities from the 150–100, 60–45, 40–30, 25–20, and 15–11 kaBP periods, respectively. The average rates of vertical activities during the Late Quaternary was 0.15–0.5 mm a^{-1}. The average rate of vertical activities during the Late Quaternary (>150 kaBP) was 0.3 $\pm$ 0.1 mm a^{-1}. From the end of the Late Pleistocene onward, the average rate has been 0.25 $\pm$ 0.10 mm a^{-1}.

A. Main boundary faults of the western side of Wenquan Basins displacing Jurassic sandstone (lens facing the north)

B. Sinter platforms at 104th point along the faults in Wenquan Basin (lens facing the northwest)

Fig. 4.7 Geological characteristics of Wenquan Basin

4.1.4 Active Faults of Lhasa Terrains

Lhasa terrains lie between the suture zones of Bangong Lake–Nujiang River and those of Yarlung Zangbo river, which mainly contain the SN-trending extensional activity faults, including the active faults of Amdo–Cuona Lake, of the western boundary of Gulu–Sangshung faults, of Bengcuo, and of Damxung–Yambajan.

4.1.4.1 Active Faults of Amdo–Cuona Lake

The extensional activity tectonic system of Amdo–Cuona Lake is mainly composed of fractures of Amdo Margin Basin and Faulted Basins. The NE-trending Amdo Basin belongs to the northern active tectonic system; however, the NNE-trending Cuona Lake Basin belongs to the southern active tectonic system. Amdo Basin lies in southern Qiangtang Plateau, which is close to the suture zones of Bangong Lake–Nujiang on the northern side and is intermittently joined by Cuona Lake Basin on the southern side. Amdo Basin is >40-km long and 5–12-km wide and has bedrock triangular facets of height 300–400 m that trend N30°–60°E, which vertically cut the water systems around the mountains. Most alluviums and diluviums are accumulated in the piedmont, and the rest are transported to middle Amdo Basin and Cuona Lake Basin, which lies in southwestern Amdo Basin. Amdo Basin is adjacent to Cuona Lake, and both these features share similar tectonic characteristics. Cuona Lake Basin is 9–12-km wide, >26-km long, and is surrounded by valleys in the SN direction. The linear bedrocks and scarps trend N5°–10°E and are >200 and 120 m high on both sides in the EW direction, respectively. Scarps are cut by the vertical basin and the water system in the EW direction at the mountain boundary, forming multilevel bedrock triangular facets. The linear scarps, north–south valleys, altitude difference between adjacent mountains, bedrock fracture zones, and cleavage belts distributed in the basin margin all evidence that Cuona Lake Basin is a faulted basin of the Quaternary controlled by extension active faults. The main active faults of Amdo Basin and Cuona Lake Basin are the fractures of the basin margin, which control the faulted basin, and includes the southern margin faults of Amdo Basin, northern margin fractures of Amdo Basin, eastern margin ruptures of Cuona Lake, and western margin faults of Cuona Lake.

Southern margin fault of Amdo Basin is >40-km long, and its overall tendency is N30°–40°E; it develops visibly into the linear bedrock fault facets, which are around 350 m higher than the adjacent valleys along the faults. Because of the fractures of the Quaternary activities, uplift of the southern faults form an obvious groundwater band-blockade, which induce the linear distribution of frost-mound groups between the fault belts and Sangqu. The northern margin fault of Amdo Basin, which trends 130°–140°, is composed of >50-km long fractures and is intermittently connected by 3 or 4 strips of faults. The northern margin fault of Amdo Basin is a fault belt of the main boundary basin that has the characteristics of the maximum intensity activities of the Late Quaternary and the obvious activities

Table 4.2 Geologic-dating results, vertical offset distances, and vertical displacement velocities of landscape bodies in Wenquan Basin during the Late Pleistocene

Observation location	Sample	Landscape (height in parentheses)	Sample properties	Methods	Age (kaBP)	Vertical dislocation (m)	Vertical displacement rate (mm a^{-1})
204th point	204-1	Top of sinter platform (~45 m)	Calcification	U-series	145.9 ± 15.1	32.1 ± 2.0	0.22 ± 0.04
208th point	208-1	Top of T6 of the tug (~45 m)	Calcareous cement	U-series	143.6 ± 13.2	31.5 ± 2.5	0.22 ± 0.04
~200 m southwest of 211th point	607-3	Bug in Buqu terraces (30–40 m)	Pebble calcic coating	U-series	127.4 ± 9.4	31 ± 3	0.24 ± 0.04
~100 m North of 210th	607-2	Lower section of the sinter platform (~20 m)	Calcification	U-series	103.0 ± 12.6	20–40	0.17–0.44
210th point	210-1	Bug in the top of alluvial fan (20–25 m)	Calcareous cement	U-series	59.3 ± 6.1	18.5 ± 1.4	0.31 ± 0.06
205th point	607-5	Upper section of sinter platform	Calcification	U-series	53.6 ± 6.1	20.8 ± 3.2	0.40 ± 0.11
~100 m north of 210th point	607-1	Top of sinter platform (~20 m)	Calcification	U-series	45.2 ± 4.6	18.1 ± 0.8	0.41 ± 0.06
203th point	203-2	Upper section of T5 with a bug (20–25 m)	Calcareous cement	U-series	31.0 ± 2.9	16.7 ± 1.1	0.54 ± 0.09
~500 m south of 210th point	607-7	Upper section of fault scarps slope (~16 m)	Pebble calcic coating	U-series	>25.3 ± 1.3	14.6 ± 1.0	<0.58 ± 0.07
203th point	203-1	Top of T4 with a bug (10–15 m)	Calcareous cement	U-series	21.3 ± 2.6	9.0 ± 1.3	0.44 ± 0.11
201th point	201-2	Upper section of T3 with a bug (6–10 m)	Silt layer	OSL	22.6 ± 2.2	5.0 ± 0.5	0.23 ± 0.44
211th point	608-3	Upper section of T2 with a bug (5–8 m)	Pebble calcic coating	U-series	21.4 ± 1.5	2.5 ± 0.3	0.12 ± 0.02

(continued)

Table 4.2 (continued)

Observation location	Sample	Landscape (height in parentheses)	Sample properties	Methods	Age (kaBP)	Vertical dislocation (m)	Vertical displacement rate (mm a^{-1})
211th point	211-1	Upper section of T2 with a bug (5–8 m)	Silt layer	OSL	12.1 ± 1.1	2.5 ± 0.3	0.21 ± 0.04
209th point	209-1	Upper section of T2 in fault fall plate with a bug (4–5 m)	Calcareous cement	U-series	18.1 ± 2.1	3.0 ± 0.1	0.17 ± 0.02
209th point	607-4	Upper section of T2 in fault fall plate with a bug (4–5 m)	Pebble calcic coating	U-series	11.4 ± 1.2	3.0 ± 0.1	0.27 ± 0.03
203th point	607-6	Terrace fault scarps slope (6–10 m)	Pebble calcic coating	U-series	>7.2 ± 1.0	9.0 ± 1.3	<1.25 ± 0.31
197th point	197-2	Upper section of T1 with the bug (~1.5 m)	Peat	^{14}C	1.15 ± 0.07	0	0

of the Holocene. The Qinghai–Tibet Railway crosses the fault near Duopuer Meander Mountain Pass. The fracture constitutes bedrock areas of average altitude >5000 m on the northwestern side and the boundary of alluvial and flood fan plain mountains of average altitude >4620 m in the southeastern side. Surfaces from the Late Pleistocene onward remain on the majority of normal fault features, including in Langlong, along the southwest of Zhasha, and in the Yingri zones to the northeast of Rilongma; 120 ± 20 m fault facets can be found in the Balong Ditch and Dingqin zones, including those from the Late Quaternary in the cutting facets ditch and multiple fault scarp landscapes formed by the reiterative breaking and vertical dislocation of alluvial and flood fans. The latter serves as important surface markers for determining the average and minimum vertical activity rate, which has been calculated to be 0.41 ± 0.22 and 0.24 ± 0.02 mm a^{-1}, respectively. The eastern margin of Cuona Lake Fracture is >25-km long and trends N70°–80°E. The fracture is composed of two strips of secondary normal faults. Fault facets of height 150–220 m control the outcrop of underground water in the uplifted and downthrown sides along the eastern fault. The fault of the western margin of Cuona Lake intermittently connects the western sections of the northern margin fault of Amdo Basin, which is mainly composed of a east-dipping normal fault and facets of height 100–150 m; obvious fault scarps are also present in this landscape.

4.1.4.2 Faults of the Western Boundary Area of Gulu–Sangshung Basin

The faults of the western boundary areas of Gulu–Sangshung Basin (namely the faults of the western boundary of Yambajan–Damxung–Gulu graben), which mainly trend NE and run for >200 km in the southeast, belong to the northern sections of the faults of Nyainqentanglha Mountain. The Gulu–Sangshung Basin trends N10°E and is 6–8-km wide and 50–60-km long. The western basin connects the Nyainqentanglha Mountain and the northern side of the basin, which lies in Qingtang Plateau; it has relatively gentle terrains of height >4900 m. The eastern and southern sides of the basin have alpine gorge landscapes crossed by valleys. Gulu–Sangshung Basin mainly includes the active faults of the main boundary of the western basin, of the inner basin, and of the eastern basin boundary. However, the active ruptures of the main boundary of the western basin are the largest scale and the most active and thus strongly control the formation and evolution of the basin.

The active fractures of the main boundary of western Gulu–Sangshung Basin are mainly of 100°∠55°–65° and control the western boundary of graben basin of the Quaternary in Gulu–Sangshung, which extends 54 km in the SN direction; recent tensional, compressional, torsional, and multiphase fault activities with different stages and properties have occurred here. The faults cut off slates of the Permian, ice water of the Middle-Late Pleistocene, and the drift sheets, alluvial fans, and regionally shifting alluvial fans of the Holocene (Fig. 4.8a). Loose fault gouges and generous fault breccias are present in the faulted structure surface, constituting the

A. Active fractures of the main boundary of southwestern Gulu Basin incising and displacing the river terraces of the Llate Pleistocene and Holocene (lens facing the north)	B. Structural geomorphological features of the active faults of the main boundary of Gulu–western Sangshung Basin (lens facing the southwest)

Fig. 4.8 Structural geomorphological features of the active faults of the main boundary of Gulu–western Sangshung Basin

boundary of the basin–mountain landscapes and forming high and multilevel fault scarps (Fig. 4.8b).

Figure 4.9 clearly shows the multiphase features of the various activity periods of the piedmont fault as well as the multilevel fault scarps, fresh fault mirrors, and dextral strike-slip characteristics. From the Pleistocene onward, the vertical displacements of the faults are 6–50 m. Some horizontal and vertical displacements since the Holocene are 40 and 80–120 cm high, respectively. Some fractures cut off the moraine soil of the Middle Pleistocene and form an indentation in the sections of the lower plates, with vertical and horizontal displacements of 60 and 36 cm, respectively. Some ruptures result in contact between the Jurassic strata and the drift sheet faults of the Middle Pleistocene, forming Gaotai stations of height 38 m and fracture zones of a width exceeding 6 m, which have since developed into fault

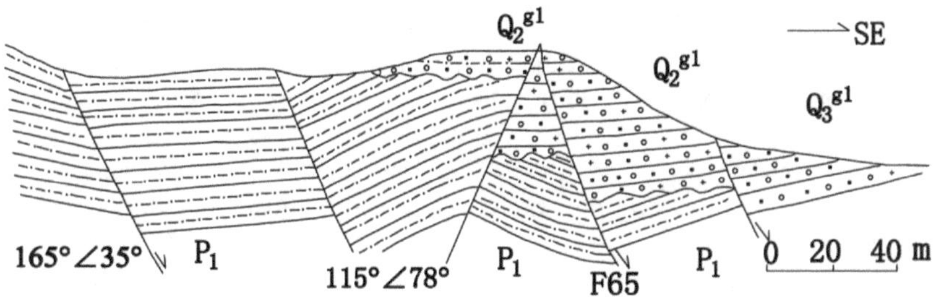

P_1: Slates and phyllite of the Permian, Q_2^{gl}: Drift sheets of the Middle Pleistocene, and Q_3^{gl}: Ice–water deposition layers of the Upper Pleistocene

Fig. 4.9 Geological profile of the western boundary of Gulu Basin gouges among the gray-black metamorphic sandstone, with the 95°∠48° faults. Seismic reflection explorations have revealed that in the F65 faults, two strips of piedmont fractures cut off the drift sheets and ice–water layers of the Middle-Late Pleistocene, one of which is under hidden state and is covered by the Holocene layers. From the Holocene onward and to date, the faulted zone has been strongly active, as evidenced by the large difference in the vegetation characteristics of the upper and lower plates, the linear Wenquan groups, fresh fault polish, loose fault gouges, and right-dislocated multilevel scarps, which is associated with the Ms8.0 earthquake of Bengcuo (1951) and the Ms7.5 earthquake of Jiuzin (1952). The cutting fractures indicate that the activity rates of the vertical and horizontal displacements are 3.3–4.7 (downslide) and 4.1–6.1 (left-slide) mm a^{-1}.

4.1.4.3 Active Faults of Bengcuo

The active faults of Bengcuo mainly trend N35°–45°W and are 100–120 km long. It starts at Jiangcuo areas in Bangoin and runs through Lanagou, which itself is at the middle section of the right-lateral strike-slip faults of Karakoram–Jiali, the largest and most active fracture in the inner Tibetan plateau; the active faults of Bengcuo passes through Naka, Naduiduo, Bengcuo, Dagze, and Dasha toward the southeast, crosses into the Nyainqentanglha Mountains, and terminates northwest of Guolongkongma in Gulu. Luokequ, Xingqu, and Darixiongqu Rivers and Jiangcuo, Pengcuo, and Cuoe Lakes are present along the faults, and modern glaciers are present in the eastern sections of the faults at Nyainqentanglha Mountains, which have formed alluviums and proluvials, lake sediments, and glacial deposits of different periods; these sediments also form river terraces, lake terraces, ancient shorelines, and glacial valleys. These sediments and geological landscape bodies are right-laterally dislocated by the faults of Bengcuo; therefore, determining the formation time of the corresponding sedimentary strata and geological landscapes is the foundation for judging the ages of the fault activities as well as the activity rate. The active faults of Bengcuo are mainly composed of three secondary faults of Naka–Weiduiduo, of Dagze–Dasha, and of Siligongnan–Guolongkongma. The faulted areas shift a series of gullies, ridges, riverbeds, river terraces, and swamps, which control the southern and northern boundaries of Bengcuo. Because of the pull-apart effect of the strike-slip fractures, the distributed areas parallel to the secondary faults form rectangular or diamond-shaped pull-apart basins, such as those at Bengcuo and Longqing–Siligong. The differences in the displacement at each section of the active faults of Bengcuo (Table 4.3) result in the following features: first, obvious segmentation, such as the maximal displacement of 1–1.5 km in Dagze–Dasha sections, multistage maximal displacement of 750 m in the Naka–Weiduiduo. Second, the vertical displacement is lower, and the distribution is partial. Fault scarps and facets only have developed on both sides of two pull-apart of Bengcuo and Dasha. Third, most of the displacement range is proportional to the length of faults, such as the maximal displacement (>1 km) only

Table 4.3 Segmentation profiles of the active fractures of Bengcuo

Segmentation		Segment characteristics							
Section	Subsection	Strike	Length (km)	Mode of motion	Displacement amplitude		Average slip rate (mm a^{-1} since 1000a)	Number of movements	Connection structure
					Horizontality	Verticality			
Naka–Naduiduo		N60°W	60–70	Light-lateral strike	750	2–3		3–4	
Bengcuo		N45°W	26	Normal light-lateral strike	0.8	1.5		1	Pull-apart basin
Dagze–Dasha	Dagze–Nagaer, Nagaer–Dasha	N60°W	60–63, 20–22	Light-lateral strike	1000–1500	20	10–11	5–6	Uplift bulge
Dasha		N43°W	4	Normal light-lateral strike	20	2–3		2	Pull-apart basin
Siligongnan–Guolongkongma		N50°W	10–13	Light-lateral strike	20–30			2–3	

occurring in the fracture of length 80 km in Dagze–Dasha. Fourth, due to multi-stage movement, the number of distinct movements differs, with Dagze–Dasha section having the most (5–6). Fifth, the rate of the horizontal displacement is largest in the Dagze–Dasha section (12–15-m horizontal displacements in the front margin scarps of the secondary terrace of the northwestern Dagze, and the 45–55-m vertical displacements in the gully of southwestern Balie); hence, since 2700a, the average rate of horizontal slip has been between 4.4–5.6 and 16.7–20.4 mm a^{-1}. The maximal displacement shows that the fault activity is equivalently strong.

4.1.4.4 Active Faults of Damxung–Yambajan

The active faults of Damxung–Yambajan lie in the Nyainqentanglha Mountains in the southeastern side, whose main attitude is 115°–180°∠50°–66°, are mainly composed of normal faults and left-lateral strike-slip fractures, and are typical tenso-shear left-laterally active faults. Movement from different activity periods can be found along the active faults of Damxung–Yambajan. These faults are composed of many faults trending NNE, NE, NEE, and NW in the developed basins. The fractures have complicated space segmentation, with obvious ruptures in the direction of parallel strike and verticality. From Nyainqentanglha Mountains to Damxung–Yambajan Basins, the vertical faulted strike is divided to the stepladder fractures of the main boundary of western basins, the active faults of western basins, and the active faults of eastern basins.

The faults of the main boundary of western basins are composed of many secondary fractures of right column juxtaposition, forming the boundary between Nyainqentanglha bedrock areas and basins of accumulation areas of the Quaternary. Fault facets of height 600–1000 m, stepped fault scarps, and bedrock fault friction surfaces have developed along the fractures (Fig. 4.10a, b). The active faults of western basins mainly have development along the drift platforms of piedmont of the Early Middle Pleistocene and consist of the right column distribution of secondary faults, which displace different ages of the Quaternary strata and are the boundary faults forming high moraines of piedmont or ice–water platforms, such as high moraines of height >200 m or ice–water platforms that are mainly distributed in the upper fracture disks. The active fractures of the eastern basin mainly lie in the middle section of eastern basins, which are intermittently distributed in the various secondary fault basins, mainly composed of three main ruptures of Horse Farm, Laduogang, and Qudengluo with worse continuity in the NE direction and some small-scale secondary fractures, and the faults cut off sedimentary of the Late Pleistocene and then form fault cliffs and scarps (Fig. 4.10c, d). Main fractures often form the boundary of the middle-low lying marshes of the basin, which control the distribution of Wenquan with a fracture in the NW direction, such as Yambajan, Laduogang, and Ningzhong.

A.Northwestern boundary fault of Damxung Basin displacing Middle Pleistocene moraine and then forming terrance fault scarps (lens facing the southwest)

B.Northwestern boundary fault of Galuogunba in Yambajan Basin cutting off bedrock and forming cliffs and mirrors (lens facing the southwest)

C.Eastern boundary fault of Damxung Basin cutting off Late Pleistocene sedimentary and forming fault cliffs (lens facing the northwest)

D.Northern strike-slip fault of northeastern Wuma ponds in Damxung Basin producing seismic scarps and sag ponds (lens oriented 70°)

Fig. 4.10 Landscape characteristics of Damxung–Yambajan active faults

4.2 Temporal and Spatial Distribution of Seismic Belts

The collision of Indian plate, which is moving EN at a stead pace, forms active faults in the Qinghai–Tibet Plateau. Consequently, the Qinghai–Tibet Plateau is currently the continental plateau with the strongest tectonic activity on Earth. Strike-slip motion rates of parts of the faults are 10–20 mm a^{-1}, and absolute motion velocities of the lithosphere are 25–40 mm a^{-1}, 5–10 times that in mid-eastern China and in the European continent. Uneven stick-sliding movement of the active fractures can gestate earthquakes of different intensity. Examples of Earthquakes stronger than Ms7.5 along the Qinghai–Tibet Railway are the Ms8.0 Yambajan earthquake (1411), Ms8.0 Bengcuo earthquake (1951), Ms7.5 Jiuzina earthquake (1952), Ms7.9 Nima earthquake (1997), and Ms8.1 Kunlun Mountain earthquake (2001). Thus, active fault belts strongly influence earthquakes, and many active faults in the Qinghai–Tibet Plateau are regionally active earthquake belts.

The Qinghai–Tibet Railway crosses seismically active areas of the Qinghai–Tibet Plateau. Seismotectonic zones that have a larger effect on the Qinghai–Tibet Railway involve twelve seismotectonic zones with different classes, which from north to south are as follows: seismotectonic zones of Kunlun Mountains, seismotectonic zones of Hoh Xil area, seismotectonic zones of Wulanwula Mountain–Fenghuo Mountain, seismotectonic zones of Tongtianheyan, seismotectonic districts of Yanshiping–Munai Mountain, seismotectonic regions of Tanggula Mountains, seismotectonic districts of Touerjiu-Suoxian, seismotectonic zones of Bengcuo, seismotectonic regions of Namucuo, seismotectonic zones of Gerencuo, seismotectonic districts of Gulu–Sangshung, and seismotectonic regions of Damxung–Yambajan. The diagonal distribution of secondary seismotectonic regions of Tongtianheyan, of Yanshiping–Munai Mountain, of Tanggula Mountain form seismotectonic districts of northern Tanggula Mountain. The diagonal distribution of three secondary seismotectonic regions of Bengcuo, of Namucuo, of Gerencuo forms seismotectonic zones of Jiali–Bengcuo. Two secondary seismotectonic zones of Damxung–Yambajan, Gulu–Sangshung produce seismotectonic regions of Yambajan–Damxung–Gulu (Fig. 4.11 and Table 4.4). This section introduces the relationship of earthquakes and active faults, especially in the seismotectonic zones of Kunlun Mountains, Bengcuo, and Yambajan–Damxung–Gulu.

4.2.1 Relationship Between Earthquakes and Active Fault Belts

Earthquake epicenter distribution data show that the vast majority of strong shocks occur in the active faults or the fault basins of the lithosphere, crusts, or bedrock strata of some stable block edges, and strong shocks occur very rarely in the inner stable blocks. Tarim blocks, Junggar blocks, Ordos blocks, are stable blocks, around which active fractures or faulted basins of the Quaternary are strong earthquake belts.

In China, 18 earthquakes of strength Ms8.0+ have occurred along the lithosphere faults that extend more than tens of hundreds of kilometers or in the fault basin; 63 earthquakes of Ms7–7.9 have occurred in the Chinese mainland, the vast majority of which occur in the crustal active faults of length >100 km or in the fault basins controlled by those fractures. More than 90% of the epicenters of 329 Ms6–6.9 earthquakes are associated with controlled fault basins and active fractures whose length are more than tens of kilometers; the relationship of the rest of the earthquake epicenters and active fault belts is unclear.

Approximately 15% of the total earthquakes related to fractures occur in the strong activity segments of actively deep-large fractures. The age of fault activity is closely related to strong earthquake activities. In China, new active faults are divided on the basis of the obvious activity actors as since Quaternary, since Neogene, and since Cenozoic. Ms6.0+ earthquakes in the continental areas in these three types of faults have a 7:2:1 proportion.

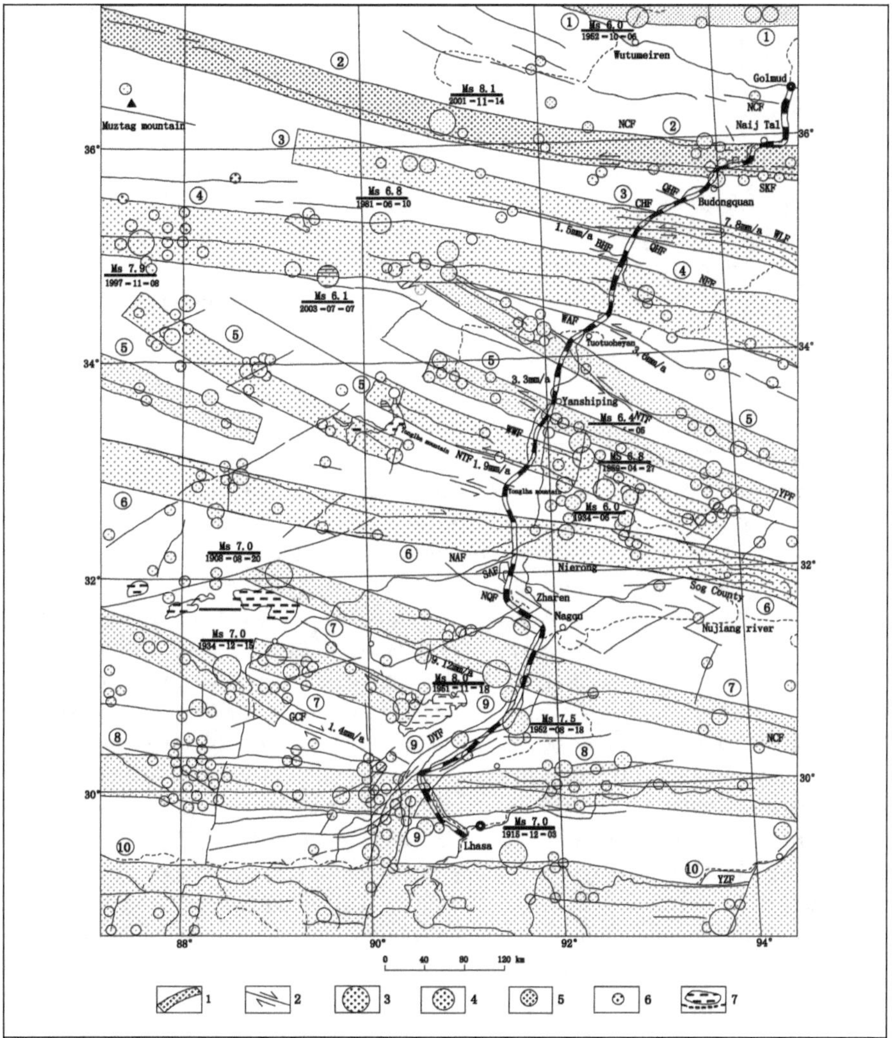

1: Regional seismic belts, 2: Active fault belts, 3: Ms7.0+ earthquakes, 4: Ms6.0–6.9 earthquakes, 5: Ms5.0–5.9 earthquakes, 6: Ms4.0–4.9 earthquakes, 7: Modern lakes and main rivers

① seismotectonic zones of northern Qaidam Basins in Wutumeiren, ② seismotectonic regions of Kunlun Mountains, ③ seismotectonic belts of Hoh Xil areas, ④ seismotectonic districts of Wulanwula Lake–Fenghuo Mountain, ⑤ seismotectonic regions of northern Tanggula Mountains, ⑥ seismotectonic districts of Bangong Lake–Nujiang, ⑦ seismotectonic zones of Jiali–Bengcuo, ⑧ seismotectonic regions of Gongbo'gvamda County–Songduo, ⑨ seismotectonic districts of Yambajan–Damxung–Gulu, ⑩ seismotectonic regions of Himalaya

Fig. 4.11 Seismotectonic map along the Qinghai–Tibet Railway

4.2.2 Seismotectonic Zones of Kunlun Mountains

Large-scale seismotectonic belts of Kunlun Mountains are important active fault belts and seismotectonic belts on the northern Qinghai–Tibet Plateau, whose length exceeds 1500 km and mainly trend NWW. According to fault structures, structural types, and spatial occurrence, from west to east, the active fault belts are divided into six secondary seismic faulted structures along Kunlun Mountains, as faults of Kusai Lake, of Xidatan, of Alake Lake, of Tuosuo Lake, of Dongqinggou, and of Maqu. In the western sections of the active fractures of eastern Kunlun, faults of Kusai Lake and of Xidatan have a juxtapositional and the composite relationship. Faults of Kusai Lake are located in southern Kunlun Mountains, whose structural activities are currently highly intense. The Ms7.0 earthquake of November 4, 1902, in the western section of Alake Lake; and the Ms7.5 earthquake of January 7, 1937, in the valleys of Huashixia; the Ms7.0 earthquake of April 19, 1963, in the valleys of Huashixia; the Ms6.8 earthquake of March 14, 1971, in the Tuosuohe areas; the Ms8.1 earthquake of November 14, 2001 in the Kunlun Mountains; and numerous ancient seismic events have occurred along the earthquake zones of Kunlun Mountains. The epicenter of Kunlun Mountains was located in the vicinity of Kusai Lake, and the instrument recording the epicenter is located in Bukadaban Feng. The strong Ms8.1 Kunlun Mountains earthquake produced earthquake rupture zones of length 426 km on the surface, cutting through the Qinghai–Tibet Highway, the communication cables, the Qinghai–Tibet Railway pioneer roads, and other engineering facilities; triggered avalanche in Kunlun Mountains; and destroyed houses in Xidatan, causing concerns among both domestic and overseas geologists. Numerous ancient earthquake ruptures with echelon distribution and earthquake bulges can be seen in the southern Xidatan Basin. Cut by active faults, weakly deformed gravel rocks of the Late Pleistocene are covered by approximately horizontal sand layers in the Holocene. Sand layers develop into new seismic ruptures in the overlying Holocene. Since seismic activities generate the earthquake faults of approximate length 100 km, with great deformation (i.e., horizontal and vertical distances and bulge length), as explained in Table 4.5. Two hundred years ago, $\geq$Ms7.5 earthquakes, which are rare, with horizontal displacement rate of 2.5 cm a^{-1} had occurred in Xidatan areas. According to earthquake trend analyze, the activity intensity of each segment varies with the geological period across the East Kunlun fault belts. After the Ms8.0 East Huashixia earthquake in 1313 turned west during the beating, earthquakes of magnitude Ms7.5 and 7.0 occurred in 1937 and in 1963, respectively, and an Ms8.1 earthquake occurred in the westernmost section of the northern Bukadaban Feng on November 14, 2001, with migration time of 30–40 years. Smaller earthquakes of around Ms5.0 occur more frequently in the fault zones. The ground stress is mostly concentrated, and earthquakes are more likely to occur in sections that are bifurcating or are the intersection, turning, or ending points. Thus, these fault zones have a high probability of being the epicenter of the devastating earthquakes (at least 1000a) of magnitude Ms8.0; these zones are also the longest and the most obvious potential strong earthquake belt crossing the central section of Qinghai. Therefore, railway crossing the section must be designed with these considerations.

Table 4.4 Main seismotectonic belts and characteristics along the Qinghai–Tibet Railway

Serial numbers and designations	Fault belts of main activities	Fault property	Earthquakes of magnitude ≥ Ms6.0
1. Seismotectonic zones of Wutumeiren in the north of Qaidam Basin	Faults of buried activities in the north of Qaidam Basin	Right-lateral strike-slip normal fault	Earthquake of Wutumeiren on October 6, 1952 (Ms6.0)
			Earthquake of Xitie Mountain on January 19, 1977 (Ms6.3)
2. Seismotectonic zones of Kunlun Mountains	Active faults of Naij Tal	Left-lateral strike-slip	Earthquake of Kunlun Mountains on November 14, 2001 (Ms8.1)
	Active faults of Xidatan		
	Active faults of the southern margin of Kunlun Mountains		Earthquake of Huashixia on April 19, 1963 (Ms7.0)
3. Seismotectonic zones of Hoh Xil	Active faults of Chuermahe	Left-lateral strike-slip	Ancient earthquake of Ms7.0–7.5 with coseismic displacement of 4–5 m
	Active faults of Wudaoliang		
	Faults of the north of Hoh Xil Mountains		≥ Ms6.0 Earthquakes since 1950
	Active faults of the south of Hoh Xil Mountains		
4. Ulan Ul Lake seismotectonic zones of Fenghuo Mountain	Active faults of Beilu river	Left-lateral strike-slip	Earthquake of Nima on November 8, 1997 (Ms7.9)
	Active faults of Fenghuo Mountain		Earthquake of Xijir Ulan Lake on June 10, 1981 (Ms6.8)
	Active faults of the north of Erdao Ditch Basin		Earthquake of Ulan ul on May 2, 1920 (Ms6.2)
	Active faults of the south of Erdao Ditch Basin		Earthquake of Dogai Coring on July 7, 2003 (Ms6.1)
5. Seismotectonic zones of the north of Tanggula Mountain, include a series of oblique secondary seismotectonic zones, such as Tongtianheyan, Yanshiping–Munai	Active faults of Tongtianheyan	Left-lateral strike-slip normal fault	Eastern earthquake of Tanggula Mountain on June 5, 1975 (Ms6.4)
	Faults of Yanshiping–Munai Mountain		Eastern earthquake of Tanggula

(continued)

Table 4.4 (continued)

Serial numbers and designations	Fault belts of main activities	Fault property	Earthquakes of magnitude ≥ Ms6.0
Mountain, and Tanggula Mountains			Mountain on April 27, 1959 (Ms6.0)
	The active faults of northern margin of Tanggula Mountain		Eastern earthquake of Tanggula Mountain on June 23, 1934 (Ms6.0)
6. Seismotectonic zones of Bangong Lake–Nujiang	Active faults of Touerjiu Mountain	Right-lateral strike-slip	No ≥ Ms6.0 earthquakes have occurred since 1950
	Active fault of the north of Suoxian County		
	Active faults of northern hills of Amdo		
7. Seismotectonic zones of Jiali–Bengcuo, include a series of oblique secondary seismotectonic zones, e.g., Bengcuo, Namucuo, Gerencuo	Strike-slip faults of Jiali–Bengcuo	Right-lateral strike-slip	Earthquake of Bengcuo on November 18, 1951 (Ms8.0)
	Transverse rupture of Namucuo-Ningzhong		
	Active faults of Gerencuo		Earthquake of Xungmei on December 15, 1934 (Ms7.0)
8. Seismotectonic zones of Gongbujiangda-Songduo	Oblique distribution of Gongbujiangda fault with the direction of nearly SW in southern Lhasa block	Right-lateral strike-slip	No ≥ Ms6.0 earthquakes have occurred since 1950
	The active faults of Songduo		
9. Seismotectonic zones of Yambajan–Damxung–Gulu, include secondary seismotectonic zones of Gulu–Sangshung, Yambajan–Damxung	Active fault of main boundary of Yambajan–Damxung Basin	Stretching structural normal fault	Earthquake of Yambajan on September 28, 1411 (Ms8.0)
	Faults of the boundary of Gulu–Sangshung		
	Faults of Cuona Lake–Amdo		Earthquake of Jiuzina on August 18, 1952 (Ms7.5)
10. Seismotectonic zones of Himalaya	Thrust faults of Himalaya Mountain	Right-lateral strike-slip thrust fault	Eastern earthquake of Qushui on December 3, 1915 (Ms7.0)
			Southern earthquake of Milin on July 23, 1947 (Ms7.7)
	Strike-slip faults of Yalu Tsangpo River		Earthquake of Chayu on August 15, 1950 (Ms8.5)

4.2.3 Seismotectonic Zone of Jiali–Bengcuo

Seismotectonic zone of Bengcuo mainly trends NW–NWW and belongs to the important diagonal seismic belts of Jiali–Bengcuo, which is controlled by the strike-slip faults of Jiali–Bengcuo and is composed of the echelon distribution of the active faults of NW–NWW. Southeast through Jiali, Bomi intermittently extends to Chayu and to the northeastern side of Selincuo along the northwest, and its main body is distributed west of Gulu–Sangshung Basin. The seismotectonic zone is very strong in western sections of Gulu Basin activities, but it is weak in the Dongjiali segments of Gulu Basin activities. The seismotectonic zone was where the Ms8.0 Bengcuo earthquake of November 18, 1951, and the Ms7.5 Jiuzina earthquake of August 18, 1952, and a few eastern earthquake activities occured. Seismotectonic zones in Bengcu mainly include the active faults of Bengcuo, southern faults of Samli, and eastern faults of Jiali. Jiali faults have developed into ancient earthquake ruins, which exhibit a diagonal distribution and control some earthquake activities (Ms4.0–6.2). Bengcuo faults right-laterally displace bedrock mountains, ridges, gullies, glacial valleys, and lake terraces, forming pull-apart basins with multiperiod seismic ruptures, multistage right-lateral displacements, multilevel seismic scarps, and different sizes. Ms8.0 Bengcuo earthquake forms the surface rupture zones and the diagonally distributed beaded earthquake bulge, and the epicenter area is with the seismic displacement of approximately 8 m. According to the observation data of fault displacements and activity times, in the active faults of Bengcuo of the Late Pleistocene–Holocene, the rate of right-lateral strike-slip movement is estimated to be 7–16 mm a^{-1}, and the average rate of slip movement to be 11.5 ± 4.5 mm a^{-1}, and the recurrence period of Ms8.0+ earthquakes to be 900 ± 500a in the Holocene. In 1951, the basic characteristics of the Bengcuo earthquake were obtained by investigating some obvious marks, including a cut mass of gullies, riverbeds, floodplains, terrace scarps, and swamps boundary on the fractures in the field. The characteristics are as follows: horizontal displacement is greater than the vertical displacement, and the maximum horizontal displacement is in the middle segment of Jiangri Mountain–Nagaer (specifically, the maximum horizontal displacement of 7.3 m is in the west of Dagze and the average horizontal displacement is 3.5 m). The maximum vertical displacement of only 1.5 m appears in the internal tension–shear fracture segment of Bengcuo. If an earthquake occurred in 1951, counting up to the 1990 survey, with 39-year terms, the maximum horizontal slip rate is 18.9 cm a^{-1}, and the average slip rate is 9 cm a^{-1}. By contrast, the vertical slip rate is only 0.9 cm a^{-1}. Second, the maximum horizontal displacement occurs mainly in the shear zone, and the maximum vertical displacement is mainly in the tension area. In the maximum horizontal displacement occurrence section, the vertical displacement is relatively small, but the maximum of the horizontal and vertical displacement never appears in the same place.

4.2.4 Seismotectonic Zones of Yambajan–Damxung–Gulu

Seismotectonic zones of Yambajan–Damxung–Gulu mainly include the seismotectonic zones of Gulu–Sangshung, seismotectonic zones of Yambajan–Damxung.

The overall seismotectonic zone of Gulu–Sangshung trends nearly SN for more than 250 km, lies in the northern seismotectonic zone of Yambajan–Damxung–Gulu, and is an important section of the stretching structural system of Yadong–Yambajan–Gulu. North through Cuona Lake Basin and Amdo Basin, it intermittently extends to the east side of Tanggula Mountain, after which it forms a small pull-apart basin. Extending southward, it connects with the seismotectonic zone of Yambajan–Damxung. The main controlling fault of the Gulu–Sangshung seismotectonic zone is the fault of the Gulu–Sangshung basin boundary. According to historical macroearthquake records, the largest earthquake was the Ms7.5 earthquake in Jiuzina, whose epicenter was in the southwest of Gulu in 1952. The macroscopic intensity of the magistoseismic areas is up to 10°, with a long-axis length of >40 km. Occurring in the surface seismic rupture zones and extending from Yanta areas in northwestern Gulu Basin to Zhuogaruo areas in northeastern Damxung Basin, its overall length is more than 60 km, forming 3–6 strips of earthquake scarps in the western side of Gulu (Fig. 4.12a–c) and forming pull-apart basins, earthquake bulges, and reverse scarps in variable fracture occurrence areas (Fig. 4.12d). The cumulative vertical displacement of the surface rupture zones is 0.6–5 m, the maximum of which is 4–5 m, and the average is 2–3 m. Several ancient earthquakes have occurred along the fault zones since the Late Pleistocene. In northwestern Samli, many ancient earthquake activities have formed multistage fault scarp landscapes, with a total cumulative displacement of 16–17, 14–15, 11–12, 8–9, and 4–5 m in the five most recent activities, respectively. The minimum displacement was formed in the Jiuzina Ms7.5 earthquake in 1952, but the highest fault scarp displaces the fluvioglacial terrace whose age is 11–12 ka. Based on the projections, the repetition period of ancient earthquakes stronger than Ms7.0 is roughly 2.2–2.4 ka. In addition to the landscape logo, stratigraphic markers on ancient earthquake can be observed along the faults. Preliminary studies have found the last two sets of colluvial wedges caused by ancient earthquakes, which had been formed before 1952. The most typical faulted colluvial wedge profile is more than 4 km from the Sanglixi area, because of the displacement caused by the surface rupture zones of Jiuzina Ms7.5 earthquake in 1952, suggesting that there had been two ancient earthquakes before 1952. The ages of the upper colluvial wedges collected from silty soil layers which are formed by the earliest ancient earthquake are respectively 4.18 ± 0.07 kaBP and 3.2 ± 0.3 kaBP in ^{14}C and TL samples. This shows that including the earthquake in 1952, at least two earthquakes of magnitude exceeding Ms7.0 have occurred along the faults since 4.18 ± 0.07 kaBP or 3.2 ± 0.3 kaBP, and the corresponding paleoseismic periods should be 2100–1600a. In summary, ancient earthquake periods is 2400–1600a and the average repeated cycles is 2000 ± 400 along seismotectonic zones of Gulu–Sangshung. Because the last >Ms7.0 earthquake occurred more than 50 years

Table 4.5 Active faults of ancient earthquake events in Dongxidatan

No.	Time before present (a)	Magnitude	Epicenter locations			Horizontal offset (annual average rate)	Vertical offset (annual average rate)	Bulge length (m)
			Northern latitude	Eastern longitude	Site			
1	11,000	Around 7.5		94°03′	Jingxiangukou			
2	10,000	Around 7.5		92°00′	Northern Kusai Lake			
3	8317	8	35°44′	94°53′	Dongdatan	45 m	4 m	23–36
4	5650	≥ 8.0	35°43′	94°08′	Xidatan	18 m (0.32 cm a^{-1})	4.5–8 m (0.1 mm a^{-1})	31–34
5	2700	≥ 7.5	35°43′	94°14′	Xidatan	8 m (0.3 cm a^{-1})	1.1 m (0.04 cm a^{-1})	11–20
6	1000	≥ 7.5	35°43′	94°42′	Dongdatan	14.5 m	2–8 m	5–7
7	200	≥ 7.5	35°43′	94°14′	Xidatan	5 m (2.5 cm a^{-1})	2	7

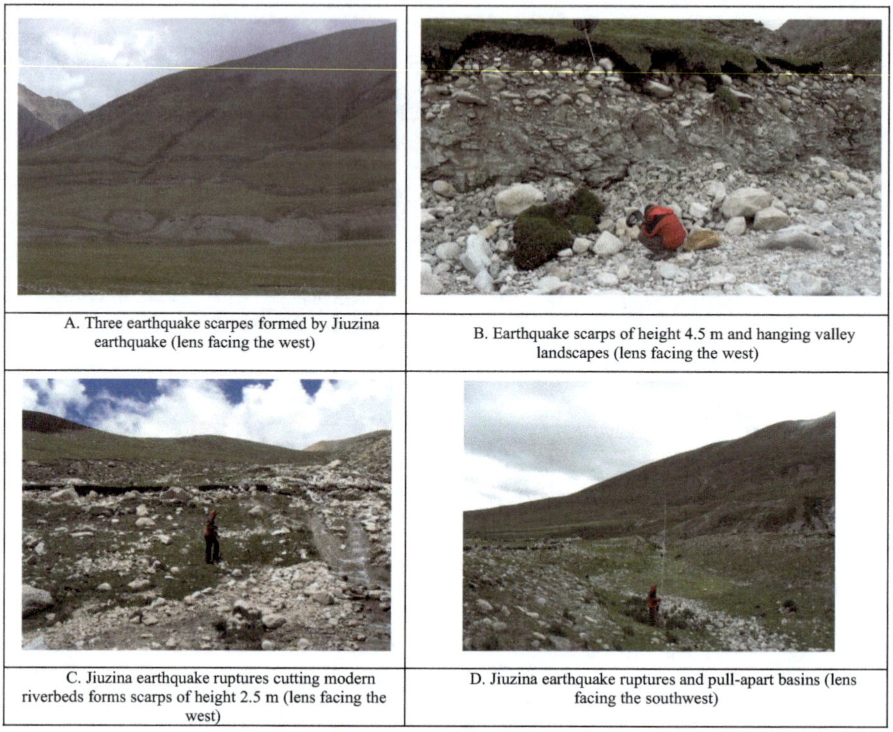

A. Three earthquake scarpes formed by Jiuzina earthquake (lens facing the west)

B. Earthquake scarps of height 4.5 m and hanging valley landscapes (lens facing the west)

C. Jiuzina earthquake ruptures cutting modern riverbeds forms scarps of height 2.5 m (lens facing the west)

D. Jiuzina earthquake ruptures and pull-apart basins (lens facing the southwest)

Fig. 4.12 Landscapes formed by Jiuzina earthquake

ago, >Ms7.0 earthquakes are less likely to occur along the fracture for a long period.

The seismotectonic zone of the overall Yambajan–Damxung is in the NE direction, which is adjacent to the seismotectonic zone of Gulu–Sangshung, consisting of the seismotectonic zone south of Yambajan–Damxung–Gulu, and extends to Nimu, and Jiangzi areas in the WS direction. The main controlling earthquakes of seismotectonic zones of Yambajan–Damxung include those along the active faults of the main boundary basin and margin branch fractures in Yambajan–Damxung and active faults within the basin. The seismotectonic belts and other configurations differ: for engineering areas, the seismic tectonic belt is parallel to the railway piers; a powerful earthquake fault would devastate the entire railway. The earthquake incidence of the seismotectonic zone of Yambajan–Damxung is much larger than that of Xidatan. Seismic records show that the vast majority of earthquakes concentrate in Nyainqentanglha Mountain Fractures in the southeast, indicating that the fault is a major controlling seismogenic structure in the areas. The strong (Ms8.0) Yambajan earthquake occurred on September 28, 1411, and a series of medium and small earthquakes of Ms4.0–5.5 have occurred in the earthquake tectonic zone. In 1411, the strong Yambajan Ms8.0 earthquake, whose epicenter was located in the southwest of Yambajan Basin, formed a multimagnitude earthquake scarp (Fig. 4.13a) in the northwestern boundary of the basin in Nimu, Jiedaguo, Yambajan, Ningzhong, and Damxung. Visible modern alluviums, proluvials, and slope wash phenomena (Fig. 4.13b) cut by the earthquake rupture zones can be seen in Hagongtang, Laqu, Gurenqu, Ranbuqu, and western Sibuqu of Jiedaguo. The vertical displacement is mostly 1–6 m, the maximum may reach 6–7 m, and the average is 2–4 m. Many ancient earthquake ruins are present along the fault zones, and fault scarp landscapes formed by numerous ancient earthquakes are the most typical in the toe of the Laqu gully of northwestern Ningzhong. East of Mizoguchi, the recent three vertical dislocations of normal faults of piedmont make river terraces with the tug of 8–10 m and with the OSL age of 10–12 kaBP form fault scarps and increasing heights of 2.8, 6.2, and 8.5 m along the fault zones, where the fault scarp formed by a recent activity is fresh. Latest faulted colluvial wedges are present in the lower sections of fault scarps, which were the products of the Ms8.0 earthquake in 1411. The OSL age of the terrace shows that the cycle of the Ms8.0 earthquake would not be greater than 3200–3700a around the areas. Northeast of Damxung County cities, two crossing sections of fault colluvial wedges reveal four and three earthquake events, respectively. They are exposed to the lower sections of the fault scarps formed by the alluvial fan with 2–15 kaBP fault vertical ages, indicating that these ancient earthquakes mainly are those since approximately 12 kaBP, that is, at least four ancient earthquakes have occurred along the fraction zones of the Holocene indicating that the maximal cycle of ancient earthquake is around 3000a. According to the average vertical displacement ($\sim$3.5 m) of the last major earthquake and the average vertical activity rate of the Holocene faults (1.5–1.8 mm a^{-1}), repetition periods of earthquakes of Ms8.0 are between 2000a and 3000a, and the average earthquake recurrence periods are 2500 $\pm$ 500a. Since the recent Ms8.0 earthquake is nearly 600a, the hazard about the earthquake zones around Ms8.0 is not very large. Because of the high activity

Fault earthquake scarp of northern boundary of Damxung Basin (lens facing the north)	A.Remained glacial deposits after avalanche ablation triggered by Gurenquogutou earthquake (lens facing the northwest)

Fig. 4.13 Landforms formed by Yambajan Ms8.0 earthquake

rate of the fault zone in the Holocene, the occurring risk of earthquakes of Ms6.0–7.0 is higher, both two earthquake ruptures were not extending to the northern area of Damxung Basin in 1411 and in 1952.

4.2.5 Other Seismic Tectonic Belts

The overall seismic tectonic belt of Hoh Xil runs NWW, controlled by the active faults of Hoh Xil. The main controlling earthquake of Hoh Xil seismic tectonic belt includes the active faults of Chumar River, Wudaoliang, northern Hoh Xil Mountains, and southern Hoh Xil Mountains. The seismic tectonic belt of Donggu–Luhuo–Kangding segment has obvious strong Xianshui River seismic belt. Strong earthquakes of Ms7.0–7.5 along the active faults of southern Hoh Xil Mountains in the Holocene, forming earthquake scarps of length >20 km and height 1.5 m. The scarps cross the Qinghai–Tibet Railway, and they cut modern turfs and sand layers of the Holocene, left-laterally dislocating modern gullies of 4–4.5 m (Fig. 4.14) along ancient earthquake ruptures west of the Qinghai–Tibet Railway. The development of earthquake ruptures and the pull-apart basins in the earthquake scarp of 5–10 m, the occurrence of the Holocene sandy tectonic wedge of earthquake ruptures in tension–torsional in the NE direction, in addition to the formation of perennial frost heave hugelboden in the axial direction of NE after the earthquake show that the strike-slip faults of Hoh Xil in the Holocene have strong tectonic activities.

The seismic tectonic zone of Tongtianheyan of length >1000 km is located in the northern Tonglha Mountains, the overall trend of which is in the NW direction; it belongs to the northern branch of diagonal seismic tectonic belt in the northern Tonglha Mountains, and it is controlled by active fault zones of Xijinwulan Lake–Tuotuo River, developing into the distribution of the linear fault scarps and sinter platforms, which cuts the sinter deposits of the Late Pleistocene–Holocene and

A. Hoh Xil earthquake rupture crossing the Qinghai–Tibet Railway (lens facing the east)

B. Ancient Hoh Xil earthquake faults (lens facing the east)

Fig. 4.14 Ancient earthquake ruins of seismic tectonic zones of Hoh Xil controls the direction of Ms4.0–6.0 earthquakes and the distribution of Wenquan under low temperature. The major controlling fault of seismic tectonic zones of Tongtianheyan is the active fault of the northern Tongtian River, followed by the active faults of Tuotuo River, Kaixinling, and Bumalangnaon, which have not highly significant effect on modern seismic activities. An Ms6.2 earthquake occurred on May 2, 1920, in Ulan Ul, and another of Ms6.8 on June 10, 1981, in Xijir Ulan Lake in the cross composite sections of active faults among the northern Tongtian River, Erdao Ditch, and Fenghuo Mountains, respectively.

Seismic tectonic zone of Yanshiping–Munai Mountain whose total trend is in the NW direction belongs to central branches of diagonal seismic tectonic zones of northern Tonglha Mountains, which is controlled by active fault zones of Yanshiping–Munai Mountain. The seismic tectonic zone merges to major active faults extending to the NW direction, is limited by active faults in the nearly EW direction, and extends to the ES direction, incorporating into the active faults of Tonglha Mountains over a total length around 300 km. The main controlling earthquake fault is the active fault of Munai Mountain. Ancient earthquakes of magnitude Ms4.2–5.4 have occurred along the active fault of Munai Mountain on both sides of east–west in the Qinghai–Tibet Highway and Railway. However, since the earthquake records began, no strong earthquakes of magnitude Ms6.0–7.0 have occurred in the seismic tectonic belt.

The major seismic tectonic zone of Tonglha Mountains is in the NW direction, controlled by the active faults of northern Tonglha Mountains, regionally extending more than 800 km, and are an important part of the diagonal seismic tectonic zone of northern Tonglha Mountains. The main controlling faults are active faults (NTF) of northern Tonglha Mountains and main boundary faults of pull-apart basin in the nearly SN direction on the eastern side of Tonglha Mountain Pass. In the eastern side of Tonglha Mountain Pass, along NTF, and main boundary faults of pull-apart basin, a Ms6.0 earthquake occurred on June 23, 1934, and a Ms6.4 earthquake of on April 27, 1959, and an earthquake of Ms6.4 on June 5, 1975, and there were many times of seismic activities of Ms4.0–6.0. The area belongs to the relatively strong tectonic belts of medium-strong earthquakes in central seismic activities of the Qinghai–Tibet Plateau. The active faults of Wenquan are present at the southern margin, crossed by the Qinghai–Tibet Railway and Highway. The current seismic activity of the fault is weak, where no clear ancient seismic remains can be found in the local area.

Seismic tectonic belt of Touerjiu Mountain–Suoxian County is located in the north of Amdo, with the global distribution in the NWW direction, regionally extending more than 700 km, which belongs to the middle of seismic tectonic zone of Bangong Lake–Nujiang River, controlled by active fault belt of Toujiu Mountain–Amdo, and after going eastward into Suoxian-eastern Baqing, it is carved up, and restricted by the active fault of Taniantaweng Mountain. The major the active faults of seismic tectonic zone of Touerjiu Mountain–Suoxian include the active faults of north of Suoxian, the active faults of north of Amdo Mountain and the active faults of Touerjiu Mountain. The active faults of Touerjiu Mountain are major earthquake faults, and there have been multiple ancient earthquakes of

Ms4.0–5.2 in the north of Amdo Basin and on both sides of the east–west of Amdo Basin. However, thus far, no Ms6.0+ earthquakes have occurred here, which belong to the weaker seismic tectonic belts of modern earthquake activities.

The overall seismic tectonic belt of Namucuo trends NWW, with smaller scales, extending <100 km, which is located in the central diagonal seismic zone of Jiali–Bengcuo, and is controlled by transverse the active faults of Namucuo-Ningzhong. In the western areas of Nyainqentanglha Mountains, fault activities are manifested more obviously, controlling zonal distributions of epicenter zones on the Namucuo earthquake at the side of northwest. East of Damxung–Yambajan Basin, faulted activity surfaces demonstrate obscure, however, medium-small earthquakes show a certain linear distribution along the faults, reflecting the buried state of the faults. The major controlling earthquake fault in the seismic tectonic belts of Namucuo is a diagonal distribution of right-lateral strike-slip faults, cutting NW–SE through the Namucuo Basins. There has been Xungmai Ms7.0 earthquake in the seismic tectonic belt, and ancient multiple earthquakes of Ms4.0–5.0 have appeared in northwestern Namucuo.

The global seismic tectonic belt of Gerencuo is distributed in the NW–NWW direction, with regional length of around 230 km, located south of diagonal seismic belts of Jiali–Bengcuo, and it is controlled by the active faults of Gerencuo and Gurenqu, extending eastward to stretching structural composite sections of Nyainqentanglha Mountains and Damxung–Yambajan, and extending to the NW-running suture zones of Bangong Lake–Nujiang. Numerous ancient earthquakes of Ms4.0–4.5 have occurred in Gerencuo and Shengzha, and there had been the Ms8.0 earthquake and a large number of ancient earthquakes of Ms4.0–5.5 in Yambajan in 1411.

In addition, seismic tectonic belts of Gongbo'gvamda County–Songduo, of Yarlung Zangbo river, and of the Himalayas are also present here. Seismic tectonic belt of Gongbo'gvamda County–Songduo runs nearly EW, extending westward to active fault west of Xietongmen, and controls the massive middle, and small earthquake swarms of Ms4.0–5.9, however, no Ms7.0+ earthquakes have occurred over the past century. Seismic structural belt of Yarlung Zangbo river belongs to the northern boundary of the seismic tectonic belt of Himalaya, the spatial location of which is close to suture zones of Yarlung Zangbo river. Several strong earthquakes of magnitude Ms7.0 have occurred, which show genetic relationship dynamics with right-lateral strike-slip of the active faults of Yarlung Zangbo river. Located on the southern side of suture zones of Yarlung Zangbo river, the seismic tectonic belt of Himalaya is an important part of seismic tectonic belt of Tetisi-Himalaya on modern earth surface, with the occurrence of strong earthquakes of Ms7.0. Strong earthquake mechanism indicates the genetic relationship dynamics with the northern dive of India subcontinent. Though the main body of these seismic tectonic belts is away from the Qinghai–Tibet Railway, strong earthquakes differentially affect the Qinghai–Tibet Railway project safety, and can induce geological disasters of landslides, debris flows, threatening the safety of railway transportation.

4.3 Effects of Active Faults and Earthquake Zones on Railway Projects

The foregoing two sections have described the active faults and seismic tectonic zones located in the Qinghai–Tibet Plateau, and this section primarily describes their effects on railway projects, such as crushing rocks and shallow groundwater distribution near the active faults, and in particular, the climate and the environment of the Qinghai–Tibet Plateau, in addition to the generation of landslides and mudslides, groundwater, and tectonic fractures that induce uneven frosts and migrating hummocks. The average movement rate of many important active fault zones is up to 4–15 mm a^{-1} and can gestate strong Ms6.0+ earthquakes.

4.3.1 Effect of Active Faults on Railway Projects

Owing to the worse rock quality on both sides of the faults, air faces are often formed by fault scarps and fault facets, where it can easily trigger geological disasters such as landslides and debris flows. In the perennial frozen-soil region of northern Qinghai–Tibet Plateau, fraction activities not only result in roadbed deformation, pavement breaking, and engineering failure, but also induce geological disasters of uneven heaving, structural fractures, and migrating hummocks, having adverse effects on line projects safety of the Qinghai–Tibet Railway and Highway and oil pipelines. The coupling effect of creep fracture movements, underwater movements, and uneven frost heaves make the roadbed loose and the pavement strongly deformed in the section of Amdo of the Qinghai–Tibet Highway, potentially threatening projects the safety of the Qinghai–Tibet Highway.

4.3.1.1 Landslides and Mudslides Caused by Active Faults

In the margin faults and strike-slip faults, with larger angle, large-scale and steep fault slopes are present along the faults between rising plates and fall plates, and with shallow groundwater, which are favorable structural positions for the formation of landslides and debris flows. For example, among the 214 disasters occurring in Huizhou section of Beijing–Kowloon line in 2005, 35 disasters belonging to cuttings or landslides and slumps have occurred, accounting for 16% of the total disasters. This is an obvious result of lacking of the knowledge on geological disasters. Each activity fault zone has dense distribution of landslides which are clearly affected by the active faults of this area, and by remote sensing and field surveys, 522 landslides whose covering area is greater than 0.01 km^2 have occurred within 50 km wide stripes in the Qinghai–Tibet Railway and Highway, which are mainly distributed in Jingxian Valleys, Fenghuo Mountain, Yanshiping–Wenquanbing stations, Tanggula Mountain, Amdo, northern Naqu, and Damxung Valleys.

4.3.1.2 Dangers of Structural Fracture Belt for Railway Engineering

The formation of erodibility gullies is under the action of active fault belts and water erosion. In constant tension–torsional movements of active fault belts, deep cracks gradually expand to the surface. The surface soil mass is loose because of tensile-tension influence and strength decrease, and a gradually widened land subsidence zone and local secondary fracture zones are present on the surface. Because of the effect of loose and secondary fracture zones or loess collapsibility, once the occurrence of surface precipitation, leakages and erosion arise, and cracks keep expending with the connection of deep main cracks along strike and in the extending direction. At this time fault belts not only become the site which is vulnerable to water erosion, but also bring a large number of surface sediments into the fracture zones and even deep fractures, or through underground cracks passages to the surface redeposition with a relatively low altitude, even into downstream regions by rivers. Thus, the occurring strong erosion function gradually erodes (headward erosion) to a central area of the fraction zone at both sides, forming a destructive gully. Active faults form numerous structural fracture zones of width >100 m and length >1000 km, which cross or miscut railways, roadways, oil pipelines along the line, and communication cables. The ability of deeply cutting roadbeds results in that thermal insulation properties of roadbeds are out of proportion with the relationship of frost heaving, yielding into pavement ruptures, and roadbed ruptures, with the effect of the project quality and the driving speed on the line, and shortens the project life. For example, three active faults in the NWW strike direction along fracture zones of Fenghuo Mountain on the northern side, whose structural fracture zones width are 150, 660, and 300 m, respectively, but also cross the Qinghai–Tibet Railway and Highway, and do great harm to the project. Pavement cracks in the profile communicate with roadbed ruptures, in the plane which are connected with structural fracture zones with several kilometers length along active fault belts. In the north of Amdo, the Qinghai–Tibet Highway is built in the Jiebuqu riverbank, in some sections, active faults in the NNE-SN direction in the Jiebuqu riverbank longitudinally cut off the Qinghai–Tibet Highway, which induces a large number of structural fractures, single fracture width of 8–30 cm, severely affecting the engineering quality of highway and driving safety. In the Qinghai–Tibet Railway, the bridge of DK1202+668 is located south of the 85 team in the Qinghai–Tibet Highway and the west of the Qinghai–Tibet Highway, the design length of which is 120 m, crossing the active faults of the northern Wuli Basin. On account of central piers of the railway bridge of DK1202+668 built in faulted and shattered zones, the pile foundation construction depth of central piers is 37 m, opening the low-temperature Wenquan channel around faulted and shattered zones; hence, it occurred into a larger migrating hummock in the winter in 2002. In the July–August of 2003 with the progress of bridge foundation construction, a greater flow of upwelling spring water was induced under the piers, with the formation of a large area of spring pits and a mass of new springs, and the winter freezing would form severe disaster risks (Fig. 4.15a, b).

A. Central bridge foundation construction of bridge of DK1202 + 668 in faulted and shattered zones of 85 beam of Wuli inducing a greater flow of upwelling spring water (lens facing the east)

B. Spring induced by bridge foundation construction in faulted and shattered zones of 85 beam of Wuli forming hummocks in late winter after underground project drainage (lens the southeast)

C. Migrating hummocks arch destroying highway bridges (lens facing the east)

D. Migrating hummocks leading to arching deformation of oil pipeline (lens facing the west)

Fig. 4.15 Disasters induced by active fault belts of Wuli

4.3.1.3 Hazards of Groundwater of Active Fault Zones on Railway Projects

An active fault is a good channel where groundwater and underground fluid migrate and accumulate. Since that active fault zones deeply cut permafrost and underground ice layer, especially in composite fracture sections and faulted and shattered zones of rich water, which form the linear spring groups in summer and moniliform deflection-structure ice cumulus. Kunlun Mountains–Tonglha section in the Qinghai–Tibet Railway is distributed in the perennial frozen soil environment. Distributed winter hummocks along active fault zones, particularly developmentally migrating hummocks along active faults, leads to railway roadbed uplift and rail deformation, with a severe threat to the railway transport safety, even resulting in major security incidents. Unevenly distributed groundwater leads to hummocks with uneven frosts in the faulted and shattered zones, which makes a railway, and highway out of shape, affecting the line engineering quality and transportation safety. Hummocks and migrating hummocks are common in the active fault zones of Wuli. From January to February in 2002, because of fault movements, and groundwater movements, the central position of hummocks of the eastern 86 team of Wuli migrates around 7 m in the SW direction, moving to the position of 4–8 m east of highway culverts, forming the larger migrating hummocks with the width of 8 m, of length 15 m, and of height 2–2.5 m. Migrating hummock arches destroy highway bridges and culverts. The migrating hummocks of 86 team cause a severe damage on the oil pipeline of Golmud-Lhasa in 2002, leading to the occurrence of arching deformation of height 2–2.5 m in the underground oil pipeline with a length of 15 m (Fig. 4.15c, d). Active faults trending NWW miscut the Qinghai–Tibet Railway and Highway, and richer underground water near the active faults form the uneven frost heaving and then aggravates faulted crushing degree. Fault fracture zones are present in the roadbed and on both sides, making roadbed embankment unevenly loose and pavement deformed and leading to an obvious wavy fluctuation deformation in the pavement of length around 300 m, which greatly shortens the life of the railway and road, affecting the traffic speed, and causing a severe damage to the transportation safety of railway and highway.

4.3.2 Effect of Seismic Areas on Railway Projects

The Qinghai–Tibet Plateau is the most intense area of modern tectonic activities and seismic activities, with the characteristics of multiple times of seismic, high frequencies, large magnitudes. In particular, in the Qinghai–Tibet Railway which crosses the Qinghai–Tibet Plateau and along the Qinghai–Tibet Highway, four strong earthquakes exceeding Ms7.5 have occurred: Yambajan earthquake of Ms8.0 in 1411, Bengcuo earthquake of Ms8.0 in 1951, Jiuzina earthquake of Ms7.5 in 1952 and Kunlun Mountains earthquake of Ms8.1 in 2001, which has yielded into large-scale seismic ruptures and earthquake disasters, causing different levels of personal casualty losses and property losses. A strong earthquake can generate a significant surface rupture, leading to the severely deformed pavement, roadbed disconnection, bridge damage, and ground collapse, which induces the unfavorable geological phenomenon the sand liquefaction, avalanches, and landslides, and worsens the disaster effect on earthquake.

Be based on the Kunlun Mountain earthquake of Ms8.1 in 2001, we discuss the hazard of the earthquake on railway projects.

The earthquake of Ms8.1 on Kunlun Mountains was called the west of Kunlun Mountain Pass of Ms8.1, occurring at 17:26 on November 14, 2001. The epicenter was located in the Bukadaban Mountain belt that was at the boundary of Qinghai and Xinjiang of longitude 90°09′, latitude 36°02′, with a focal depth of around 10 km. After the occurrence of main seismic, it occurred to a mass of aftershocks of Ms4.3–5.7 in the Xidatan–Naij Tal area which was approximately 350 km away from the epicenter, and most aftershocks having a focal depth of around 10 km. Because of the seismic areas of Kunlun Mountains were located in the alpine and no man's land of the northern Tibet with an average altitude of >4600 m, and despite the large magnitude and the longer surface rupture, no severe casualties and significant property damages were reported, and it merely occurred to some slight seismic disasters in the Qinghai–Tibet Highway and along the Kunlun pass in the Qinghai–Tibet Railway–Xidatan–Naij Tal.

The 1997 Ms7.9 earthquake of Nima occurred in the NE direction for nearly 300 km, extending the active faults of eastern Kunlun Mountains to the east and resulting in the rupture of all active faults of Taiyang Lake–Kusuihuan and Mount Buka Daban–Kusai Lake and part of those in western Kunlun Mountain Pass. Although no historical records of any major earthquakes along the fault surface, many developmental fault scarps left-laterally leap the Holocene landscapes along the faults, indicating that there were large earthquakes in the faults. The total length of the entire earthquake surface rupture zones was 426 km, with the width of from a few meters to several hundred meters, the overall trend of 90°–110° (Fig. 4.16a). Nearly linear distributed surface rupture zones indicated that the main section was the steeper angle, and most surface sections ware vertical or nearly vertical. Main earthquake rupture was composed of branch ruptures with strike-echelons (Fig. 4.16b), the strike of which was N65°–70°E and the angle was 30°–35° with main earthquake ruptures, with straight sections, with nearly vertical production, with width of 0.2–4 m and with general length of 5–50 m, and the lengths of individual branch ruptures were 200–300 m. The earthquake possessed obvious faulted segment characteristics, which was composed of five secondary rupture sections from west to east: Taiyang Lake–Kushuihuan segment, Mount Buka Daban section, western Kusai Lake section, eastern Kusai Lake section, Kunlun Mountain Pass. In addition to Taiyang Lake–Kushuihuan segment of length 25 km, the rest of four secondary rupture segments was 64–103 km in difference. Each segment consisted of several more secondary ruptures with the left-stepping or right-stepping echelons, occasionally manifested as a simple single rupture and sometimes presented as multiple braided or crossed, wide, and complicated ruptures, with the similar fractal structural features. Seismic fracture belt was based on the left-lateral strike, with a few dip-slips. The maximum left-lateral horizontal displacement of surface coseismic was 6.4 m, and the maximum vertical displacement was 4 m. The branch rupture in the NEE direction had the obvious left-lateral strike displacement, and the tension–torsional branch seismic rupture left-laterally leaping construction roads of the Qinghai–Tibet Railway, forming the pavement cracking with the width of 50–120 cm (Fig. 4.16c).

The left-lateral horizontal displacement of surface coseismic is mainly with the characteristics of a large center and two small heads. Six peaks values were observed along the strike direction of seismic rupture zones, where there is a relatively independent attenuation sequence, indicating that the earthquake has multipoint segment ruptures. Seismic-induced structural fracture width is generally 1–5 cm, locally up to 10–20 cm, however, extending up to several meters or even hundreds of meters, which is a shape of straight line or polygonal line. Seismic induces the disasters including the monument collapse of Kunlun Mountain Pass, wall collapse of Xidatan, walls, and buildings crack of Xidatan, left-lateral levorotation leap dam of main rupture (Fig. 4.16d), avalanche of Kunlun Mountains

A. Kunlun Mountains Earthquake cutting through the Qinghai–Tibet Highway (lens facing the east)

B. Seismic trunk and main rupture with Ms8.1 along the eastern side of the Kunlun Mountains in the Qinghai–Tibet Highway (lens facing the east)

C. Seismic rupture zone cutting through the construction roads of the Qinghai–Tibet Railway (lens facing the southeast)

D. Seismic fracture left -laterally leaping the western dam of the Qinghai–Tibet Highway (lens facing the west)

E. Kunlun Mountain Ms8.1 earthquake generating the Avalanche (lens facing the south)

F. Kunlun Mountain earthquake destroying a Qinghai–Tibet Highway bridge in the Kunlun Mountain Pass (lens facing the east)

Fig. 4.16 Damage of Kunlun Mountain Ms8.1 earthquake

(Fig. 4.16e), Naij Tal bubbler. However, in the cities of Golmud, Budongquan, Wudaoliang, which are away from the seismic faults of Kunlun Mountains, the earthquakes cause no significant damage on buildings. Seismic rupture is mainly along the distribution of the active faults of Kunlun Mountains, and seismic disasters are only distributed in a certain range of seismic ruptures on both sides (Kunlun Mountain Pass–Xidatan). In the areas which are away from the active faults of Kunlun Mountains, the seismic disasters are less significant. But in main shock moment, there is still intense felt in a wide large of cities of Golmud and areas of Wudaoliang.

The Ms8.1 earthquake on the Kunlun Mountains has has a certain influence on the Qinghai–Tibet Railway, damaging some temporary sheds, and resulting in that there have been rockfalls in the Kunlun Mountain Tunnel, which leaps and destroys construction roads of the Qinghai–Tibet Highway and the Qinghai–Tibet Railway (Fig. 4.16a, c, f). The earthquake and strong aftershocks have some adverse effects on the psychological construction workers, the earthquake leads to a fracture in the 2894 km place of the Qinghai–Tibet Highway, communication cable in Naij Tal–Wudaoliang, part of the pump station buildings under destruction in oil line pipe of Huamu Ditch–Golmud. But Kunlun Mountain seismic causes no significant damage to the railway roadbed along the southern Qinghai–Tibet Railway pass–Naij Tal–Xidatan, and causes no obvious crack to the Kunlun Mountain Tunnel which has been excavated. The Kunlun Mountain seismic leads to mountain crack in the eastern side of Naij Tal railway, which has made some adverse effects on the stability of the mountain, and increased aftershocks, and the risk of rockfalls and slides in rainy season.

4.4 Active Fault Zones and Line-Selection Principle and Project Settings in Meizoseismal Areas

Accordingly, active fault zones are closely related to the distribution of meizoseismal areas. This section introduces active fault zones along the Qinghai–Tibet Railway and line-selection principle and engineering measures in meizoseismal areas.

4.4.1 Active Fault Zones and Line-Selection Principle of Meizoseismal Areas

Railway-line selection is away from active fault zones and meizoseismal areas as far as possible. If the railway must be built in the active fault zones and meizoseismal areas, the regional stability is evaluated, then the project settings. According to previous experience, *Design specification of railway line (GB50090-2006)*, *Geological investigation specification of railway projects (TB10012-2001)* and *Seismic design specification of railway projects (GB50111-2006)* formulate the followed line-selection principle:

1. Avoiding the high-level of active faults as much as possible, the low-level are selected.
2. When the fractures are parallel to line, if the faults are the thrust fault or normal fault types, the strong surface deformation and the hanging walls of the branch fault and secondary fracture are avoided as much as possible (uplifted side of the reverse fault and downthrown side of the normal fault).
3. The line is oblique with active faults in the vertical or large angle to the greatest extent, resulting in that the effect which active faults have on the roadbed is limited in the crosscutting point and its vicinity, with a small range of effect. If the line is parallel or intersecting with faults in a small angle, the influence surface is the mutually parallel or nearly parallel distribution of banded range.
4. Risk zoning evaluation of construction sites with active fault belts should be made to rationally deploy buildings based on the regional risk size and the importance of construction sites. Predicting the possible surface deformation and its scale is the foundation of risk evaluation. The deformation includes the surface distortion which is the result of surface faults and horizontal displacements, inclination caused by vertical deformation. According to the previous deformation, the possible type, and size of surface deformation are predicted.
5. Taking into account the predominant period of foundation earth-rocks and natural vibration period of buildings, the resonance between constructions, and foundation earth-rocks is avoided as soon as possible. That is to say that the buildings possessing a long natural vibration period are not built on the deep and soft sediments, and the rigid structures are not built on the foundation with short predominant period.
6. Three is a large cave in superficial underground of karst areas. When the occurrence of the earthquake, the area of possible subsidence is inappropriate to be a site.
7. The line should keep away from active fault belts, when it is difficult to shy away, and passes through the narrower fault belts in the large angle of simple projects, and it is inadvisable to set up high fill and deep cut buildings which are difficult to repair, which include large, and medium bridges, high bridges, tunnels.
8. The line should bypass the unfavorable seismic sections including the severe hillside deformation, underground hollow holes which are easy to collapse, debris flow areas, unstable hanging shores, and coulees, high-rise isolated hills, and the high bridges and high fill and deep cut roadbeds, semifilling, and semiexcavating roadbeds should not be on the loose hillside accumulation horizon. The line selection should be in the open and flat or gentle slope locations with good engineering-geological conditions.
9. The line should be chosen in the nonliquefied soil zone or the liquefied soil zone with deeply buried areas and the scope of smallest areas. The line should avoid the sand layers that may produce seismic liquefaction or strongly deposited silt layers and thick reclamation soil layers or the foundation which

may generate uneven sedimentation. The liquefacient soil layer should not serve as the bearing stratum of buildings.

10. The shed gallery should take measures to prevent falling beams in the seismic zone. When fortification intensity is 8° and 9°, the cantilever shed tunnel should not be used.
11. When bypassing the soft areas including the liquefying soil and the soft soil, the line selection should be on the surface with thicker nonliquefied sand layers or overlying crusts, and the low embankment should be installed. There should not be a deep and long cutting on the soft soil surfaces or rock crushing and unfavorable geological structure surfaces.
12. The line should keep away from the unstable cliff locations and steep sections, when it is difficult to avoid, by the tunnel.
13. When located in hillside location, the tunnel should be ingression. The tunnel pass should not be located in the adverse geological areas which include collapses, landslides, scattered earth-bodies, when it occurs to the seismic.
14. The bridge location should be located in a good foundation and a stable riparian area. When it is difficult to avoid the liquefying soil and the soft soil foundation, the middle bridge route should be orthogonal with the river.
15. The roadbed is located on the semifilling and semiexcavating stable slope and the embankment is built on the stable slope whose lateral ground slope is greater than 1:5. The bench should be dug in the original ground, whose width is not <2.5 m, and the drainage works should be done well. When it is necessary, the slippery measure including supporting constructions should be set up.
16. On the rocky cutting, when rock breaking or weak sandwiches, critical rocks on the hillside or upper overburdens which are easy to collapse, are present the protection and reinforcement measures including the clearance, anchor, supporting, and retaining should be taken. Light and flexible protections or galleries should be set up on the I and II grade railway.
17. For the retaining walls which are located on the liquefying soil layers and soft soil roadbeds, some measures should be taken, including a composite foundation and a pile foundation, to reinforce the foundation. The pile toe should insert the stable soil layer.
18. When the bridge must pass through the seismic faults, the simply supported beam bridge with a small span and a low pier should be adopted. The bridge opening should be arranged at equal spans. The bridge pier should avoid suffering from partial pressure. The abutment should adopt the T-shaped or U-shaped abutments.
19. For the abutment pier on the liquefying soil layer or soft soil foundation with perennial rivers, the deep foundation should be used, including the pile, and the caisson, and the pile toe and the caisson bottom are buried within the stable soil layer of not <2 m. When the horizontal force is larger, the inclined pile should be set up in the pile foundation abutment and other reinforcement measures are taken.

20. Foundations of the super large bridge embankment and large–medium bridge embankment are liquefying soil layers or soft soil. When the possession of the I, II grade railway conditions with fortification intensities of 8° or 9° and with the abutment embankment whose height is greater than 3 m, the liquefied soil layers, or soft soil are reinforced by vibro-density, sand piles, sand drains, gravel piles, and changing and filling below the embankment base within the range of 15 m behind the abutment.
21. Located on the liquefying soil layers or soft soil foundation, the abutments of super large bridges and large–medium bridges should be set up on the stable riverbanks. The bridge pier should not be set up on the mutation places at the boundary of the main channel and the river shoal.
22. When the bridge crosses the fault zone, the pier foundation should not be set up on the severe fracture zones.
23. Bridges are located in the seismic zones. Some antiseismic measures should be taken on the upper structures of bridges, including preventing falling beams, and setting up stops. When the condition is permitted, lead rubber bearings, or other shock absorption and energy dissipation devices are adopted, for the sake of reducing the seismic effects of beams.
24. For the important bridges in the III, IV sites where the fortification intensity is greater than or equal to 7° and 6°, taking the prevention of beam measures, as follows:

    a. Longitudinal reinforcement of beam end or longitudinal supporting and retaining of beam end are adopted in the simply supported beam. For the continuous beam, longitudinal and horizontal supporting and retaining is set up in the horizontal wall on the piers, and the horizontal wall is locally reinforced. For the reinforced concrete t-beams, transverse connection between beams, and beams is strengthened.
    b. For the bridges which possess high piers, large spans, and deep water environment are difficult to repair, moderately widened energy dissipation facilities are present atop the hats of piers and abutments.
    c. When the use of lead rubber bearings, the connection, or the supporting and retaining should not be set up along the bridge in the axial direction. The supporting and retaining structure should be set up on the beam-end along bridges in the transverse direction. Cushion materials should be packed between the supporting and retaining and the beam.

25. The antiseismic should be fortified in the tunnel portal, the shallow-embedded section, and the eccentric compression section and the fault fracture zone section. The lining structure is strengthened. When it is necessary, according to the actual situation, the section clearance can be reserved along the active fault fraction zone section. On the basis of the terrain condition and the geological condition and fortification intensity, the length of structural opening fortification segment is ascertained, not <2.5 times of the structural span.

26. The tunnel of the antiseismic fortification segment should adopt the composite lining structure and curved wall lining section with inverted arches. The deformation joint is installed in the fortification segment. For the tunnels on the shallow and bias geological sections, the lining should be disposed by grouting.

4.4.2 Engineering Measures for Active Fault Zones and Meizoseismal Areas Along the Qinghai–Tibet Railway

According to the aforementioned principles, the DK1001+750 (Budongquan)–DK1419+300 (Tonglha Mountains) sections are the seismic intensity areas of 7°, and the effect which the seismic fault dislocation has on the construction of the buildings can be ignored. However, in view of the importance of the Qinghai–Tibet Railway engineering, in the design work, by the following principles, the active fault zones are optimized by the line program and the project settings.

According to the geological line-selection principle of the seismic zones of 7° or more, the line should avoid active fault belts, when it is difficult to bypass, pass active fault belts by right of a single project and a large angle on the shallow sites. The high fill dig large buildings which are difficult to repair, such as large–medium bridges, high bridges, tunnels, should not be set up in the active fault zones. Accordingly, the characteristics of new structural movements in the Qinghai–Tibet Plateau decide most strikes of active fault belts along the line vertically intersect or by a large angle with the line. For most faults distributed along the line, all of the routes are difficult to pass round. During the various stages of the investigation and design work in northern areas of Tonglha Mountains, for the geological route selection and the engineering setting involved in the active fault zones, we conduct a careful investigation, the line scheme comparison, and engineering setting research. In the final construction drawing design, when by means of the main active faults, the principle of high fill dig buildings which are difficult to repair and passed by a large angle and where large–medium bridges, high bridges, and tunnels try not to be set up is satisfied. When via the general active faults, the project principle try not to be avoided, which include takahashi, compression span and shortening the bridge length, bridges changing culverts. For the individual working points which affect line programs and projects setting, the geological work should be deepened, to ensure the optimum line programs and reasonable project measures when via active faults.

4.4.2.1 Engineering Measures of Passing Main Active Faults

The study on active faults and seismic zones of northern Tonglha areas show that main active faults of the Holocene are based on the lateral compressional and torsional movements, which is accordant to the direction of the modern tectonic stress field, that is, the present tectonic stress field is mainly in the extrusion direction of NNE or NE. The lithology in this area is based on the shale, sandstone, and marl in the Paleogene–Neogene, with softer rocks, intense weathering and fragmented rocks, and the structural, and freezing and thawing action is severely affected by rocks, adverse to storage, and concentration of ground stress. The route in the northern Tonglha is based on the roadbeds and bridge projects, with few tunnel engineering; hence, the ground stress state does not strongly affect the projects.

For the seismic belt route of 7° or more, on the basis of the line section of active faults and engineering setting principles, the project of the line via the deep-large active faults should be comparatively studied. In the final line programs, by nine deep-large active faults and eighteen faults, the route only passes the fifth Buqu Bridge (5–24 m bridge beams, bridge height of <10 m) and the Buqushunhe bridge (8–32 m bridge beams, bridge height of <15 m), with the angle of 60° and 40°, and the line is orthogonal with one beam bridge of 1–32 m, with the mileage of DK1443+920. The route passes other deep-large faults with a single project and with a large angle (Table 4.6).

4.4.2.2 Engineering Measures of Passing General Active Faults

When the line gets through the general active faults, on account of the line which is difficult to pass round, and it is difficult for the project settings to pass the simple engineering, such as roadbeds, and bridges, during the schematic studies of the line selection and project settings, we mainly consider some engineering measures, such as trying to cut across the active faults with a large angle and compress the span and shorten the bridge length, avoiding setting up the high bridges, optimizing the abutment location, avoiding the abutment location on the fault belts, conducting the bridge-changing-culvert.

The fracture of the large bridge of bridge-instead-of-embankment is surveyed, where five bridge-changing-culverts, the mileages are the following, respectively: DK976+062, DK976+168, DK1196+030–DK1196+130, DK1198+510–DK1198 +860, DK1214+900–DK1215+050. One bridge-site realignment and reaming, is with the mileage of DK1367+870. Table 4.7 summarizes the relationships among the general active faults, the lines, and the project settings from the southern mountain pass of DK975+100 to the northern Tonglha Mountains of DK1404+250, the routes get through one hundred and forty-two general active fault zones. The lines pass the general active faults by the 37 super large bridges and large bridges, three intermediate axles, and one tunnel, and others get through the general active faults via the simple projects, such as roadbeds, culverts, and small bridges.

Table 4.6 Relationship among deep-large active faults, the lines, and project setting

Fracture name	Fault occurrence	Fault property	Crossed mileages with lines	Angle with lines	Line project settings
Fault of southern margin of Qaidam Basin	SW∠40°	Insidious and reverse fault	K840 +000	60°	Existing railway, low embankment
Middle fault of Kunlun	190°∠85°	Insidious and left-lateral extrusion	DK871 +000	85°	Low embankment
Fault of eastern Kunlun	SW∠60°–80°	Insidious and left-lateral strike-slip	DK963 +000–DK964 +000	10°	Low embankment, culvert, small bridge
			DK967 +600–DK967 +700	50°	Low embankment, culvert
Fault of Kunlun Mountain Pass	190°∠90°	Left-lateral strike-slip	DK979 +700–DK979 +900	50°	Low embankment, culvert
Fault of Wudaoliang	10°–20°∠65°–85°	Left-lateral strike-slip	DK1086 +680–DK1099 +320	50°–90°	Low embankment, culvert
Fault of Wulan–Yushu	200°∠70°–85°	Left-lateral strike-slip	DK1147 +450	80°	Low embankment
Fault belts of western Wenquan Basin (five parallel faults)	132°∠80°–85°	Insidious, normal fault	DK1360 +000	40°	Low embankment
	110°∠60°	Insidious, normal fault	right side of DK1494 +000	not through	Low embankment
	125°∠87°	Insidious, reverse fault	DK1394 +415–DK1394 +430	60°	The fifth bridge of Buqu (beam of 5–24 m, <10 m bridge height)
	128°∠80°–85°	Insidious, reverse fault	DK1396 +300–DK1396 +400	40°	Maleic river bridge of Buqu (beam of 8–32 m, <15 m bridge height)
	129°∠80°–85°	Insidious, reverse fault	DK1398 +400	40°	Low embankment

(continued)

Table 4.6 (continued)

Fracture name	Fault occurrence	Fault property	Crossed mileages with lines	Angle with lines	Line project settings
Fault belts of the northern slope in Tonglha	201°∠79°	Insidious, reverse fault	DK1358 +900–DK1359 +020	45°	Low embankment
Fault belts of northern margin in Tonglha (six parallel faults)	210°∠70°–80°	Reverse fault	DK1336 +730	70°	Bridge of Zhaqinqu (beam of 9–24 m)
		Reverse fault	DK1337 +920	70°	Low embankment, culvert
		Reverse fault	DK1340 +490	50°	Low embankment
		Parallel fault	DK1342 +290	90°	Transitional sections of filling and digging roadbeds, culvert
		Reverse fault	DK1348 +050	90°	Low embankment
		Reverse fault	DK1350 +230	90°	Transitional sections of filling and digging roadbeds, culvert

4.5 Major Scheme Comparison and Selection of Active Fault Belts and Meizoseismal Areas

4.5.1 Scheme Comparison and Selection of Kunlun Mountain Tunnel

The rocky ditch of the northern slope in Kunlun Mountains is the tension–torsional fault cleft on the northern side of Kunlun Mountain Pass, with complex terrains, and zigzag and shallow gully beds, possessing the ditch width of 20–90 m, the naturally longitudinal slope exceeds 25%. The ditch pass of height >4600 m is the only way to the Tibet, which the Qinghai–Tibet Highway passes. No large side channel is for exhibition line. Both sides of the mountain are steeper, with the natural slope of 25°–40°. The spring water development and broken slopes, and the stone bar, and sizes of rocks are distributed in the slopes and toes. The rocky ditch belongs to permafrost regions, with the unique geological phenomenon, such as hummocks, ice cones, thick layers of groundwater, and possesses the adverse geological development including faults. The segment belongs to the seismic intensity zone of 8°.

The topography and geology of the section is highly complex, which belongs to permafrost regions. According to the topographies, geologies, and various opinions,

Table 4.7 Relations among general active faults, lines, and projects setting

Fault numbers	Fault strike	Intersection mileages with the line	Angle with the line	Setting types on the line engineering
F6′	EW, left-lateral displacement	DK975+100	60°	Bridge of 6–32 m, <15 m bridge height
F6-1	N80°W, left-lateral displacement	DK975+480	80°	Bridge of 12–32 m, <15 m bridge height
F6-2	N80°W, left-lateral displacement	DK975+670	80°	Bridge of 12–32 m, <15 m bridge height
F6-3	N80°W, left-lateral displacement	DK976+050	80°	Roadbed
F6-3	N78°W	DK976+105	80°	Culvert
F6-3	N78°W	DK976+140	80°	Roadbed
F6-3	N78°W	DK976+160	80°	Culvert
F6-4	N75°W	DK976+890	85°	Kunlun Mountain Tunnel
F7-1	N67°W	DK978+130	55°	Bridge of 2–24 m + 4–32 m, <20 m bridge height
F7-2	Speculate N85°W	DK978+880	85°	Super large bridge of 22–32 m, <7 m bridge height
Crushing belt of earthquake	N50°W	DK979+510	50°	Low roadbed
F7-3′	N75°W	DK981+700	40°	Bridge of Kunlun Ditch of 15–32 m, <22 m bridge height
F7-3	N60°W, left-lateral	DK981+870	50°	Bridge of Kunlun Ditch of 15–32 m, <22 m bridge height
F7-4	N70°W, left-lateral	DK982+330	80°	Low roadbed
F7-5	N70°W	DK983+180	50°	Bridge of 3–24 m + 11–32 m, <15 m bridge height
F7-6	N55°W, left-lateral	DK984+700	85°	Culvert
F8	N55°W	DK997+750–+770	85°	First bridge of 5–32 m in Kunnan
F8-1	N55°W	DK999+370	85°	Culvert of 1–1.5 m
F8-2	N70°W	DK1000+970	60°	Low roadbed
F8-3	N70°W	DK1001+960	40°	Low roadbed
F8-4	N55°W	DK1004+680	85°	Low roadbed
F9-1	N55°E	DK1005+500–	Parallel	Bridge-instead-of-embankment of large bridge in Budongquan
FD7	Geophysical speculation	DK1005+170–+300	50°–70°	Low roadbed

(continued)

Table 4.7 (continued)

Fault numbers	Fault strike	Intersection mileages with the line	Angle with the line	Setting types on the line engineering
F10	N70°W	DK1018+130	30°	Bridge-instead-of-embankment of Baladacaiqu of 473–8 m, <5 m bridge height
F10-2	N65°W	DK1021+400–+550	35°	Low roadbed
F11-2	N45°W	DK1037+050–+230	90°	Bridge-instead-of-embankment of Qingshuihe of 1366–8 m, <5 m bridge height
F11	N45°W	DK1037+600	90°	Bridge-instead-of-embankment of Qingshuihe of 1366–8 m, <5 m bridge height
F11-3	N65°W	DK1038+600	70°	Bridge-instead-of-embankment of Qingshuihe of 1366–8 m, <5 m bridge height
F12-1	N50°W	DK1041+700	65°	Low roadbed
F12-2	N75°W	DK1042+180–+460	45°	Low roadbed
F12-3	N70°W	DK1043+460	45°	Low roadbed
F12-4	N60°W	DK1044+770	50°	Low roadbed
F12	N60°W	DK1047+005	60°	Small bridge of 1–8 m
F13-1	N75°W	DK1048+265	50°	Small bridge of 1–12 m
F13-1′	N50°W	DK1048+715	75°	First bridge of 9–16 m in Chubei, <15 m bridge height
F13	N75°W	DK1051+650	40°	Low roadbed
F13-2	N40°W	DK1053+620	65°	Low roadbed
F14-1	N70°W	DK1056+900	30°	Low roadbed
F14	N50°W	DK1058+300	50°	Culvert of 1–1.5 m
F14-2	N60°W	DK1061+270	60°	Low roadbed
F15-1	N65°W	DK1064+080	60°	Low roadbed
F15-2	N65°W	DK1067+630	60°	Low roadbed
F15-3	WE	DK1065+220	40°	Low roadbed
F15	N80°W	DK1067+260	70°	Super large bridge of 21–32 m in Chumaer river
F15-4	N80°W	DK1067+305	70°	Super large bridge of 21–32 m in Chumaer river
F15-5	N80°W	DK1067+970	70°	Low roadbed
F15-6	N80°W	DK1068+120	70°	Bridge-instead-of-embankment of 78–32 m in Chumaer river
F16-1	N10°E	DK1083+390	55°	Low roadbed
F16-4	WE	DK1087+500	80°	Culvert of 1–1.5 m
F18-2	N70°W	DK1102+860	75°	Low roadbed
F18-1	N65°W	DK1103+860	65°	Low roadbed
F18-4	EW	DK1105+750	80°	Low roadbed

(continued)

Table 4.7 (continued)

Fault numbers	Fault strike	Intersection mileages with the line	Angle with the line	Setting types on the line engineering
F18	N70°W	DK1107+790	85°	Low roadbed
F19	N50°W	DK1108+670	75°	Bridge of 6–32 m in Qubei
F19-1	N40°W	DK1110+760	70°	Low roadbed
F19-2	N50°W	DK1112+640	55°	Bridge of 9–32 m in Hongliang river
F19-2′	N50°W	DK1112+760	55°	Bridge of 9–32 m in Hongliang river
F19-3	N50°W	DK1113+350	45°	Low roadbed
F19-4	N60°W	DK1114+080	85°	Low roadbed
F23-2	N40°W	DK1118+400	30°	Low roadbed
F20	N65°W	DK1126+130	70°	Samll bridge of 1–16 m
F20-1	N70°W	DK1127+040	80°	Bow roadbed
F21	N60°W	DK1132+420	80°	Bridge of 12–32 m in Beiluhe, <10 m bridge height
F22-2	N70°W	DK1138+770	70°	Bridge of 3–32 m
F22-1	N55°W	DK1139+580	65°	Low roadbed
F22	N60°W	DK1140+280–DK1142+560	80°	Low roadbed
F22-3	N70°W	DK1143+300–+530	75°	Low roadbed
F23-7	N70°W	DK1145+550	50°	Low roadbed
F24-1	N80°W	DK1148+270–+450	85°	Sirst bridge of 12–32 m in Zuomaoxikongqu, <10 m bridge height
F24-2	N80°W	DK1149+800	80°	Second bridge of 35–32 m in Zuomaoxikongqu, <18 m bridge height
F24-3	N80°W	DK1150+060	80°	Second bridge of 35–32 m in Zuomaoxikongqu, <18 m bridge height
F24-3′	N80°W	DK1150+130	80°	Second bridge of 35–32 m in Zuomaoxikongqu, <18 m bridge height
F24-4	N70°W	DK1151+160	80°	Low roadbed
F24-5	N70°W	DK1151+390–+440	85°	Bridge of 6–32 m, <15 m bridge height
F25-1	N80°W	DK1151+810	75°	Low roadbed
F25-2	EW	DK1152+590–+620	65°	Low roadbed
F25-3	N60°W	DK1153+865	80°	Low roadbed
F25-4	N75°W	DK1154+100–+150	85°	Xidagou Bridge of 20–32 m, 35 m bridge height
F25-5	N20°W	DK1154+700	30°	Low roadbed

(continued)

Table 4.7 (continued)

Fault numbers	Fault strike	Intersection mileages with the line	Angle with the line	Setting types on the line engineering
F25-10	N20°W	DK1155+345–+410	60°	Low roadbed
F25-13	N70°W	DK1158+010–+160	60°	Low roadbed
F26-1	N80°W	DK1158+410–+605	90°	Low roadbed
F26-2	EW	DK1158+690–+780	65°	Low roadbed
F26-3	N30°E	DK1163+040	20°	Low roadbed
F26-3	N35°W	DK1168+370–+500	25°	Low roadbed
F27-2	N80°W	DK1174+670	50°	Southeastern overpass of 23–32 m, 8 m bridge height
F26-4	N25°E	DK1174+790	30°	Low roadbed
F27-2	N80°W	DK1176+620	45°	Southeastern bridge of 5–32 m, 6 m bridge height
F27-2	N70°W	DK1178+310	70°	low roadbed
F26-5	N60°E	DK1178+460–+730	30°	Yamaer river bridge of 12–32 m, <13 m
F27-2	N70°E	DK1178+760	70°	Yamaer river bridge of 12–32 m, <13 m
F27-3	N70°W	DK1181+200	70°	Low roadbed
F27-3	N70°W	DK1182+000	80°	Low roadbed
F27-5	EW	DK1183+305	80°	Low roadbed
F27-5	EW	DK1183+940	80°	Middle bridge of 3–24 m with zigzag, 7 m bridge height
F27-5	EW	DK1184+270	90°	Low roadbed
F27	N80°W	DK1187+000	80°	Low roadbed
F29-1	EW	DK1192+840	90°	Low roadbed
F29-1	EW	DK1193+170	90°	Low roadbed
F29-1	EW	DK1193+330	90°	Low roadbed
F29-1	EW	DK1193+440	90°	Low roadbed
F29	N60°W	DK1196+030–+130	60°	Culvert of 1–4 m, low roadbed
F29-2	N70°W	DK1197+260	80°	Bridge of 4–32 m, <8 m
F29-2	N70°W	DK1198+510	70°	Low roadbed
F29-2	N70°W	DK1198+860	50°	Low roadbed
F29-3	N50°W	DK1200+080	60°	Low roadbed
F29-5	N55°W	DK1202+030	90°	Intermediate axle of 1–24 m, <8 m bridge height
F29-4	EW	DK1202+600–+690	70°	Bridge of 2–24 m + 2–32 m, <9 m bridge height
F31-1	N70°W	DK1203+000	30°	Low roadbed
F29-6	N65°W	DK1206+850	40°	Bridge of 8–32 m, <11 m bridge height
F31-1	EW	DK1208+500–+800	Parallel	Low roadbed

(continued)

Table 4.7 (continued)

Fault numbers	Fault strike	Intersection mileages with the line	Angle with the line	Setting types on the line engineering
F30-1	EW	DK1209+800–DK1210+000	50°	First number of overpass of 4–24 m + 5–32 m in Wuli
F30-2	EW	DK1211+000	60°	Low roadbed
F30-3	EW	DK1211+700–DK1212+120	50°	Low roadbed
F30-4	N70°E	DK1214+900–DK1215+050	50°	Low roadbed, culvert of 1–1.5 m
F30-5	N75°W	DK1214+900–DK1215+050	80°	Third number of overpass of 5–32 m in Wuli, <15 m bridge height
F31-5	EW	DK1226+620	55°	Bridge of 5–32 m, <6 m bridge height
F32-1	N55°W	DK1228+880–DK1229+330	80°	Low roadbed
F32-2	N55°W	DK1230+880–DK1231+280	60°	Tuotuo River overpass of 14–32 m
F32-3	N65°W	DK1233+140–+430	80°	Bridge of 12–32 m, <12 m bridge height
F32-4	N50°W	DK1234+860–+930	90°	Low roadbed
F32-4	N80°W	DK1235+580	60°	Low roadbed
F32	N75°W	DK1236+595–+640	70°	Low roadbed
F33	N70°W	DK1245+080	80°	Low roadbed
F33-2	N70°E	DK1248+020	80°	Low roadbed
F33-1	N80°W	DK1252+925	80°	Low roadbed
F33-3	N60°E	DK1253+230	55°	Low roadbed
F34-4	EW	DK1261+950	80°	Middle bridge of 2–32 m, <9 m bridge height
F34	N55°W	DK1263+100–+500	55°	Low roadbed
F34	N55°W	DK1264+000	55°	Low roadbed
F34-5	N50°W	DK1265+540	55°	Low roadbed
F36-1	N60°E	DK1280+200	55°	Kongjiezhaqu bridge of 7–32 m, <7 m bridge height
F36	N80°E	DK1282+900	60°	Super large Tongtianhe bridge of 20–32 m, <20 m bridge height
F36-2	EW	DK1284+000	50°	Low roadbed
F38	N60°W	DK1327+550	80°	Low roadbed, culvert of 1–1.5 m
F38	N60°W	DK1327+760	80°	First Buqu Bridge of 3–32 m + 2–24 m, <12 m bridge height

(continued)

Table 4.7 (continued)

Fault numbers	Fault strike	Intersection mileages with the line	Angle with the line	Setting types on the line engineering
F39-5	N65°E	DK1331+400–900	Parallel	Buquhefang Bridge of 9–32 m
F40	N60°W	DK1341+100	70°	Second Buqu Bridge of 19–32 m, <10 m bridge height
F41	N15°E	DK1358+100	30°	Thrid Buqu Bridge of 18–32 m, <17 m bridge height
F41-1	N50°W	DK1363+560	85°	Culvert of 1–2 m
F41-2	EW	DK1367+870	50°	Bridge of 14–16 m, <12 m bridge height
F41-2	EW	DK1368+090	50°	Low roadbed

two local programs are mainly formulated. First, the long tunnel scheme (CK scheme): the route is drawn from CK971+800, for the use of a powerful line of terrain exhibition in the rocky ditch pass, by means of the long tunnel of 1686/1 via from the state road of 109 to CK978+800 in the western bank of the ditch. Second, the short tunnel program (CIK program): the line is drawn from CIK971+800, but also for the possession of favorable lines of location exhibition in the western ditch, after two short tunnels which are set up, earlier than CK program across the state road of 109, continuously after three short tunnels are set up, connecting the CK scheme to the end (Fig. 4.17).

The rocky ditch is located in uplift disk during two larger thrust faults. Faults are a good channel of groundwater, which is also the frequent of various adverse geological phenomena; hence, the relationship between faults and the line strike must be handled well.

The reverse fault (F1) on the northern side of the rocky ditch is located in the long tunnel line of CK976+070 and the short tunnel line of CIK976+715. The fault occurrence is 198°∠84°, with the broken belt width of 44 m. The fault fracture zone is the serpentine and slates of serpentinization, presenting the graveled, and brecciated, locally silted. The upper plates on the southern side of faults are slates and schist of triassic, the lower plates of northern side of which is the phyllite of triassic. Two schemes are substantially perpendicular to the faults, and the line in this area gets through the fractures by a culvert.

The faults (F2) on the southern side of the rocky ditch are compresso-shear faults of the Kunlun Mountain Pass, which is located in the southern side of main ridges of Kunlun Mountains, of length nearly 900 km, intersecting the two program lines by the angle of 55° in the line department of CK978+800. The fault occurrence is N20° E∠64°, with the broken belt width of 27 m. The fault fracture zone is gravels and silts, and most gravels are orientational. The hanging plates of northern side of faults are slates and schist in triassic and iced product gravel soil, the low plates on the southern side of which is Quaternary under renewal series lake product silty clays and silts. Two scheme routes in this area are across the faults by the roadbed fill.

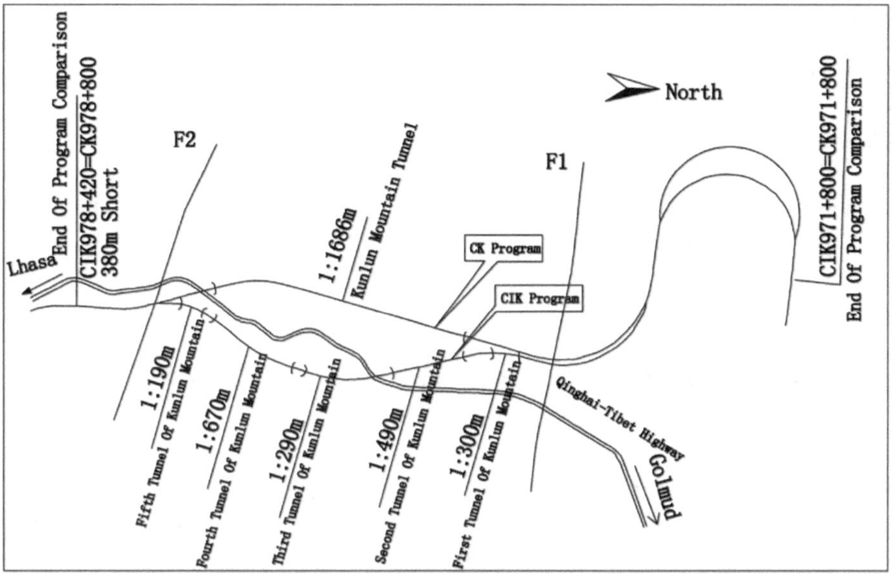

Fig. 4.17 Schematic comparison of the long and short tunnel scheme in Kunlun Mountains

The long tunnel is affected by the slope reduction, with higher line level, longer line length, and the increase of bridge projects. There is the tunnel of 1686 m with a relatively small number of frozen soil excavation works. The western bank hillsides of the rocky ditch are steeper, with a slope collapse, and stone bars and sizes of rock piles are distributed in hillsides and toes. Because of the deep burial, the long tunnel avoids a series of unfavorable geologies, such as slope collapses, ice cones, and hummocks, but also with relatively good geological conditions of the long tunnel entrance, in addition to the ice-containing soil dug outside, no rock heaps, rock glaciers, and ice cones are present.

For the short tunnel scheme, on account of the short length, the slope reduction is very long, therefore the line is 380 m shorter than the long tunnel, with decrease of bridge projects. However, increase of frozen soil excavation engineering is adverse to project permafrost regions. The scheme has five short tunnels, with the total length of 1940 m, possessing a high bridge of 54 m. Since that the rocky ditch is located in the uplift plates during the larger thrust faults, the compaction fold is very strong, with the development of joints and fractions, in addition to steep slopes, further contributing to the development of rock heaps. The slopes of the eastern bank of the rocky ditch are still steeper, with the slope collapse. The main projects are short tunnels. With the increase of tunnel entrance, but also with the deeply buried long tunnel, it is adverse to tunnel insulation. The geological conditions of the import and the export of the short tunnel are not ideal. The import and the export of the fifth short tunnel possess the distribution of rock heaps. The import and the export of the third and fourth short tunnels not only have the distribution of rock heaps, but also are affected by rock glaciers, leading to gully debris flow gasification.

By the analysis and comparison, the burial of the long tunnel is deeper than the short tunnel scheme. The surrounding rock classification of the long tunnel is superior to the short tunnel scheme, and the effect of adverse geological phenomena on the long tunnel is weaker than the short tunnel scheme. The geological conditions of the long tunnel entrance are relatively good, and it is for the roadbed project to be handled, with the better construction conditions. Therefore, we recommend the long tunnel program.

4.5.2 Scheme Comparison of Railway-Line Selection in the Wenquan Gypsum Region

The line in this region is mainly located in Buqu gorge area, and the results of field survey and geophysical data show that two faults have developed. The two faults intersect on the right side of Buqu gorge in V-shape and the right-bank region is cut into a wedge-shaped block. The fault in the western side of Wenquan Basin is a reverse fault, and the width of rupture zone is 120 m, and the occurrence of the fault is 211°–79°. Dislocated in Jurassic formation, the hanging walls of the fault are gray-white or milky gypsum layers and the footwalls are amaranthine sandstone and conglomerate. The scheme comparison region is covered with a 10–20 m thick Quaternary covered layer, the fault in the peak of right side of Buqu is largely exposed and the lithologic boundary is clear here. The fault (D12–F2) on the northern slope of Tonglha Mountains is a concealed fault with a high angle, whose strike is N42°E, which slides in Jurassic formation. The hanging wall of the fault is conglomerate (locally intercalated with limestone) and the footwall is the (Fig. 4.18). The anhydrite area of Jurassic which is controlled by the main faults of deep-large active faults in this segment, and comparison studies of two schemes in both sides of Buqu are made in the process of preliminary survey, location survey, and supplementary measurement.

The left bank scheme of Buqu (CIK): it is Qinghai–Tibet Highway in the left bank, and oil pipeline and Lanssila communication cable are distributed in the slope, which have an adverse effect on the line. Moreover, the line needs to get through the permafrost wetland in the section of CIK1358+940–CIK1359+230, crossing the Qinghai–Tibet Highway, and Buqu is passed by the type of Shunhe Bridge. The surface layer of the Buqu riverbed is the 1–2 m thick gravel soil. The bedrock is Jurassic anhydrock, the thickness of which is >50 m. The anhydrock is easy to dissolve and has high expansibility, where karst caves are also present. Controlled by the faults in both sides of Wenquan Basin and two faults of Tonglha, the line in this scheme needs to get through more areas which mainly consist of anhydrocks, and the burial is shallow and geological-engineering conditions here are poor.

The right-bank scheme of Buqu: crossing Bangerlongba, the Qinghai–Tibet Highway, and Buqu, the line mainly is in the form of fill to pass the right bank of

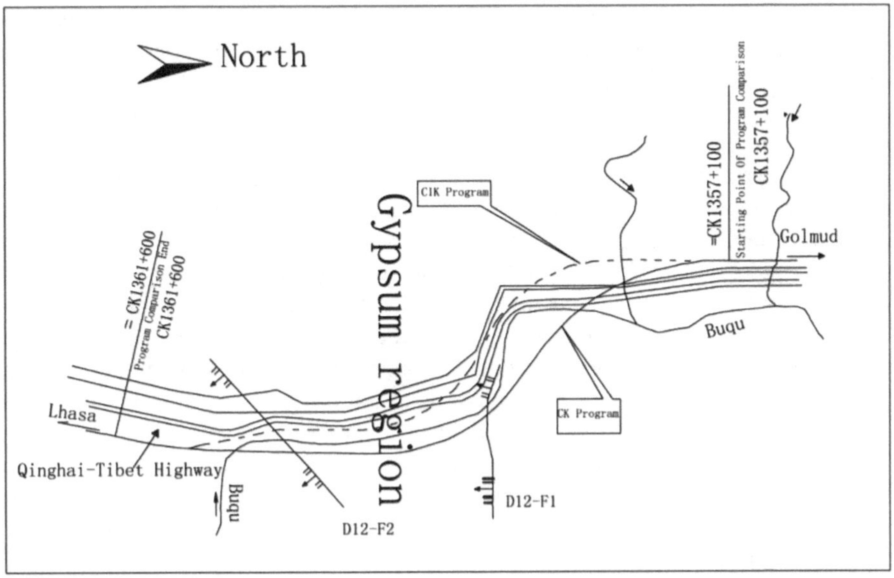

Fig. 4.18 Line program comparison schematic in Wenquan gypsum

Buqu. The thickness of cover layer of the Quaternary in the anhydrock region, which the line passed through is 10–15 m. In the initial surveyed geological-engineering exploration, the distribution of karst caves only is found in the range of CK1359+220 to CK1359+160. Compared with the left bank scheme, the anhydrite zone which the line needs to pass through in this scheme is 900 m shorter than that in left bank, and the Quaternary overburden with a large thickness, geological-engineering conditions in here are better than those in left bank. Therefore, we recommend the right-bank scheme of Buqu, and for the use of roadbed to pass through this region.

Thus, we can see that complex geological structure and anhydrite distribution are the two key points to the line's selection. Based on the geological-engineering line-selection principles, the line should keep away from these unfavorable geological phenomena. When it is impossible, by simple engineering formation such as roadbed, the major engineering project such as bridge should be avoided.

4.5.3 Bridge-Instead-of-Embankment of Large Budongquan Bridge

As a 26–32 m + 1–26 m + 179–8 m + 1–16 m + 16–32 m beam bridge, bridge-instead-of-embankment of large bridge of Budongquan is located in stake number of DK1005+350–DK1008+300 and in the central stake number of DK1006

+819. This large bridge crosses the Qinghai–Tibet Highway, Lanssila communication cable, oil pipeline, and gullies around the Budongquan. Take the marker DK1005+900 as the border, it is talik in the direction of Golmud and highly unstable temperature region in Lhasa direction. Considering these conditions, for the use of bridge in this region is better.

On the basis of the linear frost mounds along bridge site, in January of 2002, Institute of Geology and Mechanics of Chinese Academy of Geological Sciences identified a fault of strike N55°E, which intersected with the line with two small angles in markers DK1005+600 and DK1006+100, and the fault was parallel to the line. If the existence of the fault, strongly affect the project, and the location of the line plane position can be adjusted to make the bridges bypass the fault. Hereby, adding the geological work of supplementary measurement aims to find out whether the fault exists or not and that the fault's distribution forms. On the basis of the 11 exploration holes in the survey phase location, another 18 exploration holes have been set up and the depths of the hole is generally 30–35 m. At the possible location of faults, one vertical section and two cross-sections perpendicular to the line have been set up along the central line, and the geophysical exploration such as electrical methods and earthquakes are carried out here.

The investigation shows that there is no larger abnormal belt in the worksite survey range along the direction of the line, and there is no fault zone material in drilling core, either. There is an approximately 130 m wide abnormal belt with low resistivity and with low speed in the stake numbers from DK1005+170 to DK1005 +300 (lateral abutment in the direction of Golmud), and no abnormality is found in drilling core. It is speculated that there may be a hidden fault in the stake numbers from DK1005+170–DK1005+300 which is obliquely crossed with the line. According to this, the original test route plan and project settings keep invariant.

4.5.4 Scheme Comparison of Line Selection from Xiaonanchuan to Wangkun

The line from Xiaonanchuan to Wangkun is around 18 km length, and in this region, the line gets into the edge of the permafrost zone, where is an intersection area of island-shaped frozen soil and talik. On engineering geology, the preliminary design scheme can be simplified to two schemes: the northern and southern schemes take the Qinghai–Tibet Highway as the boundary. The most representative program in northern scheme is the CIK route scheme, while the CK route scheme is the most typical plan in southern scheme (shown at Fig. 4.19).

The CK route program is a main and recommended program in preliminary design stage for southern scheme. Extracted from Xiaonanchuan station, the line mainly is distributed in the piedmont alluvial fan on the left side of the Qinghai–Tibet Highway and river floodplain, and ends in Wangkun station. Except for overland flow, no adverse geological phenomena occur here. The modern mountain

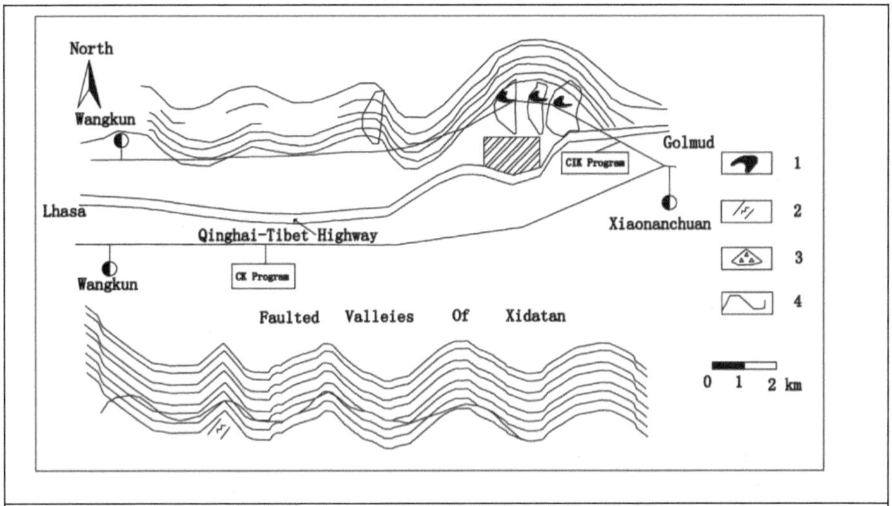

1: Crescent flowing dune; 2: Mountain glacier; 3: Rock glacier; 4: unfavorable geological boundary

Fig. 4.19 Line program comparison from Xiaonanchuan to Wangkun and geological-engineering skh map

glacier which is indeed a solid reservoir in its left side, and the main replenishment for overflow is water melted by ice and snow in warm season, meteoric water only accounting for a small section. The melt-water is mainly controlled by temperature. The current global temperature is generally increasing and the water melted by mountain snow flow increase. In view of this phenomenon, properly increasing the bridge density, pore size, and diversion dike solve the overflow problem.

The southern scheme is mainly CIK route program. Leaded from Xiaonanchuan station, the line bypassed the north side of the residential areas, and then the line which adjacent to mountain extended to Wangkun station. From the landscape, the line is mainly distributed on the northern margin of Xidatan Valley, a part of which passes through the piedmont alluvial fan and overflowing area. In all comparison schemes, this line scheme is the longest and most typical one. The line is along the compressive fault zone of Xidatan. Because of the fault is concealed and buried deeply beneath the Quaternary, it has a little effect on the line. But considering the measured areas belongs to the high earthquake intensity area; thus, the fault affects the line to some extent. Unfavorable geological conditions as follows: First, the line which in a form of cut crosses all crescent flowing dune in survey areas, and the sand dunes is developing and expanding; hence, the line would be affected; second, near the Wangkun station, mainly use of embankment, partially use of cut to pass through the front or toe areas of the large rock glacier. Because of the large scale and at high-intensity area, there is also groundwater which is mainly exposed as down spring and ice cone in the toe area of rock glacier. If selection of the CIK program, once the limited equilibrium state is destroyed by construction, there is a large slide occurring at any time, which causes a great threat to the body and machinery safety. Once the shock of the train has a positive effect on the slide of rock glacier, it be difficult to completely deal with; hence, these areas should idealy be avoided. Thirdly, there is the rockfall distribution in the toe of slope of rock glacier and in the top area of alluvial fan, which is near the line and has an adverse effect on projects. Although alluvial terrain is relatively flat, thick aeolian sand layers of thickness 0–0.5 m are the most common, which weaken the effect of rockfall and reduce the impact force generated by released potential energy when it falls. Considering this segment at the high intensity area, with addition of possible slump by rock glacier at any time, however, the rockfall comes from the rock glacier, with abundant supply, and potential geological hazards, the line selection should be appropriate to escape.

In summary, geological-engineering conditions of the area which the Qinghai–Tibet Railway pass through from Xiaonanchuan to Wangkun are poor, and the condition of the CIK route program is the worst. Once the CK route program solves the overland flow problem, its geological-engineering condition is superior to that of other schemes; hence, the CK line scheme is recommended. On the basis of fully research, utilizing of the existed geological data, and appropriate exploration, and methods, large-scale geological-engineering surveys are carried out for the geology-route selection between Xiaonanchuan and Wangkun. Through the comparison of the different schemes, the CK route scheme, under the worse geological conditions, relatively good geological-engineering conditions and the small effect of fault structure, is selected. The study has proved that the route selection should be given great attention in the complex geological-engineering conditions.

4.5.5 Scheme Comparison of Railway-Line Selection from Liangdaohe to Nagqu

Five alternative schemes are available for the Liangdaohe–Nagqu segment (Fig. 4.20). Schemes are described as follows:

C_1K scheme (along highway): Leaded from the Liangdaohe station, the line extends to the south along the Qinghai–Tibet Highway, down the Dangqingqu. The highway with two spans is collinear with the CK (scheme along Cuona Lake) behind the stake number of CK1629+400. The line extends along northern bank of Nagqu, then extends to Gangmixue along the right side of the Qinghai–Tibet Highway (100–200 m), and turns the south, entering hilly regions-Nagqu station away from highway. In comparison areas, 12 large–medium bridges of length 1.956 km are present, and the line is 77.721 km length and accounts for 2.5% of the total length. Static investment fund is 7.44 billion.

C_2K scheme: It is overlapped with C_1K from Liangdaohe to Dangqingqu. Since away from Dangqingqu station, the line crosses the Gongtuo pass, where a station is set up. After crossing of Baritanggou, there is a reserved Balu station before crossing of the Qinghai–Tibet Highway, and through the Shasiri tunnel, the line extends to Nagqu station. In comparison areas, nine bridges and tunnels of total

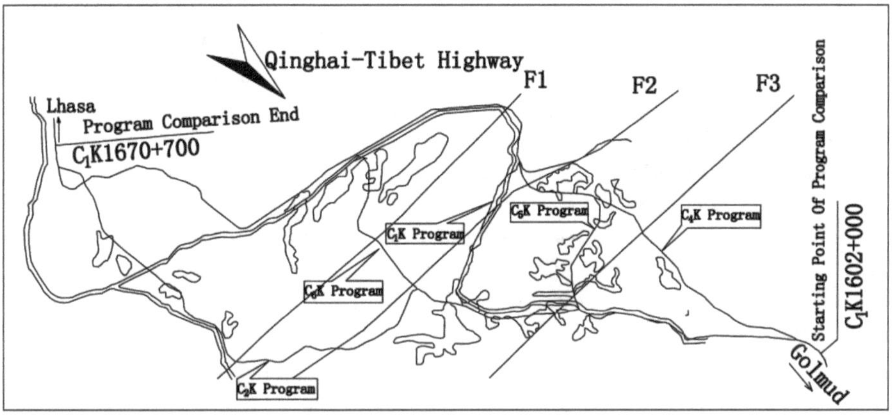

Fig. 4.20 Line program comparison schematic from Erdao river to Nagqu

length 3.869 km are present, the line is 63.052 km length and accounts for 6.1% of the total length. Static investment fund is 8.64 billion.

C_4K scheme: Leaded from starting point of comparison, the line turns to the southwest away from the Qinghai–Tibet Highway. Along the Qurugou and through the Gongna and the installment of a station, the line crosses Dangqingqu, and Nagqu. In CK1629+400, the line overlaps with CK scheme, then extends to the finishing point of comparison. In comparison areas, 13 large-medium bridges of total length 4.555 km, and the line is 66.76 km length and accounts for 6.8% of the total length. Static investment fund is 8.15 billion.

C_5K scheme: Leaded from the starting point of comparison, the line is paralleled with the Qinghai–Tibet Highway and extends to Gaqing. After crossing the highway, the line turns to southwest, then bypasses Cuobaerma Mountain where Gaqing station is set up. After passing Gongna and crossing Qingqu and Nagqu, the line is overlapped with CK scheme in the stake number of CK1629+400 until to the destination. In comparison areas, eight large–medium bridges of totallength 1.829 km are present, and the line is 72.736 km length and accounts for 2.5% of the total length. Static investment fund is 7.55 billion.

C_6K scheme: It is overlapped with CK scheme from Liangdaohe to Dangqingqu. Since away from Dangqingqu station, the line is away from the Qinghai–Tibet Highway in the stake number of CK1626+000. After crossing Sirehe and traversing the 2850 m length Hangbei tunnel, the line crosses the Qinghai–Tibet Highway again. In the stake number of CK1645+000, the line is overlapped with CK scheme to the destination. In comparison areas, six bridges and tunnels of total length 4.625 km, and the line is 66.311 km length and accounts for 7% of the total length. Static investment fund is 9.74 billion.

Five comprehensive analysis of the scheme is as follows:

C_1K scheme: Topography here is relatively wide and flat, with relatively simple engineering, but also is close to the Qinghai–Tibet Highway, with convenient construction. The line almost with a large angle passes through the fault zone, the effect of which is large, and gets through Dierma-Shaqing-Longma faults and Sheqing-Namuqing faults in the way of embankment, and the project safety is good. The defect is that the line is the longest and the freight is relatively high.

C_2K scheme: The line in this scheme is distributed in hilly regions, and topographical relief is larger. With large excavation of roadbed engineering, the environment is influenced well. The line almost is along the fault zone and the effect of the fault zone is the largest. The geological-engineering conditions are worse than the C_1K scheme. The line is far from the Qinghai–Tibet Highway and the construction is inconvenient. Although the line is smooth and short, the investment is relatively large.

C_4K scheme: Since that the line is almost orthogonal with Dierma-Shaqing-Longma faults, the range of the affected area is the smallest. Geological-engineering conditions along the line are slightly better than the C_1K scheme and the line is smooth and short. The disadvantage is: Quru Ditch which the line passes through is narrow and steep, with large bridge construction, and the line is far from the Qinghai–Tibet Highway, with inconvenient construction.

C_5K scheme: The local section of the line is far from the Qinghai–Tibet Highway, however, shortcuts that connect to the highway are present in some areas. The construction here is relatively convenient, and geological-engineering conditions along the line are similar with the C_1K scheme. Compared to C_1K scheme, the length of the line is shorter and investment is quite same, but the length of wetland that the line passes through is longer.

C_6K scheme: With a short and smooth line, Hangbei tunnel is distributed between two faults. Geological-engineering conditions are adverse, and investment is the largest.

With a comprehensive analysis, it is observed that: although the line of C_4K and C_6K scheme is relatively short, the increase of investment is larger. Despite the range of affected areas which C_6K scheme has passed through is narrow, Hangbei tunnel is distributed between two faults, with worse engineering technology conditions. C_2K scheme not only needs a large investment, but also the affected area of parallelly passing the fault zone is the longest. Compared with C_1K scheme, the investment of C_5K scheme is quite similar, while the geological-engineering conditions here are worse. In view of comprehensive comparison, although the line of C_1K scheme is the longest, the terrain along the line is flat, with the simple project, and the least investment. So we recommend C_1K scheme.

4.6 Summary

The Qinghai–Tibet Plateau has been subjected to the subduction of the India plate and has formed a lot of thrust faults and strike-slip faults. In this chapter, the spatial distribution, and fracture properties in each terrane, as well as the happening geological phenomena in fracture zone are discussed. The spatial distribution of

Kunlun Mountain seismotectonic zones, of Bengcuo seismotectonic zones, and of Yambajan–Damxung–Gulu seismotectonic zones, the faults' activity rates, and major earthquakes occurring in the seismic structural belts, are mainly introduced. The active fault zone not only affects the local production and life, but also causes greater damage to the railway engineering. The main disasters such as landslide, debris flow, structural fracture zone, and thaw slumping, solifluction, thaw lakes and ponds, frost mounds, ice cone, hummock, ice mantle, frost soil swamp, which are caused by the groundwater along active fault zones. The occurrence of earthquakes is also a great threat for the railway engineering. The destruction, which is caused by Yambajan Ms8.0 earthquake, of Ms8.0 Bengcuo earthquake, of Jiuzina Ms7.5 earthquake and of Ms8.1 Kunlun Mountain earthquake, has been mainly introduced.

Not limited to specification requirements, the line selection, and engineering settings of high intensity earthquake zone and active fault zones along the Qinghai–Tibet Railway, mainly aimed at meeting the need of the safety and the project reliability, and for the subject of building the first-class frozen earth railway on the plateau.

Per the principles of engineering-geological line selection in high-earthquake-intensity regions, railway lines should be located away from active fault zones unless unavoidable, in which case the line should be designed as a simple structure, with a large angle at its narrowest section. In addition, large and high bridges and tunnels should be avoided in active fault zones because these structures are difficult to repair when damaged. The strike of most active fault zones along the Qinghai–Tibet Railway either is perpendicular to the railway line or intersects at a large angle, depending on the neotectonic movement characteristics of the Qinghai–Tibet Plateau. In fact, most faults along the planned route could not be avoided. During the engineering investigation and design stage, all involved active faults were comprehensively evaluated and the different design schemes compared. In the final construction-drawing design stage, the aforementioned line-selection principles were followed along the main active fault zones, whereas for relatively unimportant active fault zones, efforts were made to avoid high bridges, to shorten bridge spans and lengths, and to replace bridges with culverts. At certain construction sites that influenced the route scheme and engineering setting, additional geological investigations were performed to ensure that the final line scheme and engineering measures were optimal. The line should get through these faults by simple roadbed engineering with vertical or large angles. When the bridge and tunnel engineering is necessary, some measures should be taken, such as decrease of the bridge height as soon as possible, use of a small span beam, designing a structural engineering which has good antiearthquake function and is easy to repair.

Careful investigation, for construction of the railway in active fault zones and strong earthquake zones, is necessary. Constructing a railway line which is a reasonable economic need to find out active fault zones along the railway and carry out seismic intensity zoning. The line-selection principle and engineering measures should be set up in active fault zones and meizoseismal areas.

References

1. Cao, D. Y., Lei, Y. D., Deng, X. F., et al. (2007). General engineering geologic conditions along the Naij Tal-Tuotuo River section of the Qinghai-Tibet Railway. *Qinghai Environment, 17*(2), 78–80. (in Chinese).
2. Chen, R. H., & Wang, Z. T. (2003). Engineering-geological route selection of the gypsum section from Buqu to Wenquan of the Qinghai-Tibet Railway. *Journal of Glaciology and Geocryology, 25*(1), 14–16. (in Chinese).
3. Chen, J., Chen, Y. K., Ding, G. Y., et al. (2003). Surface rupture zones of the 2001 earthquake Ms8.1 west of Kunlun Pass, northern Qinghai–Tibet Plateau. *Quaternary Sciences, 23*(6), 629–640. (in Chinese).
4. China Railway First Survey & Design Institute Group Co., Ltd. (1994). *The manual of routes* (2nd ed.). Beijing: China Railway Press. (in Chinese).
5. China Railway First Survey & Design Institute Group Co., Ltd. (2001). *Feasibility study report of Tibet Railway*. Internal information (in Chinese).
6. China Railway First Survey & Design Institute Group Co., Ltd. (2002). *Preliminary design of the Tonglha–Lhasa section of the Qinghai–Tibet Railway (Geology)*. Internal information (in Chinese).
7. Chinese Geological Atlas Editorial Board. (1996). *Geological atlas of Chinese*. Beijing: Geological Press. (in Chinese).
8. Geologic Institute of State Seismological Bureau. (1991). *Active faults in central Tibet*. Beijing: Earthquake Press. (in Chinese).
9. Geological Survey of China. (2004). *Geological map of China 1:2500000*. Beijing: China Cartographic Pressing House. (in Chinese).
10. Ministry of Railways. (2006). *Code for design of railway line (GB50090-2006)*. Beijing: China Plan Press. (in Chinese).
11. Seismological Bureau of Qinghai Province. (1999). *China crustal stress institute active fault zone in east Kunlun*. Beijing: Seismological Press. (in Chinese).
12. State Seismological Bureau of China. (1991). *Earthquake intensity zoning map of China (1990)*. Beijing: Earthquake Press. (in Chinese).
13. Wei, G. J. (2003). Geological line selection of the Xiaonanchuan-Wangkun unfavorable geological section of the Qinghai-Tibet Railway. *Geotechinical Engineering World, 6*(7), 41–45. (in Chinese).
14. Wu, Z. M., & Cao, Z. Q. (1991). Review of intensity of Damxung earthquake (Ms = 7.5) in 1952. *Acta Geophysica Sinci, 34*(2), 64–72. (in Chinese).
15. Wu, Z. H., Hu, D. G., Wu, Z. H., et al. (2006). Pressure ridges and their ages of the Xidatan strike-slip fault in south Kunlun mountains. *Geological Review, 52*(1), 15–25. (in Chinese).
16. Wu, Z. H., Wu, Z. H., Hu, D. G., et al. (2006). Holocene seismogenic faults along the Tanggula-Lhasa section of the Qinghai-Tibet Railway. *Geological Bulletin of China, 25*(15), 1387–1401. (in Chinese).
17. Yin, A. (2001). Geologic evolution of the Himalayan-Tibetan orogen. *Acta Geophysica Sinci, 22*(3), 193–238. (in Chinese).
18. Yuan, D. Y., Zhang, P. Z., Liu, X. L., et al. (2004). Tectonic activity and deformation features during the Late Quaternary of Elashan Mountain active fault zone in Qinghai Province and its implication for the deformation of the northeastern margins of the Qinghai-Tibet Plateau. *Earth Science Frontiers, 11*(4), 393–402. (in Chinese).
19. Zhao, G. G. (2006). Quaternary faulting in north Qinghai-Tibet Plateau. *China Earthquake, 12*(2), 107–117. (in Chinese).
20. Zhao, J. C., Liu, S. Z., & Ji, S. W. (2001). Engineering-geological problems in railway engineering construction. *The Chinese Journal of Geological Hazard and Control, 12*(1), 7–9. (in Chinese).

Chapter 5
Railway-Line Selection in Slope Areas

Abstract The main mountain areas along the Qinghai–Tibet Railway are Kunlun Mountains, Hoh Xil Mountains, Fenghuo Mountain, Kaixinling, and Tonglha Mountains, along the slopes of which are distributed various unfavorable geological hazards. Route selection and optimization are crucial in minimizing their adverse effects on the construction of the Qinghai–Tibet Railway.

Keywords Railway-line selection · Geological hazard · Precautions
Preventions · Scheme comparison

5.1 Distribution of Slope Geological Hazards

5.1.1 Fundamental Slope Characteristics

Slopes can be natural (e.g., riverbanks, ditch banks, and reservoir banks) or artificial. Rocks and soil on slopes, particularly in mountain areas, may tend to be displaced because of gravity or factors such as weathering, seismic, human activity, freezing, and thawing. This displacement is called slope deformation.

5.1.1.1 Slope Classification

Table 5.1 presents slope classification on the basis of the cause of the slope.

1. By material composition

On the basis of the material composition of the slopes, slopes can be classified as follows:

a. Geological slopes: cohesive soil slopes, gravel soil slopes, Loess slopes
b. Rock slopes
c. Mixed rock and soil slopes.

J. Li et al., *Geological Line Selection for the Qinghai-Tibet Railway Engineering*,
https://doi.org/10.1007/978-3-662-55572-9_5

Table 5.1 Slope classification by cause

Classification		Cause		
Natural slope	Denudation slope	Mainly due to denudation caused by the rising of Earth's crust	Linear slope	When the intensity of crust rising equals that of denudation
			Concave slope	When the rising of the crust is less intense than denudation
			Convex slope	When the rising of the crust is more intense than denudation
			Alternately concave and convex slope	When the rising of the crust and denudation are of different intensities and are alternately dominant in the movement of rocks
	Accumulation slope	Accumulation of fragmented materials produced by physical weathering		
	Erosion slope	Surface water erosion; this type can be further classified as bank erosion and gully erosion		
	Slumping slope	Natural factors such as landslides and collapses		
Artificial slope		Human activities such as excavation or accumulation (e.g., mine slopes and cut slopes of railways and highways)		

5.1.1.2 Major Types of Slope Deformation and Failure

Figure 5.1 presents the major types of slope deformation and failure.

5.1.2 Major Geological Hazards Along the Qinghai–Tibet Railway and Their Distribution

Major geological hazards along the Qinghai–Tibet Railway line include dangerous rocks, rockfall, landslides, collapse, landslides, and debris flow; in addition, solifluction, thaw slumping, and slope wetlands occur frequently in permafrost slope areas.

5.1.2.1 Dangerous Rocks, Rockfall, and Collapse

A dangerous rock refers to a rock mass or an individual block of rock on steep slopes and having a potential to collapse or fall, and rockfall refers to the

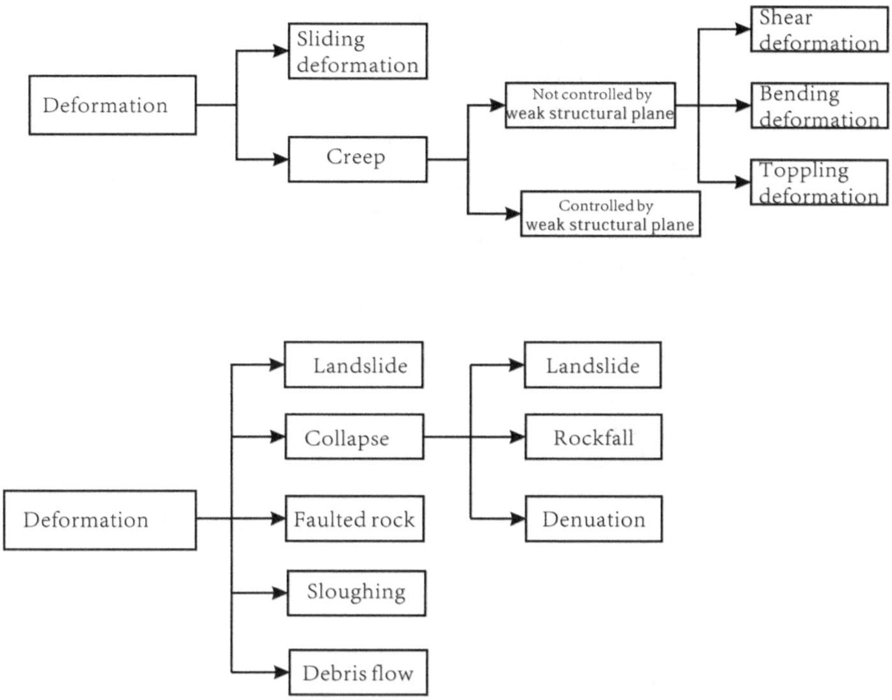

Fig. 5.1 Major types of slope deformation and failure

phenomenon wherein an individual rock block suddenly falls or rolls down a steep slope under the effect of gravity. Collapse refers to the phenomenon wherein a rock mass suddenly falls down a steep slope under the effect of gravity or other forces (e.g., toppling, caving, and rolling). Collapse occurs when the shear stress of the rock mass exceeds the strength of any of its weak structure planes (e.g., joint surfaces, bedding planes, schistosity planes, and magmatic intrusive contact zones). Rock and soil movement does not occur along a fixed surface or belt, and their vertical displacement is significantly higher than the horizontal displacement.

1. Distribution

   a. Topography and geomorphology
   Collapse occur mainly along the steep (inclination >55° and height >30 m) slopes of rock canyons and concave banks of mountain valleys; these slopes generally have a rugged surface, with steep upper sections and gentle lower sections.

   b. Lithology

      i. The steep slopes composed of hard and brittle rocks (e.g., thick limestone, granite, and basalt) contain tectonic and unloading joints and deep and steep tensile cracks running parallel to the slope; these features increase the risk of collapse and rockfall (Fig. 5.2). Along the Qinghai–Tibet Railway, such slopes are mainly distributed on the slopes of Doilung Qu Gorge Terrace area.

ii. The steep composed of alternating soft and hard bedrock (e.g., sandstone interbedded with shale and limestone interbedded with marl). Soft and hard rocks differ in their weathering-resistance properties and thus undergo weathering at different rates. Consequently, such slope surfaces often have concave soft rocks and convex hard rocks and thus are susceptible to collapse and rockfall. Along the Qinghai–Tibet Railway, such slopes are mainly distributed in sections of Kaixinling and Yanshiping.

iii. Some steep slopes are composed of hard rock in their upper sections and soft rock on their lower sections. Here, cracks in hard rocks develop further the under tensile stress due to soft rock erosion and deformation. The cracks may eventually form a continuous, penetrating fracture and can trigger large collapses. Such hazards are relatively rare along the Qinghai–Tibet Railway and mainly occur in the northern foot of the Kunlun Mountain area, which the railway line bypasses (Fig. 5.3).

c. Geological structure

i. Collapse and rockfall severely affect a railway line when the line runs closely and parallel to the regional tectonic belt and when the design entails large cuts.

ii. Collapses are more likely at the intersection of several tectonic lines.

iii. If various weak structural planes of a rock are distributed in the lower section of the slope, the slope is more likely to collapse.

Fig. 5.2 Collapse of hard and brittle slopes

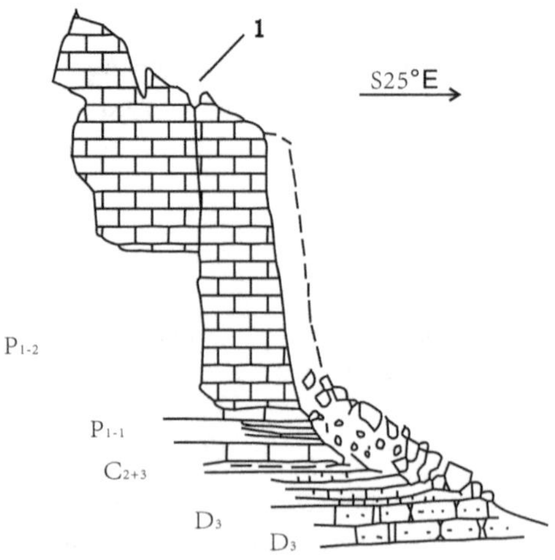

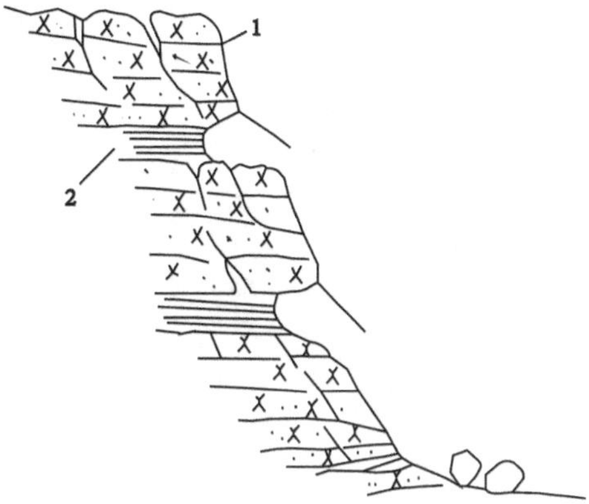

Fig. 5.3 Local collapse and fall of slopes composed of alternating soft–hard bedrocks

d. Other factors

i. Collapses are strongly dependent on hydrological factors: the vast majority of collapses occur in the rainy season or after a rainstorm. Hydraulic actions increase the geotechnical mass and hydrostatic pressure and dissolve or soften the fillings in the fractures within the mass; consequently, the weak structural plane in the mass is destroyed, leading to collapse.
ii. Strong seismic activity, large-scale blasting, or dynamic loading can contribute to or induce a collapse or rockfall.
iii. A large day–night temperature difference and seasonal temperature changes accelerate rock weathering. In addition, intrusion by plant roots, frost action of the fissure water, and the erosive action of ice all contribute to rock-mass weakening and collapse. Moreover, creating very high or steep artificial slopes can destroy the stability of mountains, which too may lead to collapse.

2. Distribution of dangerous rocks, rockfall, and collapse along the Qinghai–Tibet Railway

Along the Qinghai–Tibet Railway, dangerous rocks and rockfall mainly occur in the front pebble layer and rock slope areas of Doilung Qu Gorge Terrace. Pebble soil-containing boulders aggregate into isolated stones under the action of temperature, fissure water, and rain erosion. These isolated stones can easily roll off the slope. Rockfall is influenced by numerous factors, such as steepness of the slope. Furthermore, any exposed andesite becomes weak under the action of structural joints (e.g., at the intersection of the Mozhugongka EW-trending tectonic belt and the western end of the Nyainqentanglha NE–SN-trending tectonic belt). The frost action of fissure water, the erosive action of ice, denudation, and other exogenic forces also worsen the rockfall conditions.

Fig. 5.4 Rockfall at DK1938+700

Dangerous rocks and rockfall are common near Naij Tal (south bank of Kunlun River) and Yambajan tunnels because the geological conditions in these areas are severe and complex, rendering the weak. Dangerous rocks and rockfall that affect the railway are mainly distributed in the southern hillside of Naij Tal, specifically at sections DK901+300–DK902+050, DK1936+270–DK1937+450, and DK1972 +760–DK1972+880 (Figs. 5.4 and 5.5).

Fig. 5.5 Dangerous rocks at DK1938+900

5.1.2.2 Rock Heap

Gravel is formed through physical weathering and is transported downhill by gravity or precipitation, and a rock heap is an accumulation of loose gravel.

1. Distribution
   Rock heaps are mainly distributed on steep mountain areas or slope toe areas where the climate is dry, the day–night temperature difference large, the physical weathering strong, and the tectonic uplift intense.
   Geological conditions strongly influence the distribution of rock heaps:
   a. Rocks heaps are often distributed on one side of a river valley when the river valley runs along a fault belt.
   b. They are often distributed symmetrically on both sides of a river valley when the river valley runs along an anticlinal axis, where many joints are often present.
   c. They are mainly distributed behind the valley formed by joint cutting when the river valley is parallel to the rock strike.
2. Distribution of rock heaps along the Qinghai–Tibet Railway

Along the Qinghai–Tibet Railway, rock heaps are mainly distributed on both sides of Kunlun Mountain Luanshi Ditch and the slope toe of both sides of Doilung Qu.

Kunlun Mountain Luanshi Ditch is located on the uplift plate between two thrust faults, namely Xidatan Fault and Kunlun Mountain Pass Fault. In this strong compressive fold area, rock heaps of various sizes and shapes are well developed because of the development of joints and cracks and steep slopes. They are mainly distributed in the lower section of steep slopes or on the both sides of large gullies. Rock heaps in this area are mainly composed of loose slate and schist fragments, and the slope toes of both sides of Doilung Qu are mainly composed of weathered rock and ice–water bodies.

Some rock heaps are distributed in the steep slope areas of the side opposite to Naij Tal (Fig. 5.6).

5.1.2.3 Landslide

A landslide, also known as a landslip, is a geological phenomenon encompassing a wide range of ground movements with gravity as the primary driving force.

For a mountain railway line, a landslide is a disaster that causes the most damage. Small landslides can cause railway subgrade upwarping, sinking, or translation, whereas large landslides can bury the railway lines, destroy the railway subgrade, and damage railway bridges, tunnels, and other engineering structures. Landslides that occur during railway construction slow the construction work, and

Fig. 5.6 Rock heaps distributed in the steep slope areas opposite to Naij Tal

landslides that occur during railway operation can interrupt train traffic or even cause significant loss of life and property.

Although a few landslides have occurred along the Qinghai–Tibet Railway, they have not significantly affected the railway line; hence, they are not discussed in detail herein.

5.1.2.4 Debris Flow

Debris flows are geological phenomena in which water-laden masses of soil and fragmented rock rush down from mountain sides under the influence of gravity. Debris flow is dependent on the conditions of the terrain, geology, hydrology, meteorology, vegetation, earthquake, and human activities. The source area of a debris flow must have (1) a very steep slope, (2) an abundant supply of loose debris, and (3) a sudden flow of water from heavy rain or snowmelt.

1. The features of debris flow are as follows:

   a. They are mainly active in the mountainous area and piedmont.
   b. They occur suddenly, last for a short time, and move at high speeds.
   c. They vary widely in density (1.2–1.3 to 1.8–2.3 g cm^{-3}) and particle size range (clay to boulder).
   d. They have large inertia and entrain objects in their paths.
   e. They move rapidly and cause tremendous damage.

2. Effects of debris flows on railway

Hydrological Specifications for Survey and Design of Highway Engineering (TBJ17-86) classify debris flows according to their danger level as shown in Table 5.2.

Debris flow can damage a railway line in two ways.

a. Damage due to erosion

In sections with debris-flow paths, the railway line is most likely to be damaged by the erosion of debris flows in two ways.

First, the debris flow directly hits the buildings, thus thrusting or shearing the structures or even washing away bridges. For example, on July 9, 1981, in the Chengdu–Kunming line, the Toshiko Wajda debris flow hit bridge pier 2 and destroyed the bridge, causing cars on the bridge to crash fatally; on August 21, 1981, in the Baoji–Tianshui line, the Xiaoqiaogou Debris Flow directly washed away a 10-m-long concrete beam.

The second damage mechanism is erosion. Some debris flows undercut ditch beds or hollow out the bridge foundation abutments and slopes. For example, in August 1981, at the Shanggada Bridge along the Chengdu–Kunming railway, debris flow undercut the ditch bed 7–13 m deep, covering the pier and almost destroying the bridge. In addition, some debris flows laterally erode riverbanks; moreover, debris-flow tributaries can erode or bury roadbeds, thus blocking or slowing traffic. For example, on August 21, 1981, debris flow eroded the concave bank and buried the railway embankment of the ditch at K1305+862 along the Baoji–Tianshui Railway. Debris flow can also block rivers, leading to river bursting and consequently strong erosion on both sides of the river bank. For example, in the Chengdu–Kunming Railway line, on July 1, 1984, debris flows blocked the Niuri River, which then burst and washed away a 13-m wide terrace near the exit of the Lianghong tunnel, leaving the foundation of the tunnel hanging in the air.

b. Damage due to accumulation

In open gentle slope terrain regions, debris flows often present a loose flow with a decreasing flow rate until it eventually stops. This phenomenon can easily lead to the siltation of railway bridges and culverts in debris-flow alluvial or flood-fan

Table 5.2 Debris-flow classification according to their danger level

Danger level	Severe	Loose solid material is abundant in the river and is accumulated heavily on the riverbed, which has little vegetation coverage; the soil erosion is severe. The collapse area proportion is >10%
	Medium	Loose solid material is abundant in the river and is accumulated at some sections of the riverbed, which has low vegetation cover; the soil erosion is severe. The collapse area proportion is 5–10%
	Slight	Slight erosion in the watershed, with little accumulation in the channel. The vegetation coverage is good. The collapse area proportion is <5%

fronts, such as at Chengdu–Kunming Railway Machang Ditch and Baoji–Chengdu Railway Huanglongyu Ditch, and the accumulated silt must be cleaned every year. In particular, railway stations located in flood debris-flow alluvial and flood fans are more likely to be affected by geological disasters if no adequate preventive measures are implemented (e.g., in case of weak drainage, inadequate bridge and culvert pore size, and inappropriate ditch conditions), which is likely I the debris-flow severity is underestimated. For example, four debris-flow gullies are present around Baoji–Tianshui Railway Yuanlong Station, and debris-flow deposition under the bridges here need to be cleaned to ensure adequate drainage, and Xintiecun Station, Aidai Station, and Lianhexiang Station of the Chengdu–Kunming Railway and Lijiahe Station and Honghuapu Station of the Baoji–Chengdu Railway have been affected by debris flow.

3. Distribution of debris flows along the Qinghai–Tibet Railway

Region highly prone to debris flow must be avoided in line selection. Debris flows that affect the lines are mainly distributed between Gangou and the inlet of Kunlun Mountain Tunnel as well as the along the tributaries on both sides of Doilung Qu Gorge, where the total railway line is 1.412 km long. In these regions, rocks on the slopes, which are steep, are severely weathered, and loose debris is abundant (Table 5.3).

The slopes on both sides of Doilung Qu Gorge are susceptible to debris flow because of strong rivers incision, large seasonal changes in water flow, and weathered rock bed (which provides an abundant supply of loose debris), and the abundant moisture during the rainy season. For example, debris flows have occurred on the right side of the natural slope at DK1927+720 and on the right side of the valley at DK1933+078–DK1933+084. A small debris flow occurred on July 13, 2001, at the Qinghai–Tibet Highway Yambajan Canyon section, blocking traffic for 1 h (Figs. 5.7 and 5.8).

5.1.2.5 Solifluction

Soil becomes saturated as water that becomes abundant during thawing fails to filter through the frozen underlying ground. Solifluction, a type of creep, refers to the slow slope movement of saturated soil, which occurs when the structure of the soil and the upper cover (vegetation) are destroyed because of recurrent freezing and the thawing of the ground.

Solifluction rates vary on the order of centimeters per week to centimeters per year depending on the season. Solifluction mainly occurs in the slope areas of Kunlun Mountain Luanshi Ditch, which the Qinghai–Tibet Railway line bypasses. Thus, it does not affect the railway.

Table 5.3 Distribution characteristics of debris flows along the Qinghai–Tibet Railway

No.	Distribution		Distribution characteristics	Engineering advice
	Section beginning and ending	Distance (m)		
1	DK899+960–DK899+970	10	Flood fan; composition: thick layer of gravel soil	Build supporting and retaining structures
2	DK900+580–DK900+640	60	Developing debris-flow valleys; composition: gravel soil	Use bridges
3	DK901+215–DK901+275	60	Developing debris-flow valleys; composition: gravel soil	Use bridges
4	DK1933+078–DK1933+084	6	Developing debris-flow valleys; composition: gravel soil	Apply comprehensive treatment
5	DK1936+280–DK1936+670	390	Flood fan; composition: thick layer of gravel soil	Build supporting and retaining structures
6	DK1937+420–DK1937+960	540	Flood fan; composition: thick layer of gravel soil	
7	DK1938+510–DK1938+690	180	Developing debris-flow valleys; composition: gravel soil	Use open-cut tunnels
8	DK1938+924–DK1939+090	166	Developing debris-flow valleys; composition: gravel soil	
	Sum	1412		

Fig. 5.7 Right-side slope debris flow at DK1927+720

Fig. 5.8 Debris flow blocking traffic

5.1.2.6 Thaw Slumping

Thaw slumping refers the phenomenon wherein the thermal equilibrium state of the slope containing underground ice is destroyed, and the frozen soil melt because of external influence; consequently, the overlying soil layer slumps along the thawing–freezing interface under the effect of gravity. On the basis of the stage of development and the degree of harm to the project, thaw slumping can be classified as stable and active. Stable thaw slumps are those slumps that no longer expand, whereas active thaw slumps are those that do. All active thaw slumps become stable slumps when the slumping layer reaches the upper edge of the underground ice.

Little slump occurs on hillsides with a transverse slope less than 3°; on hillsides with a transverse slope of 3°–5°, long strip-shaped slumping is common, whereas, on hillsides with a transverse slope exceeding 5°, long strip-shaped tractive slumping may occur. Thaw slumps are initially crescent-shaped and then gradually develop upwards, changing into a long strip or twig-shaped development.

Fully developed thaw slumping beds can be divided into three zones. The upper section is the flow zone; the steeper the natural cross-slope is, the longer is the flow zone. The central section is a plastic deformation zone, where one or more tongue-shaped terraces may be present, with one tongue-shaped terrace forming each year. The lower section is the stable zone. As slump, development varies each year, so do the position of these three zones. Generally, thaw slumping is faster in steep (transverse) and high-temperature slopes with more ice.

Thaw slumping can undermine the stability of building foundation or subgrade slopes and lead to buildings being buried by the thawing mudflow. Nevertheless,

Fig. 5.9 Thaw slumping in Fenghuo Mountain area

because of the slow and small lateral development, thaw slumping hazards tend to be less severe (Fig. 5.9).

Excavation or construction activities that change the thermal equilibrium state of the ground may induce secondary thaw slumping near such activity zones. Along the Qinghai–Tibet Railway, thaw slumping mainly occurs in Kunlun Mountains and Fenghuo Mountain.

5.1.2.7 Slope Wetlands

Slope wetlands can be divided into two categories: nonfrozen-soil slope wetlands and permafrost slope wetlands.

Excessive water on the ground surface cannot filter through the frozen ground underneath; thus, surface water accumulates in low-lying areas, gradually forming swamp wetlands and swamp permafrost. The structure of a swamp permafrost is similar to that of a sandwich, with an ice layer and layered humus or peat. Hence, the structural features of soft soil and a swamp differ significantly. Most slope wetlands are meadow swamps and meadow wetlands, where pimple-like surface grass, called "tower head grass", grows. Grass roots crowd together, forming turfs of thickness 0.1–0.3 m. Wetlands commonly contain frost fissures, where water gathered in potholes in the summer dries or leaves less ice in winter. The main component of the topsoil here contains humus clay, and the weak surface layer is generally 0.2–2-m thick.

The harmful effects of permafrost swamp wetlands are as follows. First, the topsoil has a weak structure after summer melting. Topsoil of wetlands has characteristics such as high moisture content, high void ratio, uneven composition of organic ingredients, low permeability, high sensitivity, high compressibility, low strength, rheology, and physical and mechanical indexes less than those of soft soil. Second, the surface water adversely affects ground consolidation, which is essential for projects. Third, the thermal disturbance caused by engineering activities can induce the thaw subsidence of the underlying permafrost layer. Hence, without appropriate measures, permafrost swamp wetlands can be a strong hazard to engineering projects.

Slope wetlands are likely to form where groundwater discharges onto the land surface or where the overflow saturates the soil because of the lack of a drainage channel. They normally occur on sloping lands, both slight and steep. Further, wetlands are likely to become thawing mud flow and thaw slumping areas.

5.2 Influence of Slope Geological Hazards on Railway Engineering

5.2.1 Dangerous Rocks, Rockfall, Collapse, and Rock Heaps

5.2.1.1 Golmud–Amdo Section

In the Golmud–Amdo section, the terrain is relatively flat, with only one section—the slope toe of the south bank of Kunlun River, north of Naij Tal Station (DK903+450–DK903+850)—with dangerous rocks and rockfall along the Qinghai–Tibet Railway line. The terrain is steep, with the mountain sloping 70°–80°, and 0.5–1.0-m-thick Paleozoic limestone strata are bedded around N80°E/60°N. Five groups of joints with a joint spacing of 0.1–0.3 m and joint density of 4–7/m^2, and length of 1–3 m can be found at N82°E/74°N, SN/75°N, N85°W/71°N, N20°E/31°S, and N12°W/70°S. The rock at the intersection of these five groups is cut into pieces. Weathering has expanded the fissures to widths of 1–5 cm and even 20 cm in some cases. The rock body is in a critical state and is likely to form dangerous rocks. At DK903+760, many dangerous rocks of diameter 1.3–1.7 m, which has the potential to damage the railway, can be found on the left-side hilltop. The rock heaps at the foot of the mountain are approximately 120-m high and 250-m wide, with a surface slope of nearly 30°; the toe is approximately 60 m away from the line.

5.2.1.2 Amdo–Lhasa Section

In the Amdo–Lhasa section, most of the terrain is relatively flat. Dangerous rocks, rockfall, and collapse mainly occur in the rock slope areas of Doilung Qu Gorge Terrace (DK1927+000–DK1941+200), located in the high mountainous area of the Damxung–Yambajan tectonic basin edge, where the seismic intensity is 9°. The exposed rock is Yanshan–Himalayan granite. The slope surfaces are highly physical weathered under the action of external forces such as temperature, fissure-water frost, and ice splitting. The section has erosion features, deep valleys, and unstable slopes, which are conducive to dangerous rocks and rockfall; some moderately large rock heaps are distributed at the foot of the slope. Rockfalls and dangerous rocks near the exit of Yambajan tunnel 1 and at DK1972+760–DK1972+880 threaten the safety of the line. Some dangerous rocks are distributed on the hilltop near the exit of Yambajan tunnel 1, and the hillside slope varies from 40° to almost 90°. Multiple groups of joints have developed in the exposed bedrocks, and the rocks are cut into blocks and scattered at the toe.

5.2.2 *Debris Flow*

5.2.2.1 Golmud–Amdo Section

Debris flows in the Golmud–Amdo Section are mainly distributed in the Gangou–Xidatan section. The hillsides on both sides of Kunlun River Bank between Gangou (DK870) and Xiaonanchuan (DK927) are steep and have low vegetation coverage; the slopes here are undergoing intensive physical weathering, and some loose spall soil has accumulated on the slope surface and ditch bed. The flow moves loose debris downstream, creating debris flows in the tributaries. Small alluvial fans are distributed at the toe of some gullies, and the phenomenon of the river mainstream being squeezed and offset can be seen in this area. The relatively large alluvial fans are distributed at the toe of Caiyuanzi, Moshi, Gao, Duan, Vampo, and Wumin Valleys. Table 5.4 lists the characteristics and dangerousness of each valley, and Table 5.5 presents the same information for the four valleys across the river at Naij Tal.

The section of the line between boundary markers DK944 and DK965 lies in Xidatan Basin, passing through the leading edge area of north Yuzhu Mountains' ice alluvial fan and alluvial fan, 2–3.5 km from the foot of the slope. Modern glaciers have developed in Yuzhu Mountain area, where the melting snow provides water for the formation of debris flows. The north hillside is steep and produces ten short steep V-shaped gullies in an approximately 20-km section; three of these gullies extend to the toe and serve as flood catchment areas. The hillsides have almost no vegetation and have undergone severe weathering, and loose deposits are present on the slopes and ditches. The railway lines pass through the debris-flow accumulation area, where ground gullies have developed and the accumulation

Table 5.4 Characteristics and dangerousness of each valley

No.	Valley name and railway mileage	Catchment (km^2)	Main valley shape	Average gradient of the main groove	Characteristics and area of the flood fan (km^2)
1	Wumin Valley 1, Dk889 +500	12	V-shaped; two tributary gullies are present upstream; valley toe is 700-m wide	11.3° (20%)	Ancient flood fan; the main valley is ~15-m deep and has an area of ~4 km^2
2	Wumin Valley 2, DK890 +500	23	V-shaped; valley toe is 750-m wide	10° (18%)	Ancient flood fan squeezing the main river; main valley is ~5-m deep and has an area of ~2.2 km^2
3	Wumin Valley 3, DK894 +000	42	Two main valleys that converge to one after they leave the mountain area	10°–12° (18–22%)	Ancient flood fan squeezing the main river; main valley is ~20-m deep and has an area of ~16 km^2
4	Caiyuanzi Valley, DK897 +500	44	V-shaped; valley toe is 500-m wide	10° (18%)	Ancient flood fan squeezing the main river; main valley is ~5–10-m deep and has an area of ~6 km^2
5	Moshi Valley, DK905 +500	100	V-shaped; three tributary gullies are present upstream, and valley toe is 800-m wide	10° (18%)	Ancient flood fan, squeezing the main river; modern deposits are present on the frontier lines of the fan; main valley is 3–5-m deep, with an area of ~6 km^2
6	Gao Valley, DK908 +000	27	V-shaped; valley toe is 600-m wide	12° (22%)	Ancient flood fan, squeezing the main river; modern deposits are present on the frontier lines of the fan; main valley is 3–5-m deep, with an area of ~6 km^2

(continued)

Table 5.4 (continued)

No.	Valley name and railway mileage	Catchment (km^2)	Main valley shape	Average gradient of the main groove	Characteristics and area of the flood fan (km^2)
7	Duan Valley, DK915 +000	24	V-shaped; valley toe is 300-m wide	12° (22%)	Ancient flood fan, squeezing the main river; modern deposits are present on the frontier lines of the fan; main valley is 3–5-m deep, with an area of ~4 km^2
8	Wanbao Valley DK916 +000	80	V-shaped; three tributary gullies are present upstream; valley toe is 700-m wide	11° (20%)	Huge ancient flood fan; the main valley south of the road is 20-m deep, with an area of ~9 km^2
No.	Loose material quality (m^3 km^{-2})	Gradient of hillside	Vegetation cover rate (%)	Frequency (times per year)	Location and dangerousness of the railway section
1	1000–3000	25°–30°	<5	1–3	Frontier line of accumulation area; severe
2	1500–4000	25°–30°	<5	1–3	Other side of the river; no harm
3	3000–5000	25°–30°	<5	1–3	Other side of the river; no harm
4	1500–4000	25°–30°	<5	1–3	Frontier line of accumulation area; severe
5	3000–5000	25°–30°	<5	1–3	Frontier line of accumulation area; severe
6	2000–4000	25°–30°	<5	1–3	Frontier line of accumulation area; severe
7	2000–4000	25°–30°	<5	1–3	Crosses the main river using a bridge at the frontier line of accumulation area; medium
8	3000–5000	25°–30°	<5	1–3	Frontier line of accumulation area; medium

Table 5.5 Characteristics and dangerousness of four valleys across the river at Naij Tal

No.	Railway mileage	Catchment (km^2)	Main valley shape	Average gradient of main groove	Flood-fan area (m^2)	Loose material quantity ($m^3\ km^{-2}$)	Hillside gradient	Vegetation cover rate (%)	Frequency (times per year)	Location and dangerousness of the railway section
1	DK901 +968–DK902 +027	1.4	V-shaped; valley toe is 18-m wide	11.3° (20%)	50000	1000–3000	25°–30°	<10	1–3	Small scale and low intensity; a bridge is used; Medium dangerousness
2	DK902 +570–DK902 +600	1	V-shaped; valley toe is 20-m wide	10°(18%)	10000	1000–2000	20°–25°	<10	1–2	Small scale and low intensity; a bridge is used; Medium dangerousness
3	DK903 +198–DK903 +240	2	V-shaped; valley toe is 18-m wide	7°(12%)	16000	1000–2000	20°–25°	<10	1–3	Small scale and low intensity; a bridge is used; Medium dangerousness
4	DK904 +016–DK904 +040	1.3	V-shaped; valley toe is 20-m wide	8.5°(15%)	5000	<1000	20°–25°	<10	<1	Small scale and low intensity; a bridge is used; Medium dangerousness

body is relatively stable. Therefore, the line may be affected by glacier debris flows. However, since the railway is located on the leading edge of the debris flows, this hazard is of relatively low severity.

5.2.2.2 Amdo–Lhasa Section

In the Amdo–Lhasa Section, debris flows occur mainly between Sangli and Lhasa. The terrain north of Sangli is relatively flat, and most rivers here have wide and shallow riverbeds with mostly wide and slow tributaries. Hence, debris flow is rare in north of Sangli. Field investigations have revealed that the environmental conditions of the tributaries on both sides of some sections of Dangqu River and on both sides of Doilung Qu Gorge facilitate debris flows; specifically, the conditions are as follows: (1) a very steep slope, and ditch beds of longitudinal gradient 10–20%; (2) abundant supply of loose debris due to strong weathering, intense tectonic movement, and low vegetation coverage; and (3) sudden flow of water due to heavy rain or snowmelt. According to meteorological observations in Damxung and Lhasa, the maximum annual and monthly precipitation in these two districts are 685.8–796.6 and 253.2–313.5 mm, respectively, with the latter being concentrated in July–August; the daily maximum precipitation is 41.6–50.4 mm. Solids accumulated in gullies are mainly gravel soils, and the main types of debris flows here are mudrock and water rock.

Along the Qinghai–Tibet Railway, ancient proluvial fans and flood fans are distributed in the area between Wumatang and Longren, where many gullies and recent deposits are distributed on the surface. Table 5.6 summarizes the characteristics and dangerousness of each debris-flow valley.

The hillsides on both sides of Doilung Qu Gorge are steep and have low vegetation coverage on the east; the western side, by contrast, has high vegetation coverage. The slopes in this area are subject to intensive physical weathering, with some loose spall accumulating on the slope surface and ditch beds. Sudden rainstorms induce flood flows in the tributaries, which carries loose debris downstream, resulting in debris flows that accumulate at the toe of gullies, forming alluvial fans or cone-shaped accumulations, which squeeze or redirect the mainstream.

When selecting the railway route of the Qinghai–Tibet Railway, the majority of actual and suspected debris-flow areas were bypassed. During the construction period, field investigations revealed that eight debris-flow areas had the potential to severely harm railway sections of total length 1.412 km (Table 5.3).

In addition, vegetation along some sections (of total area 16,000 m^2 in) of the left bank slope of Ribaron Ditch is severely damaged, especially on the left side of the DK1810+310–DK1810+360, DK1810+366–DK1810+390, DK1810+415–DK1810+490, and DK1811+640–DK1811+760 sections. Small gullies have developed on the slopes, and small debris flows occur in the rainy season, causing sedimentation on the left side of the line; this is a safety risk for railway, and protective measures—such as ensuring adequate stabilization, interception, and drainage—must, therefore, be implemented before an actual debris flow occurs.

Table 5.6 Characteristics and dangerousness of each debris-flow valley in the Sangli–Lhasa section

No.	Ditch name and railway mileage	Catchment (km^2)	Main valley shape	Average gradient of main groove	Characteristics and area of flood fan (km^2)
1	Mailang, DK1820 +200–DK1822 +300	52	V-shaped; valley toe is 700-m wide, and tributaries are present upstream	11° (20%)	Ancient flood fan; main valley depth is 5–10 m, and area is ~16 km^2
2	Yangrialang, DK1822 +600–DK1823 +500	90	V-shaped; valley toe is 900-m wide, and tributaries are present upstream	10°–12° (18–22%)	Ancient flood fan; main valley depth is 5–10 m, and area is ~14 km^2
3	Lurenlang, DK1826 +800–DK1827 +700	85	V-shaped; valley toe is 1000-m wide, and tributaries are present upstream	10°–12° (18–22%)	Ancient flood fan; main valley depth is 5–10 m, and area is ~12 km^2
4	Dabuqu, DK1932 +650–DK1933 +500	32	V-shaped; main valley is 12-km long, and valley toe is 400-m wide	12° (22%)	Modern flood fan, squeezing the main river; area is ~4 km^2
5	Dulang, DK1938 +000–DK1941 +200	30	V-shaped; main valley is 10-km long, and valley toe is 500-m wide	10° (18%)	Modern flood fan, squeezing the main river; area is ~2 km^2
6	Nongbalang, DK1950 +100–DK1951 +500	100	V-shaped; main valley is 19-km long, and valley toe is 700-m wide	10°–12° (18–22%)	Ancient flood fan; main valley depth is 3–5 m, and the second- and third-level terraces are farmed. Area is ~16 km^2, and modern deposits are present on the frontier lines of the fan, squeezing the main river
7	Mengapu, DK1952 +100–DK1952 +500	24	V-shaped; valley toe is 700-m wide	12° (22%)	Ancient flood fan; main valley depth is 1–3 m, and the second- and third-level terraces are farmed. Area is ~4 km^2, and

(continued)

Table 5.6 (continued)

No.	Ditch name and railway mileage	Catchment (km^2)	Main valley shape	Average gradient of main groove	Characteristics and area of flood fan (km^2)
					modern deposits are present on the frontier lines of the fan, squeezing the main river
8	Chamulang, DK1953 +300–DK1954 +300	80	V-shaped; main valley is 16-km long, valley toe is 700-m wide, and tributaries are present upstream	11° (20%)	Ancient flood fan; main valley depth is 3 m, and the second- and third-level terraces are farmed. Area is ~5 km^2, and modern deposits are present on the frontier lines of the fan, squeezing the main river
9	Duocuogou, DK1959 +000–DK1960 +300	18	V-shaped; main valley is 6-km long, and valley toe is 800-m wide	10° (18%)	Ancient flood fan; main valley depth is 3–5 m, and the second- and third-level terraces are farmed. Area is ~2 km^2, and modern deposits are present on the frontier lines of the fan, squeezing the main river
10	Puyonggou, DK1960 +200–DK1961 +300	24	V-shaped; main valley is 14-km long, valley toe is 2000-m wide, and tributaries are present upstream	12° (22%)	Ancient flood fan; main valley depth is 1–3 m, and the second- and third-level terraces are farmed. Area is ~8 km^2, and modern deposits are present on the frontier lines of the fan, squeezing the main river
11	Cunmailang, DK1961 +200–DK1962 +500	70	V-shaped; main valley is 14-km long, valley toe is 900-m wide, and tributaries are present upstream	10° (18%)	Ancient flood fan; main valley depth is 3 m, and the second- and third-level terraces are farmed. Area is ~5 km^2, and

(continued)

Table 5.6 (continued)

No.	Ditch name and railway mileage	Catchment (km^2)	Main valley shape	Average gradient of main groove	Characteristics and area of flood fan (km^2)
					modern deposits are present on the frontier lines of the fan, squeezing the main river
12	Xiaduipu, DK1980 +500–DK1981 +500	30	V-shaped; main valley is 8-km long, valley toe is 800-m wide, and tributaries are present upstream	10° (18%)	Ancient flood fan; main valley depth is 3–5 m, and the second- and third-level terraces are farmed. Area is ~4 km^2, and modern deposits are present on the frontier lines of the fan, squeezing the main river
No.	Loose material quality (m^3 km^{-2})	Hillside gradient	Vegetation coverage (%)	Frequency (times per year)	Location and dangerousness of the railway section
1	1000–3000	20°–30°	<30	1–3	Frontier line of accumulation area; medium dangerousness
2	2000–5000	25°–30°	<30	1–3	Frontier line of accumulation area; medium dangerousness
3	1500–4000	25°–30°	<30	1–3	Frontier line of accumulation area; medium dangerousness
4	1500–4000	25°–30°	40	1–3	Middle of fan; severe dangerousness
5	2000–5000	25°–30°	40	1–3	Middle of the fan and top section of accumulation area; severe dangerousness
6	2000–5000	25°–30°	>50	1–3	Middle of the fan and top section of accumulation area; severe dangerousness

(continued)

Table 5.6 (continued)

No.	Ditch name and railway mileage	Catchment (km^2)	Main valley shape	Average gradient of main groove	Characteristics and area of flood fan (km^2)
7	1000–2000	25°–30°	<5	1–3	Other side of the river; debris flow is dangerous for the highway
8	1000–3000	25°–30°	>50	1–3	Middle of the fan and top section of accumulation area; severe dangerousness
9	1000–3000	25°–30°	>50	1–3	Middle of the fan and top section of accumulation area; severe dangerousness
10	1000–2000	25°–30°	<20	1–3	Other side of the river; debris flow is dangerous for the highway
11	1000–3000	25°–30°	>50	1–3	Middle of the fan and top section of accumulation area; severe dangerousness
12	1000–3000	25°–30°	>50	1–3	Middle of the fan and top section of accumulation area; severe dangerousness

Here, stabilization refers to the drainage of building debris dam or other such methods that enhance soil and rock stability, interception refers to the building of dams on the upper reaches of the debris flows to intercept the debris-flow solids, and drainage refers to the building of drainage canals to prevent the deposition of debris flows.

5.2.3 Thaw Slumping

Some slopes are prone to thaw slumping, which can severely harm railway lines. The Qinghai–Tibet Railway avoids all but ten thaw-slumping areas, seven, and three of which are on and near the route, respectively. Engineering treatments have been performed at the seven sites directly affecting the railway (Table 5.7).

Table 5.7 Major thaw-slumping areas along the Qinghai–Tibet Railway

No.	Railway mileage	Location	Distance (m)	Engineering measures
1	DK978+260–DK978+390	Kunlun Mountain area	130	Bridge
2	DK978+593–DK978+722	Kunlun Mountain area	129	Bridge
3	DK979+140–DK979+200	Kunlun Mountain area	60	Bridge
4	DK979+366–DK979+462	Kunlun Mountain area	96	Bridge
5	DK1420+800–DK1420+816	Tonglha Mountain area	16	Bridge
6	DK1502+240–DK1502+270	Touerjiu Mountain area	30	Air-cooled riprap subgrade or thermal-probe subgrade; foundations replaced with embankments
7	DK1503+450–DK1503+565	Touerjiu Mountain area	115	

The slope area of Kunlun Mountains is highly susceptible to thaw slumping because of the ground ice in this area. In addition, excavations and constructions in the slope areas of the Kunlun Mountains–Budong Spring, Fenghuo Mountain–Erdao Ditch, and other sections have made exposed ground ice, leading to melting, solifluction, and thaw slumping.

Thaw slumping in Tonglha area at DK1420+800–DK1420+816 is mainly due to the development of springs. Thaw slumps here are ancient and has little effect on the railway. Hence, bridges of span 1–24 m are used to pass through this area.

Thaw slumping in Touerjiu Mountains area at DK1502+240–DK1502+270 and DK1503+450–DK1503+565 is mainly due to the collapse of gullies shore. The strata are mainly formed by silt clay and are in a soft-plastic state, and the weak layer of the strata is 0.5–2.0 m thick. Thaw slumping can destabilize building basements and embankment slopes, and the solifluction can block or bury buildings. Thus, these phenomena strongly affect railway projects.

5.2.4 *Slope Wetland*

Permafrost slope wetlands—which are slopes with peat accumulations in permafrost regions that are either permanently or seasonally saturated with water,

Fig. 5.10 Slope wetland in Fenghuo Mountain area

Fig. 5.11 Slope wetland in Touerjiu Mountain area

forming a distinct ecosystem—significantly affect railway projects. The primary factor that distinguishes permafrost slope wetlands from other slopes is the characteristic vegetation of aquatic plants and peat accumulation (Figs. 5.10 and 5.11).

Permafrost slope wetlands are distributed in the Kunlun Mountains, Fenghuo Mountain, Kaixinling, Buqu Valley, Tonglha Mountains, and Touerjiu Mountain area. Much ground ice develops in the permafrost slope wetlands area, making the area prone to solifluction and thaw slumping. Liquid water is often present upstream of permafrost slope wetlands, rendering these regions susceptible to the formation of drumlins. Thus, the effects of permafrost slope wetlands on the railway in permafrost areas warrant much attention.

The Qinghai–Tibet Railway uses bridges or retaining structures to pass through steep slopes with poor stability. Fifteen permafrost slope wetlands are present along 38.60 km of the Qinghai–Tibet Railway, of which railway bridges and railway roadbeds account for 31.56 and 6.04 km, respectively.

Slope wetlands are distributed in the nonpermafrost area between Amdo and Lhasa and the mountain slopes of the Gangxiu–Nagqu, Sangli–Wumatang, and Yambajan–Angga sections; these slope wetlands typically have a low gradient; nevertheless, groundwater resources in these areas are well developed.

The presence of a transitional zone between the hard and soft layers beneath the cross-slope belt severely affects the stability of railway embankments. The slope areas were a key focus of the Qinghai–Tibet Railway Geological Survey performed to identify critical sections requiring engineering treatments. Fifteen slope wetlands are present in the nonpermafrost area, over which passes 18.4 km of the line. Of the total 55 sections in these wetlands, 18 (5.7 km) affect the railway; in these sections, subgrade antislide measures have been implemented.

5.3 Principles of Railway-Line Selection and Preventive Measures

5.3.1 Principles of Railway-Line Selection in Slope Areas

5.3.1.1 General Principles

Slope areas are characterized by complex geological and topographical conditions and large variations in elevation. These conditions strongly influence design parameters such as the railway grade, minimum curve radius, railway strike, and line standard.

Before line selection in slope areas, the engineering-geological conditions and geological hazards along the slopes should be evaluated. In such areas, the line selection of the Qinghai–Tibet Railway adhered to the following principles:

(1) Adopt precautionary measures in tectonically active areas, namely, areas with rock-mass accumulation, avalanches, dangerous rocks, falling rocks, landslides, and unstable permafrost.
(2) In elevation design, ensure that the line is higher than the design flood level, and reduce the excavation volume by not locating the line in the upper slopes. To this end, it is essential to fully utilize the favorable topographical conditions and to consider the developmental characteristics of the geological hazards. To

minimize the bridge elevation and valley embankment, the railway exhibition line should be started early. By considering the topographical, hydrological, and elevation characteristics in the design, a line with an appropriate undulation can be realized.

(3) In areas with a low slope hazard, compare the costs of the slope-avoidance and straight-line schemes by considering the costs of slope treatment. The merits and demerits of these two design schemes are as follows: (a) the slope-avoidance scheme has poor line reliability because of the unfavorable geological conditions and river erosion. (b) In the straight-line scheme, the line uses cutting slopes, tunnels, and extension lines to pass through the slope hazard areas. This approach has high reliability because the line is short and straight; however, it entails higher construction costs and time than does the slope-avoidance scheme. The straight-line scheme is preferred when the construction costs and time are similar.

(4) When embankment stability is uncertain, use tunnels to pass through slope areas with a severe slope hazard. For example, the Kunlun Mountain area and Yambajan area have many long tunnels at a shallow depth. Use bridges in areas where the slopes are so steep that adequate tunnel safety cannot be ensured. For example, bridges were designed to navigate the slopes near the exit of the Kunlun Mountain Tunnel because of the large development and wide distribution of geological hazards (e.g., thaw slumping, thaw-frozen mudflow, and rockfall) around this area.

5.3.1.2 Principles of Railway-Line Selection in Dangerous Rocks, Rockfall, and Collapse Areas

1. Avoid high mountain and steep slope areas and areas where the rocks are severely cut by joints; in addition, avoid areas with high hazard density.
2. Locate the line in areas whose stability can be ensured through engineering measures at a reasonable cost.

5.3.1.3 Principles of Railway-Line Selection in Rock-Heap Areas

1. Avoid rock-heap areas because rock heaps are loose; moreover, the near-ground and basement regions of rock heaps are very steep, and loose debris and groundwater are abundant in areas with rock heaps. Furthermore, the engineering treatment of rock-heap areas is difficult.

2. In relatively stable rock-heap areas, necessary adopt engineering measures to ensure rock-heap stability and to intercept rolling stones before they reach the roadbed. Wherever possible, locate the railway along the toe of the rock heap.

5.3.1.4 Principles of Railway-Line Selection in Debris-Flow Areas

1. Implement engineering measures in sections that are highly prone to large debris flow, and avoid locating the line where the train traffic is likely to be blocked by debris flow.
2. Use bridges, open-cut tunnels, or tunnels to cross debris-flow ditches. Ensure that the bridges do not have sharp bends; ensure that the longitudinal gradient change points have sufficient clearance and span, depending on the debris-flow volume and sediment thickness.

5.3.1.5 Principles of Railway-Line Selection in Permafrost Areas

1. Locate the line in sections with less ice, in dry areas with rocks and coarse particles, or on sunny slopes. Permafrost regions, especially ice-rich permafrost regions, should be avoided. When they cannot be avoided, implement adequate preventive measures.
2. Prefer embankments over cutting. When cutting is unavoidable, minimize the excavation depth, and ensure that the line is not too flat.
3. The design of the profiles should account for the distribution of frozen soil and the complexity of permafrost. Minimize zero sections and low embankments. Avoid subgrade engineering in slope areas, and when unavoidable, carefully consider the asymmetrical thermal conditions of the slope and their effects on roadbed stability.
4. Source and dispose roadbed filling and cutting materials carefully to prevent the development of new permafrost hazards. Consider and modify the line according to the geological environment.
5. In regions with solifluction, thaw slumping, and frozen-soil slope wetlands, locate the line at the bottom of the outer edge of the hazard area, using bridges and embankments where necessary. When embankments are used, choose the foundation and form (e.g., air-cooled riprap subgrade and thermal-probe subgrade) such that they ensure permafrost stability.
6. In frozen-soil slope wetland areas, adopt the engineering design with the minimal environmental effects.

Table 5.8 Prevention of dangerous rocks, rockfall, and collapse

Engineering measures	Applicable conditions	Specific methods and techniques
Shelter	Unstable small and medium natural slopes and steep artificial slopes	Build open-cut tunnels, hangar tunnels, or other sheltering buildings. Such methods not only shelter from collapse and rockfall on the road but also reinforce the toe of slope, increasing slope stability and sustainability
Interception	For nearly stable slopes prone to collapse or rockfall in the rainy season, and when the cost of other engineering measures are cost-prohibitive	(1) If the line is adequately far from the toe of slope and the gradient of lower section of slope is less than 30°, build dikes or grooves to stop the flow of collapsed materials. (2) If the engineering conditions do not allow such construction, consider stone walls. (3) Extend the existing retaining wall to protect the railway against collapse and rockfall. Repave the cushion layer to reduce the effect of stones. (4) Use obsolete rails or steel wires to form a fence against rockfall
Retaining structure	For nearly stable slopes with huge isolated stones and large boulders that are difficult to remove	Build a retaining structure or use anchors to stabilize the isolated stones and dangerous rocks
Protective facing	For stable slopes but with rocks susceptible to weathering and peeling	Build retaining wall to protect the surface of steep slopes, and use gunite or trowel finish to protect the surface of gentle slopes. Although such measures cannot withstand strong lateral pressures, they can provide some support and protection under its own weight and thickness
Fill cracks	For stable slope foundation, where the slope is composited of hard rock or hard and soft rock, but the cracks and holes on the surface may cause collapse or rockfall	Fill holes with slice stones and reinforce it through grouting or anchoring
Smooth surface	For slopes where maintaining stability is challenging (e.g., when dangerous rocks and isolated stones are distributed on the slope)	Slope smoothing can be attempted to reduce the slope gradient. However, note the following: (1) Avoid this method if the slope is located in a structural fracture zone or if slope has joint fissures. (2) Avoid this method at heights above 30–40 m considering ease of operation and maintenance. (3) To avoid affecting the traffic, this method should be

(continued)

Table 5.8 (continued)

Engineering measures	Applicable conditions	Specific methods and techniques
		implemented only when the line is not in operation. (4) Remove huge isolated stones and dangerous rocks by blasting (or burning or melting if the line is in operation)
Drainage	For slopes with active surface water or groundwater	According to the runoff data, build drainage structures to divert and intercept the water

5.3.2 *Preventive Measures in the Slope Area Along the Qinghai–Tibet Railway*

5.3.2.1 Preventive Measures in Dangerous Rocks, Rockfall, and Collapse Areas

Rock, rockfall, and collapse are complex phenomena that often occur suddenly and have the potential to cause large harm. In sections of the line passing through such area, effective preventive measures (Table 5.8) should be implemented to ensure safe railway operation.

5.3.2.2 Preventive Measures for Geological Hazards in Rock-Heap Areas

Use embankments to cross areas with relatively stable and static rock heaps (e.g., Figs. 5.12III and 5.13I). Avoid semifilling and semiexcavating. When the slope cutting is parallel or nearly parallel to the bedrock surface, use the slenderness ratio of the remaining soil to determine whether the remaining soil has sufficient rigidity. If the cutting slope is not parallel to the bedrock surface, the remaining soil should

Fig. 5.12 Using embankment in rock-heap areas

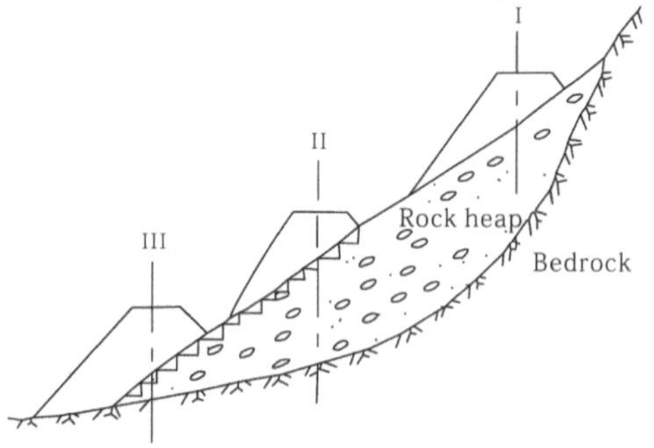

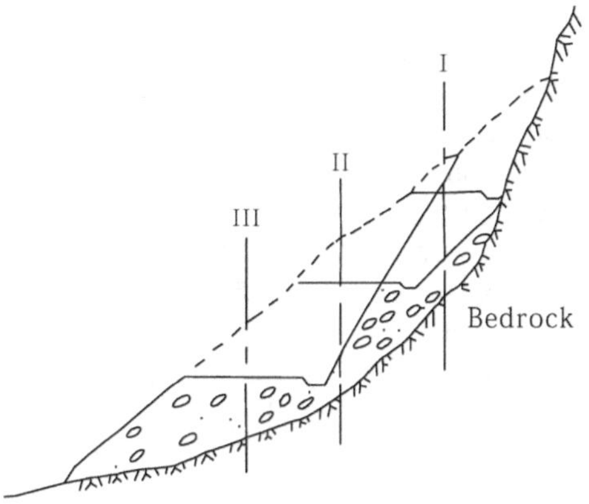

Fig. 5.13 Using cutting in rock-heap areas

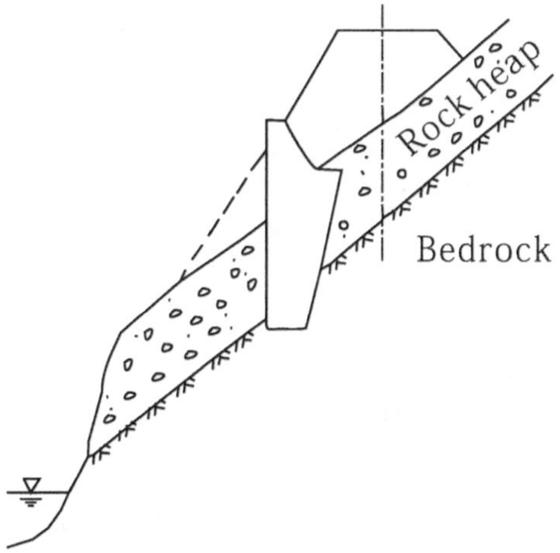

Fig. 5.14 Using embankment on steep rock heaps

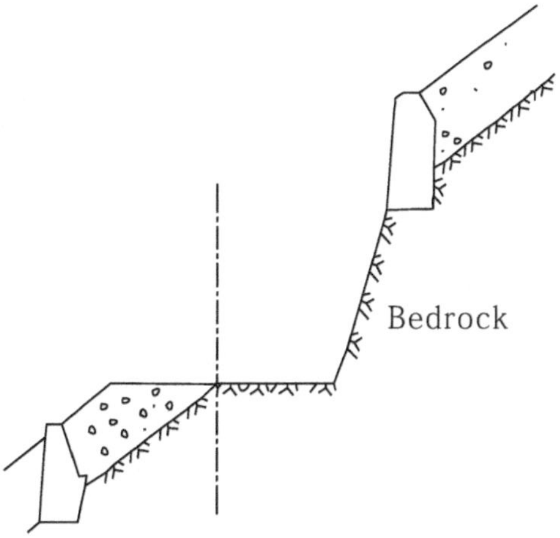

Fig. 5.15 Cutting of the whole rock heap

be designed as a wedge (i.e., wide top and narrow bottom), and appropriate treatments should also be implemented.

For embankments built on steep rock heaps, embankment walls and road epaulements should be built to prevent the sliding of the base rock heaps. Build the foundation of wall deep in the bedrock (Fig. 5.14).

If the whole rock heap must be cut, prevent sliding along the bedrock surface by building a retaining wall (Fig. 5.7).

When the line pass through rock heaps under the effect of surface water or underground water, surface-water retaining works and underground-water drainage works should be implemented to protect the roadbed (Figs. 5.14 and 5.15).

Furthermore, depending on the local conditions, the following measures also can be implemented to prevent the deformation of rock heaps: (1) Make the slope step-shaped, with each step being 5–10 m wide and 0.5–2 m high. Stepped wall can be made using dry stone or rubble stone. This is the most convenient and effective method. (2) Sow seeds—Sprinkle planting soil on the slope with rock heaps and scatter seeds over it. Grass roots can stabilize the slope surface. (3) Pressure grouting method—For loose rock heaps with large pieces, use the pressure grouting method to fill the pores of the slope, making it a single coagulate rock mass and reducing the likelihood of a collapse during slope excavation.

If the stability of rock heaps is affected by the surface water (river erosion and cutting) or groundwater activities, necessary drainage measures and riverbank protection measures should be implemented.

5.3.2.3 Prevention Measures in Geological Hazards in Debris-Flow Areas

The preventive measures for geological hazards in debris-flow areas should be a combination of precautionary and post-treatment methods.

1. Precautions

   a. In the feasibility study stage, particularly in the early testing phase, assiduously distinguish debris flows. Choose the best route according to the principles of railway-line selection in debris flows area.
   b. After the route has been selected, appropriate engineering measures should be adopted to pass through the debris-flow ditch.
   c. Biological measures should also be implemented—For example, forest protection, reasonable farming, and grazing can control surface runoff and prevent slope erosion; thus eradicating debris flow.
   d. Strengthen livestock management—the grazing area should be delineated and strictly enforced, especially in the Doilung Qu area.

2. Treatment measures

To minimize the effects of debris flow on the railway, appropriate engineering measures (e.g., landing dams, exhaust canals, bridges, tunnels, open-cut tunnels measures) should be implemented considering the local conditions, including the location of the line with respect to the slope.

a. Retaining structures

The retaining structures are used to protect riverbanks, stabilize slopes, intercept solid materials, and reduce the riverbed slope, all of which reduce the strength and the scale of debris flow.

i. Retaining wall

The dam position should be determined considering the terrain conditions, and the debris-flow formation and circulation zones or the zone with good silt storage conditions and favorable geological conditions are ideal. Dams also can be built in groove beds and hillsides that must be stabilized.

The design height of the retaining wall depends on the ditch bed topography, the geological conditions, and the particle size and velocity of the solid material. The retaining wall would halt the debris-flow upstream; hence, selecting a lower design height (2–4 m) is ideal to reduce the probability of wall damage. Conversely, a higher dam (5–7 m) is more effective in the downstream region.

ii. Silting ground

A silting ground is a flat and extensive field to stop debris-flow deposition. It can be set in the lower or middle section of broad deposition areas or in the low-lying areas between two fans. A silting ground is typically formed by building retaining structures on debris-flow fan. For example, in 1972, a flat field (area is 0.36 km^2) located in the upper section of the inlet section of Jiangjiagou drainage ditch was used as a silting ground to stop debris flow. It stopped 19.8×104 m^3 of debris, greatly reducing the siltation in the drainage groove as well as the desilting workload.

b. Drainage and diversion structures

The main role of drainage and diversion structures, such as diversion dike, torrent ditch, and beam dike, is to improve the speed and direction of debris flow.

c. Bridge and culvert structures

Railway bridges and culverts are the main methods of crossing debris-flow areas. Appropriate bridge location is crucial for reducing the effects of debris flow on bridges. The principles of bridge location selection are as follows: (1) Locate bridges where the ditch bed is stable and the local mainstream is smooth. (2) Avoid areas where the slope gradient of the ditch bed is steep. (3) Strictly avoid excavating the ground to build a bridge. (4) Avoid zones with strong debris flow. (5) In debris-flow areas, locate the bridge as a straight segment. (6) In debris-flow deposition zones, locate the bridge in the tail section and edge of the sediment fan, and avoid the fan waist and apex. Locate the railway bridge line along the contour, and multiple bridges in a section should be appropriately spaced out. If the debris-flow alluvial fan is located on the verge of the main river, the bridge should be located a safe distance from the river to prevent river erosion.

Determine the necessary scale of the bridge by considering the river flow, terrain, river width, and nature and development patterns of debris flow should all be considered.

Adequate clearance under the bridge is a major control condition in bridge design, which can ensure the drainage of debris flows. Hence, tall (high) bridges should be preferred over short (low) ones.

d. Submerged tunnel, open-cut tunnel, and flume

i. Submerged tunnel

If the railway elevation is much higher than the elevation of a debris-flow ditch or if the railway is passing through a heavy debris-flow area, which poses a severe threat to the railway safety, a deep tunnel should be chosen to cross the area.

In the area exposed to a severe threat from the other side of large debris flow, it is better to use a tunnel to pass through the area when the embankment stability is uncertain. In addition, considering river erosion, the tunnel should be located a safe distance from the river.

If the entrance of tunnel is located in the edge area of a debris-flow alluvial fan, the entrance can easily be buried by debris-flow. Hence, the entrance of the tunnels should follow the "enter early and exit later" principle, and the tunnel design should follow these principles: (1) Tunnel length should not be shorter than the full range of the debris flow. (2) The tunnel should pass through the stable bedrock beneath the debris flow. (3) The tunnel roof thickness should be adequate prevent the tunnel against erosion.

Continuous bridges and tunnels should have high clearance under the bridges for the drainage of debris flow.

ii. Open-cut tunnel

If the railway elevation is not higher than the elevation of the debris-flow alluvial fan and if the railway is passing through a debris-flow area that poses some threat to railway safety, the open-cut tunnel should be chosen to cross the area.

Design the entrance of the open-cut tunnel by considering the probability of being buried by debris flows. The design of open-cut tunnel should follow these principles: (1) Minimize eccentric compression on the tunnel. (2) Ensure that the outside of the tunnel is covered. (3) To facilitate the construction of the open-cut tunnel, the roof the open-cut tunnel should be close to the ground surface. (4) To improve the integrity, strength, and wear resistance of the tunnel, reinforce it using waterproof and drainage structures.

Under the following four conditions—(1) deep cutting truncates a dilute debris-flow ditch on slope, (2) cutting truncates a small dilute debris-flow ditch on the marginal of a table fan, (3) downstream terrain facilitates drainage, and (4) the height of the flume is adequate—use flumes to guide the debris drainage. However, for heavy debris flow and dilute debris flow with particle size >0.5 m, the design scheme with an open-cut tunnel under the flume is the ideal choice.

Flume should be connected to the original ditch on a plane; the width and longitudinal flume slope should also be adapted to that of the original ditch. Sharp turns can cause debris-flow overflow because of poor discharge. When calculating the flume depth, consider not only the debris flow but also such factors as debris-flow residual layer thickness, changes in the wave height, huge floating boulders, and mud pies, all of which increase the safe depth by 1–2 m.

5.3.2.4 Preventive Measures for Geological Hazards in Thaw Slumping Areas

Bridges with the foundation should set in the permafrost layer are ideal to cross thaw-slumping areas.

When subgrades are used, use embankment instead of cutting. Insulating berm should be set on both sides of the embankment toe, and the retaining structure should be built at the leading edge of the slumping block when necessary. Loose soil and frozen soil under the subgrade should be replaced. Moreover, locate longitudinal drainage systems upstream to prevent embankment flooding.

5.3.2.5 Preventive Measures for Geological Hazards in Slope Wetland Areas

Bridges with the foundation set in the stable layer are ideal to cross-slope wetland areas.

This approach also has the least effect on the environment.

When subgrades are unavoidable, in slope wetlands in nonpermafrost areas, to reinforce the subgrade base, replace the original soil with slice stone or use a gravel pile composite foundation with cement injection. In permafrost slope wetland areas, roadbed design should consider permafrost protection as an overriding principle; this goal can be achieved by using air-cooled riprap subgrade, rubble stone, and crushed-rock revetment or thermal-probe subgrade. Insulating berm should be set on both sides of the embankment toe, and drainage systems are also essential.

5.4 Scheme Comparison for Slope Areas

5.4.1 Comparison of the Yambajan Long Tunnel Scheme and the Yambajan Short Tunnel Scheme

5.4.1.1 Introduction to the Railway-Line Selection Scheme

1. Short tunnel scheme (C3K): This scheme passes through the Yambajan valley from the right bank using Yambajan tunnels 1 (l: 1535 m), 2 (l: 1839 m), and 3 (l: 1640 m). On the hillside are scattered Quaternary Holocene slope sediments

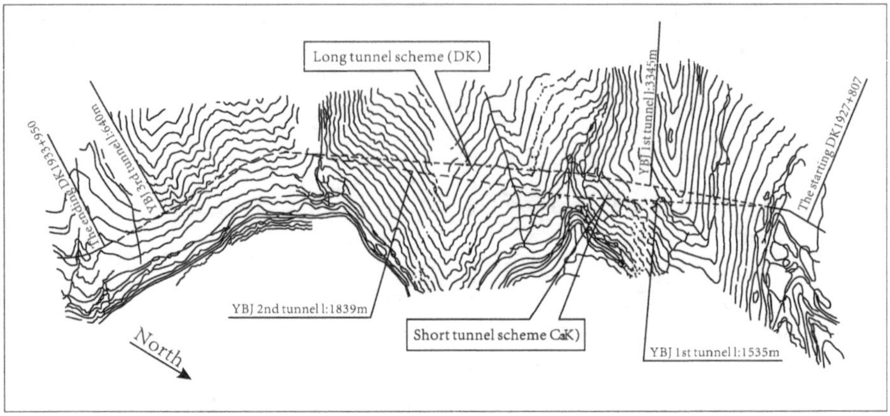

Fig. 5.16 Yambajan long and short tunnel schemes

of granular soil, and the bedrock is Yanshanian granite. Dangerous rocks and rockfall are distributed in the area between Yambajan tunnels 1 and 2 and in the area between Yambajan tunnels 2 and 3; the hazards are more severe in the area between Yambajan tunnels 1 and 2 (Fig. 5.16).

2. Long tunnel scheme (DK): This scheme is based on the short tunnel scheme (C3K), but with the tunnels located at some distance to the mountain side. Yambajan tunnels 1 and 2 are connected using an open-cut tunnel and flume to form the new Yambajan tunnel 1 (l: 3445 m). The Yambajan tunnel 3 of the C3K scheme is retained, becoming the new Yambajan tunnel 2 (l: 640 m). The geological condition of the tunnel bodies are the same as that in the C3K scheme, but it avoids the dangerous rocks and rockfall between the exit of Yambajan tunnel 1 (C3K) and the entrance of Yambajan tunnel 2 (C3K).

5.4.1.2 Evaluation of the Railway-Line Selection Scheme

Tunnel length is shorter in the C3K scheme than in the shorter DK scheme, and the C3K scheme does not require mechanical ventilation, but dangerous rocks and rock fall may endanger the safety of tunnel construction and operation. By contrast, the DK scheme has better geological conditions near the tunnel entrance. Since DK scheme connects tunnels 1 and 2 in the C3K scheme, which increases the length of the tunnel, mechanical ventilation systems must be implemented.

In summary, the C3K scheme has the following advantages: (1) short tunnel, (2) more working faces, (3) no mechanical ventilation is required, and (4) lower operating expenses. However, this scheme may be damaged by dangerous rocks and rockfall near the tunnel portal, which may endanger the safety of the tunnel construction and operation. Overall, the DK scheme has higher reliability and safety. Thus, we recommend the DK scheme.

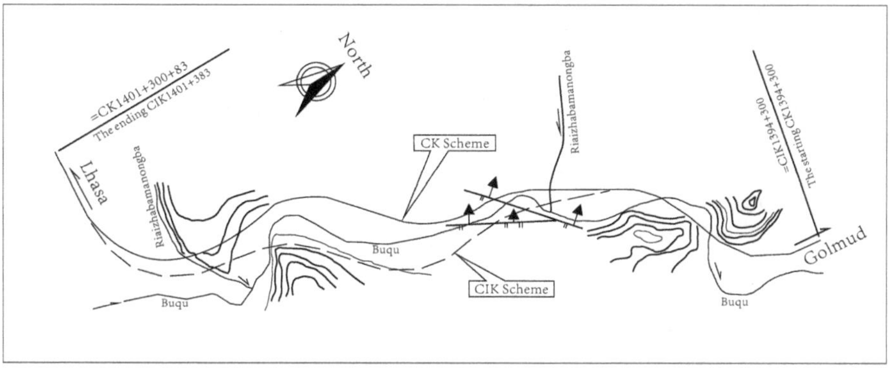

Fig. 5.17 Buqu River Valley left bank and right bank schemes

5.4.2 Comparison of the Left Bank Scheme and the Right Bank Scheme in Buqu River Source Region Valley

The Buqu River Valley segment belongs to the northern slope of the Tonglha Mountain Gorge area, and the valley is relatively narrow and with steep hillsides. The bottom of Buqu Valley is 180–200-m wide, and the average natural longitudinal gradient of riverbed is 3–7‰. The bedrock on both sides is exposed and houses permafrost wetlands. Approximately 15% of the surface is turfed. The vegetation on the slope is 5–15-cm thick, and the grass lumps are 0.2–0.5-m thick. This valley is a type of permafrost. Figure 5.17 illustrates the left bank (CK) and right bank (CIK) schemes.

5.4.2.1 Introduction to the Railway-Line Selection Scheme

1. CK scheme: After starting from the (relative) zero point and crossing the left bank of Buqu river using two connecting bridges, the line extends along the river. The CK1399+500–CK1400+000 section of the line is 500-m long, and the maximum excavation in this section is 12-m deep; the line terminates after 7 km.
2. CIK scheme: After starting from the (relative) zero point and crossing over the left bank of Buqu river by using two connecting bridges, the line extends approximately 800 m, crossing the river to the right bank. Subsequently, the line extends along the river for nearly 1.4 km and crosses the Buqu river again to the left bank. Finally, the line terminates after 7.08 km.

5.4.2.2 Geological Conditions

The surface layer of the left bank is mainly composed of thick pellet soil, pebble soil, brecciated soil, and gravel soil, and the thickness of the surface layer is 2–5 m. The main components of the hillside on the right bank are sand clay, with Jurassic limestone and sandstone underneath. The geological condition of the left bank is more favorable than that of the right bank.

These two schemes pass upstream on either side of Buqu River; the riverbed in this section is narrow, and the river S-shaped. This area has four interactive faults that affect the mountains, causing the mountains to break; thus, the internal stratum in this area is chaotic and affect bridge location and construction. For example, A bridge is located at CIK1396+950, which is also the intersection of two faults and thus is a risk to the line. Therefore, considering the geological structure and safety of the line, the left bank is more favorable than is the right bank.

The schemes pass through a permafrost area, where the permafrost table is 2.4–4.0 m, with thaw regions, highly unstable high-temperature regions, and unstable high-temperature regions. Both schemes pass through this section using embankments. The excavation of CK scheme is 750-m longer than that of the CIK scheme, and the frozen soil in the CK1298+499–CK1398+977 section has high ice content. This section should ideally be built in winter. Permafrost wetlands are scattered in this area, with a weak layer of thickness 0–1.5 m. The CK and CIK schemes pass through 620 and 850 m of permafrost areas, respectively.

The advantages of CK scheme are its relative shortness, use of fewer bridges, more favorable geological conditions, and low engineering costs. However, it requires deep excavation in the permafrost area. By contrast, the CIK scheme avoids deep excavations but is longer, uses more bridges, has worse geological conditions, and entails higher engineering cost. Hence, the CK scheme (i.e., the left bank scheme) is recommended.

5.4.3 Comparison of the Bridge Scheme and the Road Scheme in the Slope Area Near the Exit of Kunlun Mountain Tunnel

This section is located in the Kunlun Mountain area. The surface layer is brecciated soil, and the underlying layer is composed of brecciated soil and silt clay. The permafrost table is 2.4–4.0 m. The annual average ground temperature is lower than −2.0 °C, and the permafrost table in this area is frozen soil with high ice content or saturated frozen soil. The DK978+260–DK978+390 section is a thaw-slumping area of size 75 m × 250 m. Soil-containing ice is present below the permafrost table. In the DK979+500–DK979+650 section, soil-containing ice is present below the permafrost table. Kunlun Mountain area is an area with a seismic intensity of 8°.

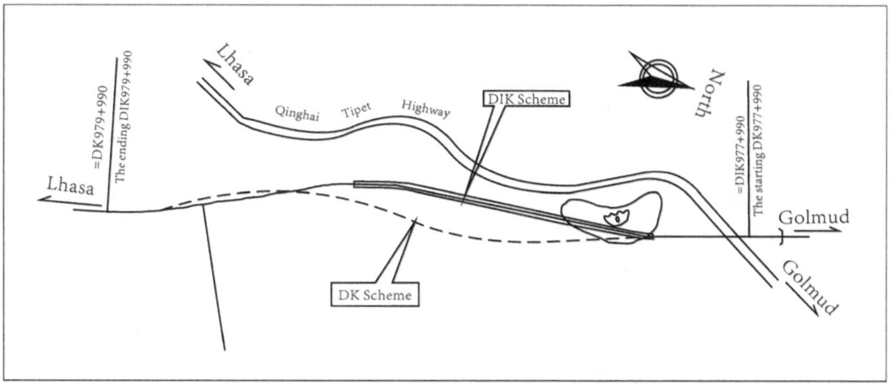

Fig. 5.18 Diagram of the bridge scheme and road scheme

Considering the topography and permafrost geological conditions in this section, we investigated the bridge scheme (DIK) and the road scheme (DK) (Fig. 5.18). In the DK scheme, the hillside near Kunlun Mountain Tunnel exit (DK978+100–DK978+800) is steep, and the thaw slumps are distributed on the slope; a 350-m-long excavation is also required on the hillside. To avoid the instability, the DIK scheme is located close to the Qinghai–Tibet Highway, and bridges are used instead of a roadbed.

5.5 Summary

The major geological disasters along the Qinghai–Tibet Railway include dangerous rocks, rockfall, landslide, collapse, talus, and debris flow; furthermore, solifluction, thaw slumping, slope wetlands occur in permafrost slope areas. Slope areas are characterized by complex geological and topographical conditions and large variations in elevation. These conditions strongly influence design parameters such as the railway grade, minimum curve radius, railway strike, and line standard.

Before line selection in slope areas, the engineering-geological conditions and geological hazards along the slopes should be evaluated. In such areas, the line selection of the Qinghai–Tibet Railway adhered to the following principles:

1. Adopt precautionary measures in tectonically active areas, namely, areas with rock-mass accumulation, avalanches, dangerous rocks, falling rocks, landslides, and unstable permafrost. The preventive measures for these geological hazards should be a combination of precautions and post-treatments.
2. In areas with a low slope hazard, compare the costs of the slope-avoidance and straight-line schemes by considering the costs of slope treatment. The merits and demerits of these two design schemes are as follows: (a) The slope-avoidance

scheme has poor line reliability because of the unfavorable geological conditions and river erosion. (b) In the straight-line scheme, the line uses cutting slopes, tunnels, and extension lines to pass through the slope hazard areas. This approach has high reliability because the line is short and straight; however, it entails higher construction costs and time than does the slope-avoidance scheme. The straight-line scheme is preferred when the construction costs and time are similar.
3. When embankment stability is uncertain, use tunnels to pass through slope areas with a severe slope hazard.
4. Avoid permafrost areas, and when unavoidable, locate the line in sections with less ice, in dry areas with rocks and coarse particles, or on sunny slopes. Permafrost regions, especially ice-rich permafrost regions, should be avoided. When they cannot be avoided, implement adequate preventive measures. Avoid cutting to the extent possible. Prefer schemes that have the least environmental impact.

Source and dispose roadbed filling and cutting materials carefully to prevent the development of new permafrost hazards. Consider and modify the line according to the geological environment.

References

1. Cao, D. Y., Lei, Y. D., Deng, X. F., et al. (2007). General engineering geologic conditions along the Naij Tal–Tuotuo River section of the Qinghai–Tibet Railway. *Qinghai Environment, 17*(2), 78–80. (in Chinese).
2. China Railway First Survey & Design Institute Group Co., Ltd. (1994). *The manual of routes* (2nd ed.). Beijing: China Railway Press. (in Chinese).
3. China Railway First Survey & Design Institute Group Co., Ltd. (1999). *The railway engineering geology manual*. Beijing: China Railway Press. (in Chinese).
4. China Railway First Survey & Design Institute Group Co., Ltd. (2001). *Line survey and instruction of Tibet railway*. Internal information. (in Chinese).
5. China Railway First Survey & Design Institute Group Co., Ltd. (2001). *Special report for nature reserve and passage for wild animals along the Qinghai–Tibet Railway*. Internal information. (in Chinese).
6. China Railway First Survey & Design Institute Group Co., Ltd. (2002). *Preliminary design of the Tonglha–Lhasa section of the Qinghai–Tibet Railway (Geology)*. Internal information. (in Chinese).
7. Chinese Geological Atlas Editorial Board. (1996). *Geological atlas of Chinese*. Beijing: Geological Press. (in Chinese).
8. Cold Building Scientific Research Institute, Heilongjiang Province. (1998). *Code for design of building foundation in frozen earth area*. Beijing: China Building Industry Press. (in Chinese).
9. Dai, J. X. (1990). *The climate of the Tibetan Plateau*. Beijing: Meteorological Press. (in Chinese).
10. Editorial Board of Engineering Geology Manual. (2007). *Engineering geology manual* (4th ed.). Beijing: China Architecture & Building Press. (in Chinese).

11. Lanzhou Glacier Frozen Soil Research Institute, Chinese Academy of Sciences. (1983). *Collected papers of Tibet frozen soil research.* Beijing: Science Press. (in Chinese).
12. Li, J. C. (2008). Frost damages in permafrost regions along the Qinghai-Tibet Railway in 2001–2006: Investigation and analysis. *Journal of Glaciology and Geocryology, 30*(1), 147–152. (in Chinese).
13. Liu, Z. P. (2008). Distribution characteristics of permafrost development along the Qinghai–Tibet Railway. *Railway Investigation and Surveying, 2,* 78–82. (in Chinese).
14. Liu, Z. P. (2008). Research on the technology of engineering geology investigation on permafrost along the Qinghai–Tibet Railway. *Journal of Railway Engineering Society, 8,* 1–5. (in Chinese).
15. Liu, Z. Q., Xu, X., Pan, G. T., et al. (1990). *Geotecture and evolution of the Tibetan Plateau.* Beijing: Geological Press. (in Chinese).
16. Liu, J. K., Tong, C. J., & Fang, J. H. (2005). *Introduction of geotechnical engineering in cold regions.* Beijing: China Railway Publishing House. (in Chinese).
17. Ma, T., Zhou, J. X., Zhang, X. D., et al. (2007). Preliminary studies on characteristics of vegetation distribution along the Qinghai–Tibet Railway. *Research of Soil and Water Conservation, 14*(3), 150–154. (in Chinese).
18. Meng, X. L. (2006). Geological prospecting for permafrost engineering in the Qinghai–Tibet Railway. *Railway Investigation and Surveying, 3,* 28–31. (in Chinese).
19. Meng, H., Zhang, Y. Q., & Yang, N. (2004). Analysis of the spatial distribution of geohazards along the middle segment of the eastern margin of the Qinghai–Tibet Plateau. *Journal of China Geology, 2*(31), 218–224. (in Chinese).
20. Ministry of Railways. (2006). *Basic specification for design of railway bridge and culvert (TB10002.1-2005).* Beijing: China Railway Publishing House. (in Chinese).
21. Papers and reports on permafrost in the Qinghai–Tibet Plateau. *Annual Report of State Key Laboratory of Frozen Soil Engineering.* (in Chinese).
22. Shen, W. T., et al. (2005). *Prediction and evaluation of the ecological impact of the Qinghai–Tibet Railway.* Beijing: China Environmental Science Press. (in Chinese).
23. State Seismological Bureau of China. (1991). *Earthquake intensity zoning map of China (1990).* Beijing: Earthquake Press. (in Chinese).
24. Sun, Y. F. (2005). Permafrost engineering in the Qinghai–Tibet Railway: Research and practice. *Journal of Glaciology and Geocryology, 27*(2), 153–162. (in Chinese).
25. The Glacier Permafrost Research Room, Institute of Geography, Chinese Academy of Sciences. (1965). *Investigation of frozen soil along the Qinghai–Tibet Highway.* Beijing: Science Press. (in Chinese).
26. Tong, C. J. (1996). *Geological evaluation and treatment offrozen soil in permafrost along the Qinghai–Tibet Highway.* Beijing: Science Press. (in Chinese).
27. Wang, Y. D. (2002). Preliminary exploration of geologic route selection in multiyear tundra of Qing–Zang Railway line. *Journal of Railway Engineering Society, 2,* 57–59. (in Chinese).
28. Wang, Z. J. (2002). Permafrost engineering in the Qinghai–Tibet railway construction. *Chinese Railways, 12,* 31–37. (in Chinese).
29. Wang, Z. H. (2003). Landslides along the Qinghai–Tibet Railway and Highway. *Modern Geological, 17*(4), 355–362. (in Chinese).
30. Wang, Z. H. (2006). Active tectonics and its secondary disasters along the Qinghai–Tibet Railway. *Journal of Railway Engineering Society, Suppl., 1,* 264–269. (in Chinese).
31. Wang, Z. H. (2007). Geological environment and disasters along the railway line in the Qinghai–Tibet Plateau. *Earth Science Frontiers, 14*(6), 031–037. (in Chinese).
32. Wu, Z. W., Chen, G. D., et al. (1988). *Roadbed engineering in frozen earth area.* Beijing: Lanzhou University Press. (in Chinese).
33. Wu, H. Z., Hu, D. G., Wu, Z. H., et al. (2005). *Active faults and induced geological disasters in central Tibetan Plateau.* Beijing: Geological House. (in Chinese).
34. Yang, W. F., & Zhou, T. (2007). Research on the geological disaster in the sideslope of tunnel at complex geological conditions. *Shanxi Architecture, 33*(4), 318–319. (in Chinese).

35. Yi, Z. M., Zhou, J. X., & Zhang, X. D. (2006). Analysis of hydrologic conditions along the Qinghai–Tibet Railway. *Technology of Soil and Water Conservation, 4,* 14–16. (in Chinese).
36. Zang, E. M., Wu, Z. W., et al. (1999). *Permafrost degradation and road engineering*. Lanzhou: Lanzhou University Press. (in Chinese).
37. Zhang, G. X., Wang, S. J., & Zhang, Z. Y. (2000). *China engineering geology*. Beijing: Science Press. (in Chinese).
38. Zhao, J. C., Liu, S. Z., & Ji, S. W. (2001). Engineering-geological problems in railway engineering construction. *The Chinese Journal of Geological Hazard and Control, 12*(1), 7–9. (in Chinese).
39. Zhou, Y. W., et al. (2000). *China frozen soil*. Beijing: Science Press. (in Chinese).

Chapter 6
Route Selection in Nature Reserves

Abstract The Qinghai–Tibet Railway passes through not only urban areas but also several nature reserves, and its construction and operation substantially affect its the social and ecological environment along the line, which renders crucial the balancing of human activities and the environment. This chapter details the route selection in the nature reserves: Sects. 6.1 and 6.2 overview the major nature reserves along the railway and the location of the railway relative to the nature reserves, respectively. Sections 6.3 and 6.4 describe the principles that should be followed in route selection in nature reserves, and Sect. 6.5 explains the effects of the railway on the environment and presents the main solutions.

Keywords Route selection · Nature reserves · Principles · Effects · Solutions

Railway construction substantially affects the ecological and social environment along the line. Environmental, soil, and water conservation; reducing or preventing environmental pollution; and rebuilding the ecological loss should therefore be comprehensively considered at the route selection stage. Because nature reserves are crucial components of the ecological environment, protecting and sustaining them greatly benefits the environment and the ecology. As an important and exemplary engineering infrastructure for nation-building, the Qinghai–Tibet Railway should ensure human–environment harmony.

6.1 Overview of Nature Reserves Along the Line

Of the 1142 km of the Qinghai–Tibet Railway, 594 and 548 km is in the Qinghai Province and Tibet Autonomous Region (TAR), respectively. These two provinces contain 24 national or provincial nature reserves, five national and four provincial nature reserves in Qinghai and nine national and six provincial nature reserves in TAR. The railway affects 11 (five existing and six planned) of these reserves. Table 6.1 overviews the nature reserves, and Fig. 6.1 indicates their distribution.

J. Li et al., *Geological Line Selection for the Qinghai-Tibet Railway Engineering*,
https://doi.org/10.1007/978-3-662-55572-9_6

Table 6.1 Overview of nature reserves along the Qinghai–Tibet Railway

Name	Administrative district	Area (km^2)	Grade	Main targets of protection	Remarks
Hoh Xil	Qinghai—Yushu	45,000	National	Wildlife (e.g., Tibetan antelope and kiang) and alpine ecosystem	Built
Sanjiangyuan	Qinghai—Yushu, Golog, Huangnan, Hainan, Haixi, and Golmud	318,000	National	Headwater environment, rare species, and alpine ecosystem	Built
Qiangtang	Tibet—Amdo, Shuanghu, Nyima, Gêrzê, Rutog, and Gê'gya	298,000	National	Rare flora and fauna and alpine ecosystem	Built
Linzhou–Pengbo Black-Necked Crane	Tibet—Lhünzhub	96.8	Provincial	Black-necked crane and its wintering habitat environment	Built
Lhalu Wetland	Tibet—Lhasa	6.2	Provincial	Urban wetland ecosystem	Built
Yarlung Tsangpo Nature Reserve	Tibet—Nagqu, Ngari, Shannan, Nyingchi, Qamdo, and Shigatse, and 41 counties in Lhasa	22,075	Provincial	Rare species (e.g., black-necked crane) and their living environment	Planned
Nam Co Lake	Tibet—Damxung and Baingoin	11,424	Provincial	Alpine lake, wetland ecosystem, and rare endangered species	Planned
Yambajan Thermophis Baileyi Nature Reserve	Tibet—Damxung	80	Provincial	Tibet Thermophis baileyi and its living environment	Planned
Lhasa Cypress Forest	Tibet—Lhünzhub and Nyêmo	75	Provincial	Large orchards and cypress forests	Planned
Dagzê Yeba Karst Landform	Tibet—Dagzê	60	Provincial	Karst landform	Planned
Marqu Nonconformable Contact	Tibet—Doilungdêgên	8.5	Provincial	Nonconformable contact	Planned

Note Data in this table are current as of 2006

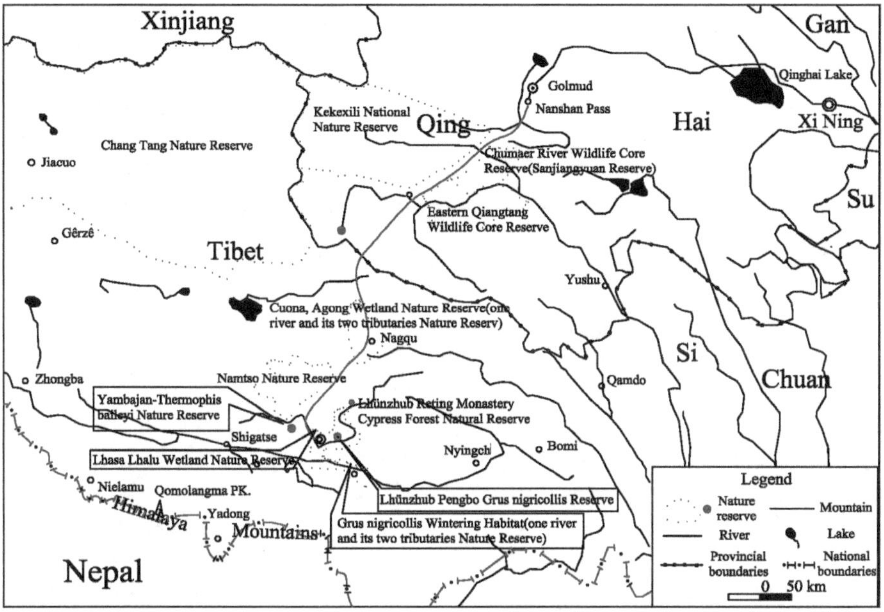

Fig. 6.1 Location of nature reserves along the Qinghai–Tibet Railway

6.1.1 Overview of Nature Reserves Along the Qinghai–Tibet Railway

6.1.1.1 Hoh Xil Nature Reserve

The Qinghai Hoh Xil Nature Reserve (89°25′–94°05′E, 34°19′–36°16′N; area 45,000 km^2) was established in October 1995, and upgraded to a national nature reserve in December 1997. It is located in the northwest hinterland of the Qinghai–Tibet Plateau, with Kunlun Mountains as the northern border; the provincial boundary of Qinghai as the western border; Tonglha, a township in Golmud, as the southern border; and expands eastward to the Qinghai–Tibet Railway. The Qinghai–Tibet Railway spans 103.4 km in the eastern buffer region of this reserve, coming as close as 29 km to the central area.

6.1.1.2 Sanjiangyuan Nature Reserve

Sanjiangyuan Nature Reserve (89°24′–102°23′E, 31°39′–36°16′N; area 318,000 km^2), which occupies 44.1% of Qinghai, was established on August 19, 2000, and upgraded to a national nature reserve in January 2003. It is located in the hinterland of the Qinghai–Tibet Plateau, which is the headwater region of Yangtze

River, Yellow River, and Lancang River. The Qinghai–Tibet Railway passes through the experimental and buffer region of Chumar River and core reserves of eastern Qiangtang. The nature reserves adjacent to the railway are Geladandong Glaciers Core Reserve and Damqu Wetland and Guozongmucha Wetland Core Reserve. The railway runs 336.6 km in total through these reserves, with 204.7 km in experimental region, 119.6 km in the buffer region of Chumar River Wildlife Core Reserve, and 12.3 km in the buffer region of Eastern Qiangtang Wildlife Core Reserve.

6.1.1.3 Yarlung Tsangpo Nature Reserve

Yarlung Tsangpo Nature Reserve (78°40′–98°20′E, 28°06′–33°40′N) was established in 1993 and includes the provincial Xainza Black-Necked Cranes Nature Reserve, which is the main breeding site of black-necked cranes and is the central reserve, and the provincial Linzhou–Pengbo Black-Necked Crane Nature Reserve, which is the main wintering ground of black-necked cranes. These two reserves were combined as a permanent reserve and a seasonal conservation region to protect the breeding site of the black-necked cranes as well as the main habitats and winter foraging sites of Tibet black-necked cranes. Administratively, it belongs to Nagqu, Ngari, Shannan, Nyingchi, Qamdo, and Shigatse, and 41 Tibetan counties. Almost 4000–5000 black-necked cranes are present in this reserve. The most important breeding site of the black-necked cranes is the Xainza Central Reserve, and Zhari Namco, Lake Palgon, Anglarencuo, and Lake Yamdrok are four satellite conservation regions. This Yarlung Tsangpo Nature Reserve also protects many other waterfowl living in this ecosystem, especially migrant fowl whose life habits are similar to those of black-necked cranes (e.g., Anser indicus, Tadorna ferruginea, and Anas strepera). Moreover, the numerous lakes, rivers, and mountains in this area form a beautiful landscape, which is also worthy of protection.

1. Central Conservation Region

Xainza Central Conservation Region (86°40′–89°30′E, 30°10′–32°10′N; area 22,075 km^2), which includes the Zhari Namco, Lake Palgon, Anglarencuo, and Lake Yamdrok divisions, has the main ridge line of Kailas Range as its southern border, the southern foot of Mugagangri Mountain as its northern border, the western watershed of Tangra Yumco as its western border, and the Xainza–Baingoin County boundary as its eastern border. Administratively, it belongs to Xainza, Baingoin, and Nyima, which in particular is the most important breeding site of black-necked cranes. The core region of the reserve is the main lake and the surrounding wetlands—where the black-necked cranes reproduce and brood in the summer—and the riverside swamp and meadow wetlands—where the lakes flow into river. The experimental region is the main tourism destination and where the local residents engage in agriculture. The rest of the reserve is classified as a buffer region.

2. Satellite Conservation Region

Satellite conservation region includes four relatively large lake and wetland reserves (Zhari Namco, Lake Palgon, Anglarencuo, and Lake Yamdrok) and some scattered wetland reserves. These four satellite conservation regions are far from the railway route.

3. Seasonal Conservation Region

The overwintering habitat of the black-necked cranes is located in the basin of Southern Tibet Yarlung Zangbo River, Nyang Qu, and Lhasa River; the river wetlands where they stay overnight and rest form the seasonal core region, whereas the farmland where they forage forms the seasonal buffer region. After the cranes and other migrant fowl depart in summer, the restrictions on the sites are removed, and the local residents are permitted to continue their agricultural activities. Winter husbandry activities are permitted in the seasonal buffer region as long as the activities do not interfere with the normal foraging activities of the cranes.

This nature reserve has three distinct regions where the black-necked cranes overwinter: Lhazê-Dagzhuka Seasonal Core Region; middle and lower reaches of Lhasa River Seasonal Core Region, and Minya Konka-Zêtang Seasonal Core Region, where 67, 26, and 7% of the cranes overwinter, respectively.

The Qinghai–Tibet Railway passes through only the middle and lower reaches of Lhasa River Seasonal Core Region ($\sim$17.1 km). The alternative project in Lhünzhub passes through the middle and lower reaches of Lhasa River Seasonal Core Region and Seasonal Buffer Region.

6.1.2 Overview of Nature Reserve Adjacent to Railway

6.1.2.1 Qiangtang Nature Reserve

Tibet Qiangtang Nature Reserve (area 29.8 km^2), established in 1993 and upgraded to a national nature reserve in 2000, is located in the northern region of Tibet. The northern region of this reserve is part of the Tibet administrative district, the western region contains the northern section of National Highway 219 and Rutog Domar, the eastern region covers the Meiduoqudong Mountain of Tonglha Mountains, and the southern region contains the Meiduoqudong Mountain–Jiaotangmayue region, including Jiangmaodong–Sari–Miduochalong–Jiarideng–Xidaer–Ajuri–Jiare–Lamada–Nazhacuo–Duomari–Laanla–Duomuzari–Tingyuanla–Qiangmai–Yalongri in the west and mountain ridges (elevation 5300 and 5618 m) reaching Rutog Domar in the northwest. This reserve involves the areas of Amdo, Shuanghu, Nyima, Gêrzê, Rutog, and Gê'gya, all of which are part of either Nagqu or Ngari provinces.

Qiangtang Nature Reserve mainly protects rare wildlife and the plateau ecosystem and is home to around 30 mammal species, 90 bird species, and many other fauna. Among them, snow leopards, kiangs, wild yaks, ibexes, golden eagles, and black-necked cranes are first-class national protected animals, and brown bears, Felis bieti, lynxes, manuls, Tibetan antelopes, argalis, and Tibetan snowcocks are second-class national protected animals. The main ecosystems are alpine desert and grasslands, mainly supporting phytocoenosiums such as Stipa purpurea, Stipa basiplumosa, Stipa glareosa, Carex moocroftii, Ceratoides latens, and Barnadesioideae. On the basis of the geographical characteristics and species distribution, Qiangtang Nature Reserve can be divided into the core, buffer, and experimental regions. The Qinghai–Tibet Railway passes within 104.9 km of the experimental region.

6.1.2.2 Nam Co Lake Nature Reserve

Nam Co Lake Nature Reserve (89°25′–91°21′E, 30°05′–31°10′N; area 11,424 km^2) is located in the center of TAR and on the north slope of Nyainqntanglha Mountains. In 1995, the People's Government of Lhasa established it as a provincial reserve, but its original range only included Damxung. Later, following *The Master Plan of Nam Co Lake Nature Reserve* (edited in 2000), the border was expanded to include the watershed of Bamucuo, the southern basin of Bengcuo, and the northern basin of Shencuo and Namtso as the northern border of the reserve and the main ridge line of Nyainqentanglha Mountains as the southern border. The southern section of the western border coincides with the county boundary of Baingoin and Xainza, which is the watershed between Zhunzangbuqu and Boqu. The middle section of the western border is the watershed of Chaguoqu, Longjuequ, and Buqu, and the northern border is the watershed between Nagqu, Xiongqu, and Gangyasangqu. The eastern border is the east watershed of Niyaqu of Moduoxiga Mountain. The administrative regions cover Damxung Namtso of Lhasa and some or all areas of Qunxue, Qinglong, Namujie, Baoji, Xinji, Deqing, and Dongerde Townships of Nagqu Baingoin County. The Nam Co Lake Nature Reserve mainly aims to protect the alpine lake and wetland ecosystem and the many endangered species within, such as Tibetan antelopes, kiangs, Tibetan gazelles, and argalis. In addition, the reserve has spectacular snow–lake landscapes bounded by Nyainqentanglha Mountains as well as Namtso, the highest large lake in the world; this region is well known for classical Tibetan cultural heritage.

The reserve is divided into core, buffer, and experimental regions, and the closest approach of the Qinghai–Tibet Railway to the buffer and core regions is 15.4 and 32.0 km, respectively.

6.1.2.3 Yambajan–Thermophis Baileyi Nature Reserve

Yambajan–Thermophis Baileyi Nature Reserve (90°12′–90°23′E, 29°38′–29°55′; planned area, 80 km^2) is located in Damxung County of Lhasa and aims to protect Thermophis baileyi, which is endemic to Tibet and its habitat.

Thermophis baileyi is the only high-altitude snake species in the world; it lives around hot springs located at elevations of 4000 m or more altitude in Tibet (e.g., Yambajan). Its habitat has been extensively damaged (e.g., disappearance of hot springs, reduction in the temperature, drying up of warm lakes, and degeneration of warm swamps) because of the development and utilization of geothermal resources and other influences of other human activities. Thermophis baileyi, which is currently an endangered species, is listed as a priority reptile species in the China National Biodiversity Conservation Action Plan.

This reserve is also home to leopards, brown bears, lynxs, Tibetan gazelles, bharals, black-necked cranes, and Anas strepera. The major wild flora includes nagamatsu, barberry root, Rhodiola rosea, cuckoo, Goldendeer muifa, and saussurea. To the northeast of the reserve lies the Yambajan Geothermal Steam Plant. The geothermal resources in this reserve are been utilized, but no modern industries are present. The Qinghai–Tibet Railway runs as close as 15.5 km from the reserve, and 21.6 km from the core region.

6.1.2.4 Lhasa Cypress Forest Natural Reserve

Lhasa Cypress Forest Natural Reserve (area, 75 km^2) was an important construction in TAR in “tenth five-year plan” period. The reserve consists of Lhünzhub Reting Monastery Natural Cypress Forest Reserve (91°29′–91°33′E, 30°18′–30°21′N; area 65 km^2) located in northeast of Lhasa and Nyêmo Natural Cypress Forest Reserve (89°52′–90°E, 29°20′–29°26′N) located at the edge of southwest Lhasa. 10 km^2 of which is composed by Lhünzhub Reting Monastery Natural Cypress Forest Reserve. This reserve is divided into key reserve (core region) and normal reserve. Key reserve area in Lhünzhub Reting Monastery natural cypress forest reserve is approximately 3 km^2, the normal one is 7 km^2, while key reserve area in Nyêmo natural cypress forest reserve is 4 km^2, and the normal one is 61 km^2. The shortest distance from railway to reserve is 25.2 km, and to core region is 26.0 km.

Lhasa Cypress Forest Natural Reserve aims at mainly protecting the large orchard cypress forest. The position of the reserve is the transition area from forest to grassland, and most of the forest have disappeared. Thus, the epibiotic ancient forest ecosystem in this area is a good science base to make scientific research and explore the evolution rule of the local wild creatures. This large orchard cypress forest protected by the reserve has both virgin forest with age of near a thousand years and the unmature forest having grown up over years, which are quite different. And the Lhünzhub Reting Monastery in reserve is a famous Tibetan Buddhism furudera, which is beautiful and quiet for tourism.

Nyêmo County, where Nyêmo natural cypress forest reserve locates, is a half farming, and half nomadism county. The reserve is located in the area of Karu Township Zelanghesangnu, and the population there is around 500. Sino-Nepal Highway goes along the edge of the reserve, and no highways have been built in the reserve. Bonismo temple and Sangnu temple are in the core region of the reserve.

6.1.2.5 Linzhou–Pengbo Black-Necked Crane Nature Reserve

Linzhou–Pengbo Black-Necked Crane Nature Reserve (90°50′–91°25′E, 29°54′–30°40′N; area 96.8 km^2) was established in 1993, located in Pengbo catchment. This reserve aims at protecting black-necked cranes and their overwintering environment. The living environment of black-necked cranes in this reserve is various and complicated, including not only river valley, farmland, river, and reservoir but also wetland such as river shoal and swamp. Wintering black-necked cranes can find the crop seeds such as highland barley and wheat in the farmland, and find root and stem of aquatic plants and farmland weeds in the swamp and river shoal where the plants are lush. Other normal waterfowl in this reserve are merganser, gadwall, barhead goose, tufted duck, and larus brunnicephalus.

Lhünzhub County is 65 km away from Lhasas, and the highways are allowed to pass through the reserve. The shortest distance (alternative project) from reserve to railway is 27.5 km.

6.1.2.6 Dagzê Yeba Karst Landform Nature Reserve

Dagzê Yeba Karst Landform Nature Reserve (91°00′–91°26′E, 29°43′–29°45′N) was planned in 2001–2010 by TAR, the main protection object is the karst landform. Dagzê Yeba karst landform (exposed area, 60 km^2) is distributed around the Yeba Temple of Dagzê County. Dagzê Yeba karst landform is a fantastic landscape resulting from the hydrochemical effects of Middle Jurassic Yeba Strata limestone, such as stone forest, stalacto-stalagmite, natural bridge, karst cave, and karst spring. Dagzê Yeba Karst Landform Nature Reserve is not only beautiful but also shares great value for research on earth evolution; there is Yeba temple in the reserve, making a blending landscape of human and nature. Thus, it is a good area for sightseeing, archaeology, tourism, and religious worship.

The shortest distance from railway to reserve is 10.2 km, and to core region is 18.8 km.

6.1.2.7 Doilungdêgên Ma Unconformable Contact Natural Reserve

Doilungdêgên Ma Unconformable Contact Natural Reserve (90°41′–90°43′E, 29°51′–29°53′N; area, 8.5 km^2) was planned in 2001–2010 by TAR, which is aimed at protecting the nonconformable contact. It is located in the west side of the Qinghai–Tibet Highway in Lhasa Doilungdêgên Ma. The underlying of unconformity surface is interbed of aubergine mudstone of Upper Cretaceous shexing group, sandstone, and gray-greenish tuffaceous sandstone. The layers have draped. Above the unconformity surface, there is Tertiary Dianzhong Strata with conglomerate at the bottom of 50–60 m in thickness. Upper coverings are andesite, dacite, rhyolite, tuffaceous sands, and glutenite, with thickness of 1500–2000 m. The contact relationship here is not only very clear but also quite spectacular. It is a good area for geological education, scientific research, and sightseeing. The shortest distance from railway to reserve is 2.5 km.

6.1.2.8 Lhalu Wetland Nature Reserve

Lhalu Wetland Nature Reserve (91°03′48″–91°06′51″E, 29°39′46″–29°41′05″N; planned area, 6.2 km^2) was established in 1999. It locates in the northwestern corner of Lhasa. High mountains are present in the northern section, which belongs to extension of Kailas Range. The northeast section connects to Liusha River, joined by Niangre cheuch, and Duodi cheuch. The eastern and western sections are connected to Lhasa Chengguan Lhalu Township and Baerku Road, respectively. The southern border is the central canal of Lhasa River diversion and the Dangre Road. It is divided into core region (2.0 km^2), buffer region (3.45 km^2, expanding to 5.75 km^2 at middle term plan), and experimental region (0.75 km^2).

Lhalu Wetland Nature Reserve is aimed at protecting the wetland ecosystem in suburb area. Lhalu wetland is called "dangba" by local people, which means wetland with bulrush, and it is the main grazing space and mowing pasture. In addition, it is also the important landscape and environmental resources of Lhasa. The shortest distance from railway to reserve is 2.0 km.

6.2 Location Relationship Between Nature Reserve and Railway

6.2.1 Situation of Railway Passing Through Nature Reserve

The Qinghai–Tibet Railway passes through Hoh Xil Nature Reserve, Sanjiangyuan Nature Reserve, and Yarlung Tsangpo Nature Reserve.

Railway intersects between Hoh Xil Nature Reserve and Sanjiangyuan Nature Reserve. The route passes through Hoh Xil Nature Reserve from Kunlun Mountain Pass to first Budongquan Bridge, from Xiushuihe station vicinity to Erdaogoubing station vicinity and from second bridge of Yamaer River to Wuli station vicinity, with a total length of 103.4 km. And the railway passes through core and buffer regions of Sanjiangyuan-Chumar River Wild Animal Reserve from Budongquan to Xiushuihe station vicinity, 119.6 km in length, and then it passes through the core and buffer region of Sanjiangyuan Eastern Qiangtang Wild Animal Reserve from Erdaogoubing station vicinity to second bridge of Yamaer River, with a length of 12.3 km. It runs through the experimental region of Sanjiangyuan Nature Reserve from Kunlun Mountain Pass to Tonglha Mountain Pass, with a length of 204.7 km. The total length of railway in nature reserves is 440 km, occupying 38.5% of the whole, with 103.4 km in Hoh Xil Nature Reserve and 226.6 km in Sanjiangyuan

Nature Reserve, respectively. Counting the 78.2 km in plan, which goes through Yarlung Tsangpo Nature Reserve, the total length reaches 518.2 km, making up 45.4% of the whole length. Therefore, the length of railway in buffer regions of Hoh Xil Nature Reserve and Sanjiangyuan Nature Reserve is 235.3 km in total (Tables 6.2 and 6.3).

Table 6.2 Summary of relative locations of the Qinghai–Tibet Railway and natural reserves

Railway location	Reserves with railway passing through	Length of railway in reserves (km)	Largest distance of railway into reserves/m	Average distance of railway into reserves/m
Kunlun Mountain Pass—first Budongquan Bridge (DK980–DK1000+500)	Buffer region of Hoh Xil Nature Reserve	20.5	400	200
First Budongquan Bridge–National Highway 109 overpass, vicinity of Xiushuihe station (DK1000+500–DK1120+100)	Buffer region of Sanjiangyuan-Chumar River wildlife core region	119.6	2400	500
DK1120+100–DK1170 +200	Buffer region of Hoh Xil Nature Reserve	50.1	2000	200
DK1170+200–DK1182 +500 (first Budongquan Bridge–National Highway 109 overpass, vicinity of Xiushuihe station)	Buffer region of Sanjiangyuan eastern Qiangtang wildlife core region	12.3	2000	200
DK1182+500–DK1215 +300 (vicinity of Erdaogoubing station—second bridge of Chumar River)	Buffer region of Hoh Xil Nature Reserve	32.8	600	200
DK979+300–DK1420 +000 (in section)	Experimental region of Sanjiangyuan Nature Reserve(passing through)	204.7		
DK1604+500–DK1608 +200 DK1641+300–DK1673+800 DK1689 +900–DK1714+800	Wetland of Yarlung Tsangpo and its two tributaries, Kalong, and Agong in plan	61.1		
DK1976+200–DK1993 +300	Seasonal reserves in Yarlung Tsangpo and its two tributaries and middle and lower reach of Lhasa River in plan	17.1	200	100

Table 6.3 Summary sheet of situation of nature reserves with railway through

Name of reserve	Length of railway through reserve (km)			
	Sum	Experimental region	Buffer region	Core region
Hoh Xil	103.4	–	103.4	–
Sanjiangyuan	336.6	204.7	131.9	–
Yarlung Tsangpo and its two tributaries	78.2	–	–	–
Sum	518.2	204.7	235.3	–

Table 6.4 Nature reserves adjacent to railway

Name of reserves	Closest distance from railway to reserves (km)		
	Border of reserve	Buffer region	Core region
Namtso	15.4	15.4	32.0
Yambajan Thermophis Baileyi Nature Reserve	15.4	–	21.6
Linzhou–Pengbo Black-Necked Crane	27.5	–	–
Lhasa Cypress Forest	25.2	–	26.0
Yeba Karst landform	10.2	–	18.8
Marqu nonconformable contact	2.5	–	–
Lhalu wetland	2.0	–	–
Qiangtang	104.9	–	–

6.2.2 *Relationship Between Railway and Adjacent Nature Reserves*

Table 6.4 lists the shortest distances between the eight reserves and the Qinghai–Tibet Railway line. Qiangtang Nature Reserve is far from railway, more than 100 km, and the rest ones are 10–30 km away from railway. Doilungdêgên Ma Unconformable Contact Natural Reserve is relatively close to the railway.

6.3 Principles of Route Selection in Nature Reserve

The Qinghai–Tibet Railway passes through Hoh Xil Nature Reserve, Sanjiangyuan Nature Reserve and some area where is under planning to be part of Yarlung Tsangpo Nature Reserve. Thus, when choosing the route, principles of "pay equal attention to exploitation and protection" should be followed throughout the procedure. High attention should be paid to the protection of nature reserves and wild animals along the railway, and the effects from railway should be reduced to the minimum level.

6.3.1 *Principles of Railway Route Selection*

When selecting the railway route, we set up high awareness of environmental protection. As for areas which are quite susceptible to environment, we have alternative projects when selecting the route. In particular, when the railway passes the Hoh Xil Nature Reserve and Sanjiangyuan Nature Reserve, we decrease the occupation of planned nature reserve area as possible.

6.3.1.1 General Principle

1. Route choosing should avoid nature reserves. If impossible, route should avoid going through too many buffer regions of nature reserves.
2. Route should pass through wetland as few as possible. When route has to pass the wetland, it should be guaranteed not to cut the surface and underground water flowing, which ensures the continuity of wetland. It is better to build bridges in wetland.
3. Route should avoid animal habitats and the susceptible zones which function as the water conservation. It should ensure the diversity of creatures and integrity of alpine ecosystem (e.g., wetland and waterhead areas).

6.3.1.2 Specific Principles in Each Reserve

1. For the overlapping sections of Hoh Xil Nature Reserve and Sanjiangyuan Nature Reserve, route selection should fully consider the properties of each reserve. For example, in the Budongquan–Wudaoliang section, route should be selected on the east side of the Qinghai–Tibet road and avoid core region of reserve.
2. The alternative project in the Damxung–Dagzê–Lhasa section has considered both the problems regarding the length of route, and operation fee, and the Linzhou–Pengbo Black-Necked Crane Nature Reserve. The final decision is Yambajan project, which avoids the nature reserve.
3. For the experimental region in Sanjiangyuan Nature Reserve where railway passes, habitats of wild animal, and the susceptible zones with function of water conservation should be protected.

6.3.2 *Setting Principles of Wildlife Passageways During Route Selection*

1. Fully consider the setting of wildlife passageways along the line and leave sufficient passageways for animal migration and communication between different population to meet the requirement of rest, reproduction, and migration.

2. In the design, the setting of wildlife passageways should have a long term plan. For example, after the wildlife is under protection, the population increase gradually, so different types of passageways should not be too narrow, and the channel clearance under the bridge should not be too low.
3. According to the wildlife passageways decided by final environmental effect assessment, passageways should be set scientifically and reasonably. The main ways include the passageways under the bridges, over the tunnel, and parallel to gentle slope of roadbed.
4. For the section where railway leaves great influence on the passageways and habitat environment, proper ecological engineering should be taken to decrease the influence.
5. For long-term consideration, fully consider the tourism potential of the Qinghai–Tibet Railway and make use of the function of sightseeing. So, along the railway, particularly in the vicinity of wildlife passageways, reengineering program should be enhanced.

6.4 Main Route Selection in Nature Reserve

6.4.1 Route Selection in Sanjiangyuan Nature Reserve

6.4.1.1 Basic Situation of Sanjiangyuan Nature Reserve

The main object of protection in Sanjiangyuan Nature Reserve is water resources and environment, including diversity of creatures, and integrity of alpine ecosystem. According to resources distribution situation, protection aim, and protection objects, the reserve is divided into four types as 25 core regions, 25 buffer regions, and one experimental region. A core region and its corresponding buffer region make up a core reserve. The area of all core reserves is 11.2 km^2 which composes 35.2% of the whole reserve, while the core regions are 6.2 km^2 and the buffer regions are 5.0 km^2. The outer contour of these 25 core reserves is an experimental region with an area of 20.6 km^2, which occupies 64.8% of the total area. It is a key national wildlife refuge, as many rare species endemic to the highlands can be seen in large numbers in the reserve.

Railway through Sanjiangyuan Nature Reserve is 336.6 km in total, among which 204.7 km is in experimental region, 119.6 km in buffer region of Chumar River Wildlife Core Reserve, and 12.3 km in buffer region of Eastern Qiangtang Wildlife Core Reserve. Railway passes through three adjacent core reserves with different nature resources and environment, resulting in differences in conservation targets.

1. Railway passes through two core reserves

a. Chumar River Wildlife Core Reserve

The Chumar River Wildlife Core Reserve (area, 27,000 km^2) in middle and lower reach of Chumar River, a tributary of Changjiang River, in Qumarlêb County., while 7158 km^2 in core region and 20,542 km^2 in buffer region; 3242 people, 203,000 livestock live here, and so do many wild animal species, mainly the Tibetan antelope, wild yak, snow leopard, kiang, and brown bear. At present, sandification is getting worse and worse, and grassland degradation is aggravating. Thus, the main task is to protect and recover the native vegetations, making them have water conservation function and provide a nice habitat for wild animals. The length of railway in buffer region of Chumar River Wildlife Core Reserve is 119.6 km.

b. Eastern Qiangtang Wildlife Core Reserve

It is in Zhidoi County and the (planned area, 22,700 km^2), among which 9786 km^2 is core region and 12,914 km^2 is buffer region. It connects to Hoh Xil in the west. The area of swamp and riverway is large with an altitude exceeding 5000 m, with 3424 people and 203,000 livestock living here. The number of wild animal is over 16,000, and species are more than 50, mainly including black-necked cranes, kiang, snow leopard, white-lipped deer, and wild yak. The main objects of protection are wild animals and wetland ecosystem. The railway in the buffer region of Eastern Qiangtang Wildlife Core Reserve is 12.3 km.

2. Railway passes through three adjacent core reserves

a. Guerra Danon Glacier Core Reserve

Guerra Danon is the headstream of Changjiang River and locates in Tonglha Township in Golmud City. It is the area where glacier concentrates, with alpine meadow in particular. Wild animals there include snow leopard, white-lipped deer, and wild yak. The main protection objects are glacier landform and rare wild animals. The shortest distance from railway to Guerra Danon Glacier Core Reserve is 50.1 km in buffer region and 54.9 km in core region.

b. Damqu Wetland Core Reserve

It is located west of Zadoi County, and many swamps, lakes, and rivers are present. The main birds and wild animals here are black-necked cranes, Larus brunnicephalus, Tadorna ferruginea, kiang, and wild yak. Main objects of protection are wetland ecosystem and wild animals. The shortest distance from railway to Damqu Wetland Core Reserve is 92.45 km in buffer region and 119.1 km in core region.

c. Guozongmucha Wetland Core Reserve

It is located in Zadoi County, and it is the headstream of Lantsang River. There is around 50,000 km^2 of rivers, lakes, and swamps. It mainly protects the wetland ecosystem. The shortest distance from railway to Guozongmucha Wetland Core Reserve is 171.1 km in buffer region and 185.3 km in core region.

6.4.1.2 Geographical Environment of Sanjiangyuan Nature Reserve

The altitude of Sanjiangyuan Nature Reserve is 4500–4700 m in general. Terrain there is even and wide. Land vegetations grow well. And thermokarst lakes are widely spreading. Sandpiles, sand dunes, and sand dune chains have developed in certain areas.

The climate in this area is cold, dry, and varied. Four seasons are not clear and the air is thin in infrabar. Annual average temperature is −6.9 to −2.6 °C, and the extreme high temperature is 28 °C, while the extreme low one is −45.2 °C. Annual average precipitation is 250–300 mm, and the annual evaporation capacity exceeds 1300 mm. The relative humidity is 49–52%. The fastest wind speed is 22–28 m/s, and the prevailing wind direction is northwest wind. Freezing period is up to 7–8 months in a year (from September to April or May). The evaporation capacity is much more than rainfall. Rainfall in high Mountain areas appears in forms of snow, frost, and hail. But it rains on the wide highland, and 60–90% of rainfall is down in thermotropic seasons. Atmospheric transparency is good, clouds are less, sunshine is strong, total radiant quantity is large, and sunshine duration is quite long, with 2600–3000 ha^{-1} on average. The average total radiant quantity on the highland with altitude under 5000 m is 600–800 kcal/(cm^2 a). It is the area with largest radiant quantity all over the nation. Because the wind on highland blows hardly, the annual sensible heat flux accounts for 60–80% of the total radiant quantity, and latent heat flux takes up 20–30%, which consumes most of the surface heat.

Surface turf is thick or inexistence with many roots. Humus layer is 5–10 cm in thickness with well-developed graininess and fine crumb aggregate structure. Organic matter content is around 1%. Under the humus layer or the transition layer, there is well-developed and well-distributed granular structure. The characters of soil are gritty soil and sandy loam.

Plant community has strong characters of cold resistance and drought tolerance, but the species are few and vegetation is sparse. The main auxiliary species are oxytropis densa, oxytropis falcata bunge, hemerocallis citrina baroni densa, and hemerocallis citrina baroni cluster. Plant coverage is 20–35%, and the rich constructive species of Stipa purpurea could reach 8–20% in general.

6.4.1.3 Relationship Between Sanjiangyuan Nature Reserve Border and Railway

Chumar River Core Reserve takes administrative boundary of Qumarlêb County and Golmud City as its northern border, the Qinghai–Tibet road as the western border, county boundary of Qumarlêb County and Zhidoi County as the southern one. Eastern Qiangtang Core Reserve takes county boundary of Qumarlêb County and Zhidoi County as the northern border (which means it connects to Chumar River Core Reserve), the Qinghai–Tibet road, and administrative boundary of Zhidoi County and Golmud City Tonglha Township as the western border, administrative boundary of Zhidoi County and Zadoi County as its southern border. In fact, from the perspective of protection function, we could regard these two reserves as an independent core reserve, with an area of 5,040,000 km^2. For better description, they are called “Chumar River–Eastern Qiangtang Core Reserve” in the following. In 2000, the border of it experienced adjustment, which took further 2 km from the railway (when they went through buffer regions) or roads (when railway did not go through buffer region) to east as a new border, while being vertical to route. And the buffer region of Chumar River–Eastern Qiangtang Core Reserve was adjusted into experimental region with an area of 868 km^2. The total range of Chumar River–Eastern Qiangtang Core Reserve was the same after adjustment, but the area of buffer region changed into 32,588 km^2, and a new experimental region was added with an area of 863 km^2 (Fig. 6.2).

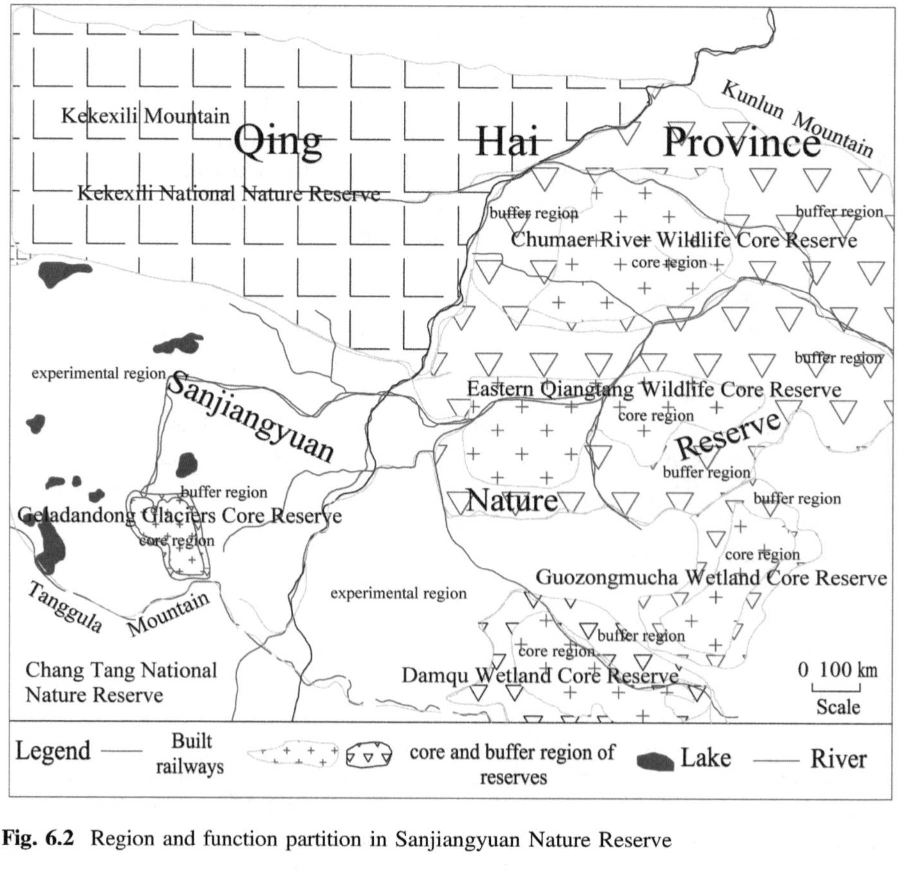

Fig. 6.2 Region and function partition in Sanjiangyuan Nature Reserve

According to the suggestion after adjustment, it is better to go in the buffer region when selecting the railway route.

6.4.1.4 Route Selection in Sanjiangyuan Nature Reserve

Changjiang source area in Sanjiangyuan Nature Reserve is from Kunlun Mountain Pass to Tonglha Mountains Pass, mainly including Chumar River Wildlife Core Reserve, Eastern Qiangtang Wildlife Core Reserve and 50,000 km^2 of buffer regions. To reduce the effects of railway on the reserve, optimization of part route selection scheme is carried out when comparing with the route selection schemes. Then, we take Tongtian River route selection scheme as an example as following to indicate the route selection situation in this reserve.

This region is alluvial plain, and it is located on the Tongtian River terrace and Buqu terrace, with altitude of 4600–4700 m. The terrain is open, riverbed is wide and shallow, river island develops well, terraces are most fixed sand and vegetation is rare. The engineering geological conditions are great. Thus, route selection mainly concerns of the influence on nature reserve.

This scheme was an alternative scheme under primary measurement. The route was along the Qinghai–Tibet road to south from Kaixinling station. In CK1258 +700–CK1310+000, two alternative schemes were put forward, one was close to the Qinghai–Tibet road while the other was far from that. This period is on the gentle slope in front of mountains and on the valley terrace (Fig. 6.3).

The close scheme (CK scheme): railway goes on the right side of the Qinghai–Tibet road with distance of 80–500 km. At CK1275+803, the railway goes over the

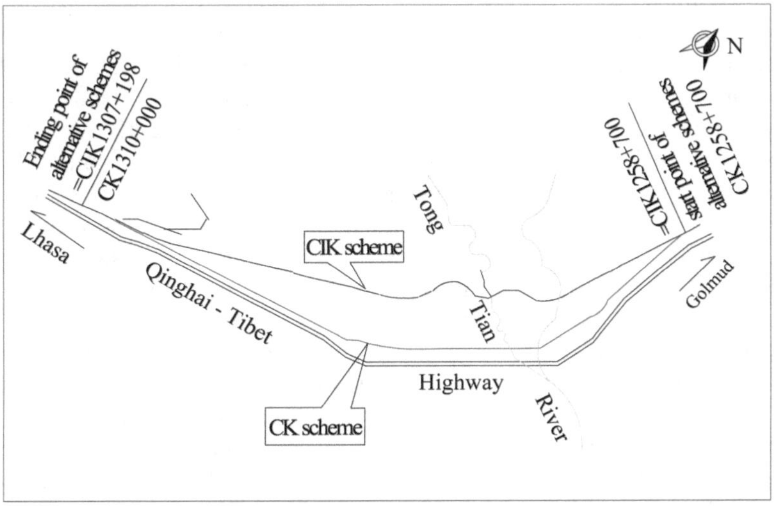

Fig. 6.3 Schematic of Tongtian River alternative schemes

Tongtian River by grand bridge (500 km away from road), and then along the Qinghai–Tibet road till the end of comparison.

The far scheme (CIK scheme): from the starting point of comparison, it goes along the gentle slope in front of Kongjie Mountain and along the Kongjiechaqu Valley terrace. At CIK1277+500, it goes across the Tongtian River by 49–24 m grand bridge (4.5 km away from road), and then along the wide valley of Buqu till the end of comparison.

Analysis from the aspect of influence on nature reserve and wildlife: the positions of two schemes are both on the river terrace, while the CIK scheme is far from road. The surface mainly consists of aeolian sand, and the small pass is in excavation section where permafrost is extremely unstable under high temperature. Although CK scheme is 2.802 km shorter than CIK and saves operation fee, it has the following disadvantages: first, it is far from road, and it involves 4.5 km of experimental region of Sanjiangyuan Nature Reserve, which is bad for border adjustment of reserve; second, Tongtian River Bridge is on reverse curve (radius 1200 m) and this is not appropriate to act as passageways of wildlife. By contrast, CK scheme is close to the road, taking the Tongtian River Bridge as the main bridge. It may leave small influence on the nature reserve. According to the wildlife habit, taking Tongtian River Bridge as passageways for wildlife is appropriate. Thus, we choose the CK scheme.

6.4.2 Route Selection in Hoh Xil Nature Reserve

6.4.2.1 Basic Situation of Hoh Xil Nature Reserve

Hoh Xil Nature Reserve is in the Qinghai–Tibet Plateau hinterland, and the main objects of protection are highland ecosystem and wildlives such as Tibetan antelope, wild yak, Tibetan gazelle, and kiang. Because of the high altitude, cold and dry climate, the vegetation forms are simple, the food condition, and shelter condition are poor and the animal fauna is simple as well. But, excepting the beasts and raptors which live alone, the ungulate has group activities and habitat. Thus, the population density is large, and the number is large, with 19 species of mammals, belonging to ten orders, 18 families and five genera, and 48 species of birds, belonging to 11 orders and 20 families, and fishes of six genera and three species. According to the primary investigation, the number of birds and beasts in the reserve is more than 100,000 (excluding murine).

The main types of ecosystem are alpine steppes and alpine meadow. Alpine periglacial plants have wide distribution, and alpine desert steppes, alpine cushion plants, and alpine desert plants spread less. Because of the high altitude, cold climate, and thin air, most areas are still no man’s lands up to now. Only the eastern edge area sees seasonal graze of herdsman. There is 240 km trunk line of the Qinghai–Tibet road east of the reserve, 160 km away from Golmud City.

The nature reserve is divided into core region and buffer region. Core region is between Hoh Xil Mountains and Wulanwula Mountain–Dongbulei Mountain. It expands eastward from western section of Gongzerima to Goulucuo, and it is west to Qinghai provincial boundary, north to the line from border of Qinghai and Tibet to the east through Heituofeng–Malan Mountain–Hoh Xil Hubei Mountain–Heishi Mountain and then turning southeast through Hoh Xil Mountains to western section of Gongzerima, and south to the line from Dongbulei Mountain to east through southern bank of Wulanwula Lake–Wulanwula Mountain to Goulucuo. It is 25,500 km^2. The outer area of core region is buffer region.

The Tibet Railway passes through the eastern buffer region of Hoh Xil Nature Reserve, with 103.4 km in length. The shortest distance from railway to core region is 29.0 km.

6.4.2.2 Geographical Environment of Hoh Xil Nature Reserve

Qinghai Hoh Xil connects to north Qiangtan Plateau in the west, and it is the core area of the Qinghai–Tibet Plateau. (area, 75,000 km^2) Terrain is wide and even. Excepting many mountains in Tonglha Mountains and Kunlun Mountains in the east and north edge where the altitudes exceed 5600 m, most massif in the middle area of Hoh Xil Mountains and Wulanwula Mountain are low and even, which extends from west to east with an altitude of 5200–5400 m. Most areas show landscape as lake basins and broad valley and plain landform. The average altitude is 5000 m. And the landscapes such as periglacial, denudation, and sand are quite common.

The climate here is typical inland plateau cold and arid and semiarid. The average temperature is below 0 °C, and it remains in low temperature throughout the year with only cold and warm seasons. The cold season is long with more than 8 months of freezing period. The warm season only lasts around 3 months, and frost and freezing are quite common, especially at night. The underground permafrost, undercooling soil, and surface periglacial are quite common. Even in warm season, the temperature of frozen earth under the ground 15–30 cm remains between −1.4–4 °C. Soil texture is not good, and most of them are original desert soil or frozen thin soil. The rainfall is little. Excepting Geladandong in the southeast and the glacial area in extremely high mountains in the northwest which rainfall could exceed 500 mm, rainfall in most areas is approximately 200 mm. Rainfall concentrates in warm season and often appears in gustiness and in solid, which is easy to evaporate and be blowed away. This area is high, even, and open, with both long sunshine duration and strong radiation (total radiation is more than adjacent Qaidam Basin). The air pressure suffers great changes, and it is easy to be affected by the high-altitude stream, which causes strong wind. Number of days with 8+ wind reaches as much as 100–200 days a year. It is the high wind speed region in the Qinghai–Tibet Plateau, even in the whole nation. Average wind speed in warm season is 4–13 m/s, and the largest wind speed reaches 24 m/s.

The nature condition in Hoh Xil is so harsh, lack of heat in warm season (average temperature is only 2.7–4.7 °C in growing seasons from June to August).

Strong ultraviolet radiation, frequent wind deflation, and low pressure are extremely harmful to the survival of normal creatures and human activities. Growth of any high mountain vegetations is highly limited. However, cushion plants develop very well in this area for their unique adjustment mechanism.

6.4.2.3 Relationship Between Hoh Xil Nature Reserve Border and Railway

The actual border of Hoh Xil Nature Reserve is decided by the considerations that Hoh Xil is no man's land in general and the administrative boundaries. This reserve is divided into core region and buffer region. Railway is far from the core region, but it involves some sections of buffer region (Fig. 6.4). The outer area of core region is buffer region of area 19,500 km^2, which is the isolated belt of core region to prevent it from outside disturbances and damages and to provide an environmental protection for wild animals and vegetation in the nature reserve, it is used to scientific research, domestication, and propagation, wildlife rescue, and ecotourism.

6.4.2.4 Route Selection in Hoh Xil Nature Reserve

Railway going through Hoh Xil Nature Reserve is from Kunlun Mountain Pass to Wuli (DK983+700–DK1218+000). The reserve is on the west side of railway. The

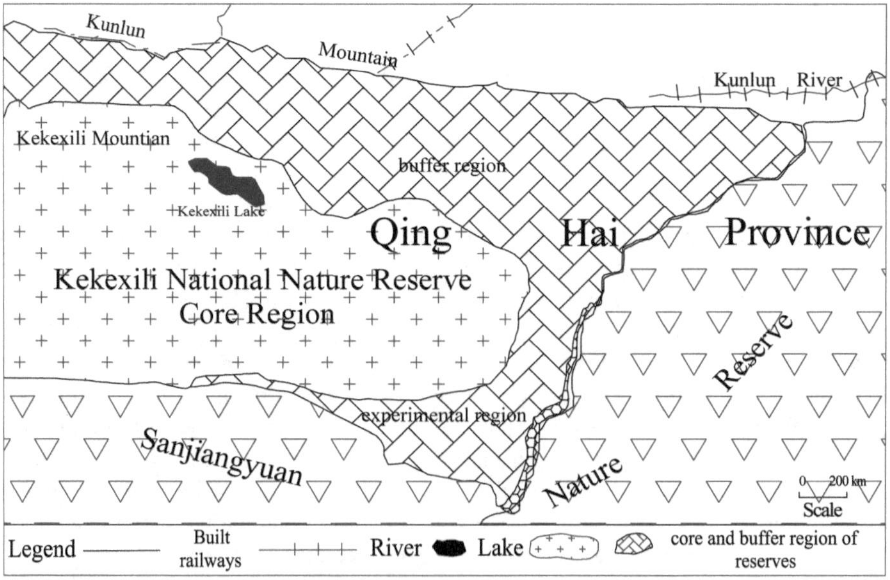

Fig. 6.4 Region and function partition in Hoh Xil Nature Reserve

first-class national protected animals in reserve are Tibetan antelope, wild yak, kiang, and snow leopard, and the second-class national protected animals are argali and Tibetan gazelle. Identified animals in the reserve are 19 species of mammals, 48 species of birds, six species of fishes, and 202 species of higher plants. Main protection objects are peculiar rare wild animals and vegetation, original nature environment, alpine grassland ecosystem, and unique plateau landscape.

As for route selection in Hoh Xil Nature Reserve, we emphasize the selection scheme in primary survey district from Budongquan to Yanshipin.

The length of railway going from Budongquan (CK 1002+100) through Xiushuihe, Wuli to Yanshipin (CK 1334+550) is 330 km. We take the east and the west schemes to make a comparison. West scheme goes largely along the Qinghai–Tibet road, while the east one goes on the east side of road in 20–45 km distance away (Fig. 6.5).

This district is at altitude of 4500–4700 m. Ridge is gentle, terrain is wide, and even, surface vegetation develops well, and thermokarst lakes spread widely. Lithology is glutenite, and sandpiles, sand dunes, and sand dune chains are present. Streams disperse in the river valley which is 300–1000 m in width, and the undercut is not clear. The general landform is combination of valley and ridge, and the geographic structure develops. It passes through permafrost district, resulting in difficulties in geological route selection and the route cannot go straightly.

According to the aspect of influence on nature reserve, the west scheme is 301 km and the east scheme is 294 km, so the west scheme is 7 km longer than the

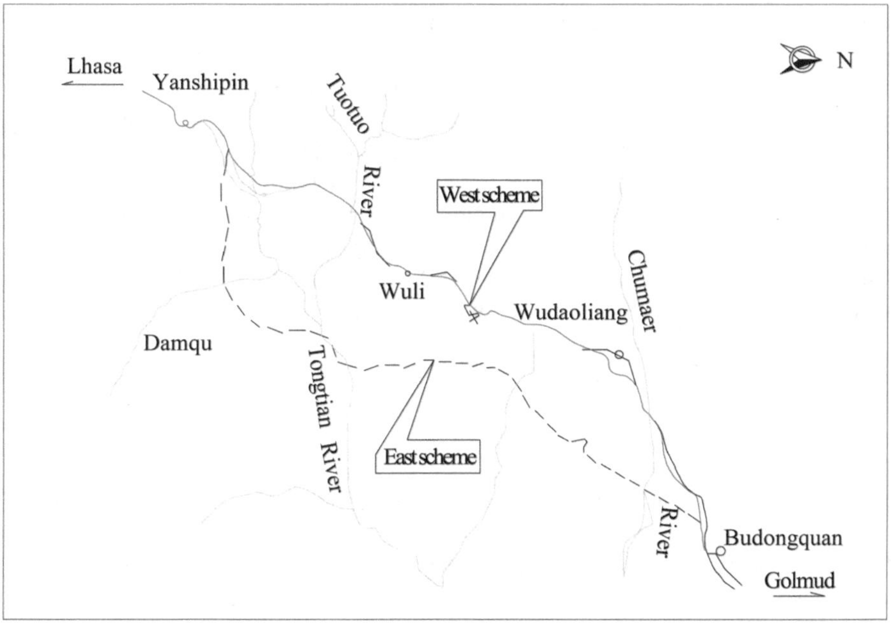

Fig. 6.5 Schematic of west and east alternative schemes from Budongquan to Yanshipin

east one. Route goes 32.8 km through Hoh Xil Nature Reserve, but all is in buffer and experimental region. The east scheme is far from road, and the deepest distance running in Sanjiangyuan Nature Reserve is 45 km, going not only through the buffer region but also through Chumar River Wildlife Core Reserve and Eastern Qiangtang Wildlife Core Reserve; the railway in wind-sand district is 39 km longer than the west one and 1 km longer than the west one in ice piton and drumlin district. Construction road is 200 km longer and the temporary ground during construction is larger, which means more influences on the reserve.

In conclusion, although west scheme has some influence on Hoh Xil Nature Reserve, it is less than the east scheme has. This is good to protect the integrity of the whole nature reserve and to adjust the function border, it reduces the engineering difficulties and saves engineering investment. Thus, we recommend this scheme.

6.4.3 Route Selections in Other Nature Reserves

Other nature reserves are represented by Yarlung Tsangpo Nature Reserve, which has great influence on the whole railway construction. Thus, it is important to handle its relationship with railway. This reserve is combined by permanent reserve and seasonal conservation region to protect the breeding site of black-necked cranes and to protect other main habitats and winter foraging sites of Tibet black-necked cranes, taking Xainza black-necked cranes breeding site as a center. It involves six regions as Nagqu, Ngari, Shannan, Nyingchi, Qamdo, and Shigatse and 41 counties in Lhasa. The whole reserve consists of central conservation region, satellite conservation region, and seasonal conservation region. The satellite conservation region consists of four large lake and swamp reserves and some scattered wetland reserves (without function division). The Qinghai–Tibet Railway passes through some scattered wetland reserves with a length of 64.1 km in the vicinity of Nagqu and Amdo (3.7 km in Kalong wetland, 32.5 km in northern section of Agong wetland, and 24.9 km in southern section of Agong wetland). Riverine wetland where black-necked cranes stay overnight and rest is divided into seasonal core region, and the farmland where they find food is classified as seasonal buffer region. The Qinghai–Tibet Railway passes through the seasonal buffer region in the middle and lower reach of Lhasa River with a length of 17.1 km. The followings are route selection scheme comparisons in certain section of the route.

6.4.3.1 Route Selection Comparison from Amdo to Gajia

Starting from Amdo station, comparison schemes are put forward in CK1560+000–CK1622+285, from Amdo to Gajia. One is along the road and the other is along the Cuona Lake (Fig. 6.6).

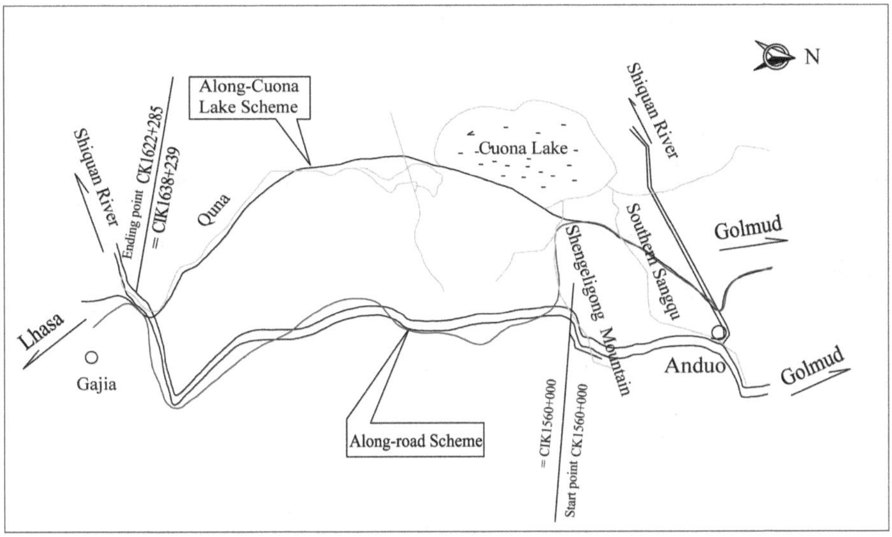

Fig. 6.6 Schematic of alternative schemes from Amdo to Gajia

Scheme along Cuona Lake (CK scheme): railway is mainly along Cuona Lake, and the comparison section is 62.285 km. The route starts from Amdo, goes down along the plain, bypasses Shengeligong Mountain with 230 m of relative height difference, passes through Southern Sangqu and wind-sand land, and then goes along the east bank of Nacuo Lake across Liantong River and along the Nagqu River Valley. It turns southeast at the intersection of the Qinghai–Tibet road and Nagqu–Baingoin Road, and then goes along the Qinghai–Tibet road to Gajia station.

According to the analysis of influence on wetland, CK scheme is relatively far from Yarlung Tsangpo Nature Reserve and covers 3650 m in Kalong wetland, CIK scheme passes through wetland with a length of 18,846 m, which means much more influence on the wetland compared to CK scheme. From the aspect of protection on Yarlung Tsangpo Nature Reserve, no matter whether it reduces the deterioration of wetland and damages on wetland, or provides a better environment for black-necked cranes and other birds to breed and rest, CK scheme is much better than CIK scheme.

That, CK scheme is 15.954 km shorter than CIK scheme without disturbance to the Qinghai–Tibet road, oil pipeline, and communication optical cable in general. Along the route, the districts with frozen earth, sand, and snow disaster are less than those in CIK scheme. It has better geological conditions comparing to CIK, saving investment of 180 million RMB. Thus, we choose the CK scheme.

6.4.3.2 Route Selection Comparison from Damxung to Lhasa

Two comparison schemes are put forward from Damxung to Lhasa (Fig. 6.7).

1. Yambajan scheme (CK)

Route starts from Damxung and goes along the Qinghai–Tibet road to south through Qudeng, Ningzhong, Duiling, and then goes over Yangba ridge through Laduogang to Yambajan. After the Yambajan station, it goes on west bank of Duilongqu through Yambajan Valley to reduce the intersection disturbance with the Qinghai–Tibet road on the east bank. The route goes over Duilongqu at the vicinity of Deqing, then parallels to the Qinghai–Tibet road on the east bank, and then goes across Lhasa River where we set Lhasa station. The route is 154.102 km in total.

2. Dagzê scheme (IVCK)

After starting from Damxung station, the route is similar to CK scheme and it goes toward southwest and turns southeast at vicinity of 149 maintenance squad along the right bank of Wululongqu and the right side of Linbao Road, and then passes through Wululongqu and Linbao Road to Panduo before Seanong. This route is disturbed by road and the terrain, and geological conditions are complicated. Proluvial fan, impact pile, and debris-flow valleys are widely distributed. The smallest curve radius is 600 m. Because of the limitation by Panduo reservoir and terrain conditions, Panduo station is located in the reservoir area, which requires more protection engineering. The later route is mainly along the I and II terraces of Lhasa River on the left bank through Alang, Zhagang, Longzhu, and Zhaxue to

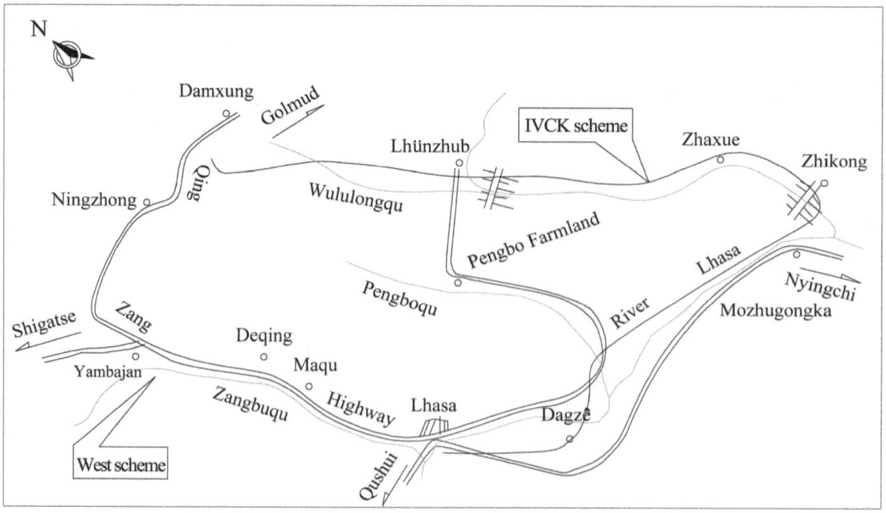

Fig. 6.7 Schematic of alternative schemes from Damxung to Lhasa

Zhikong. The Zhikong–Mozhugongka section is controlled by Zhikong reservoir plan, control water level of which is 3972.5 m. The route goes over Lhasa River at around 1 km from the lower reach of Zhikong reservoir. In the Mozhugongka–Pengbo section, the route goes along the I and II terraces of Lhasa River on the right bank. This section has good terrain and geological conditions. In the Pengbo–Dagzê section, there is a SN direction fault with a fracture zone of 600–700 m wide. And there is an ancient faulted rock mass in front of the mountain and a debris-flow valley here. The route passes through Pengbo and over Lhasa River and Sichuan–Tibet Highway through Dagzê County and then along the Sichuan–Tibet Highway to the end of this scheme, Lhasa.

3. Assessment on route selection scheme

Yambajan scheme (CK) passes through a large area of swampiness wetland in the districts such as the vicinity of Ningzhong and the entrance of Yangbajai first tunnel. Swamp is scattered at Yangbajai station, right bank of Youqu, and the vicinity of Saiqu. Rockfall, debris flow, and alluvial and flood fans are scattered on the slope between Yambajan and Saiqu. The aforementioned districts have unfavorable geological conditions while other districts are good.

As for Dagzê scheme (IVCK), well-developed modern alluvial and flood fans and debris-flow valleys can be seen between the origin and Lhünzhub. Massif from Lhünzhub to Zhikong mainly consists of igneous rock. And collapse, dangerous rock, rockfall, and debris flow are well developed. Tunnel passes through fault and the engineering geological conditions are poor. In the period from Zhikong to Lhasa, the route goes on flood plain and I terrace of Lhasa River. The route goes partly on alluvial and flood fan, and then it passes through a fault at the vicinity of Dagzê. This fault is between the lower Cretaceous schist and the upper Jurassic limestone, and the fracture zone is 600–700 m in wide, most of which is fragmental block. There is bedrock faulted rock mass on some sections of slope with poor engineering geological conditions; the route during Zhikong is affected by Zhikong reservoir. Location of Lhasa River first bridge is poor since it is affected by reservoir dam.

Dagzê scheme passes through Lhünzhub County and the edge of Linzhou–Pengbo Black-Necked Crane Nature Reserve, which affects the reserve. And it passes through the seasonal core reserve and buffer reserve at the middle and the lower reach of Lhasa River along the Lhasa River terrace, which leaves significant effect on reserves.

Comparing Dagzê scheme (IVCK) with Yambajan scheme (CK), the former route is 70.838 km longer, and has greater disturbance on Linbao Road and Sichuan–Tibet road. The areas where route passes through are under complicated terrain. Two reservoirs in planning have great influence on the route scheme. Most major projects are concentrated (19 tunnels). It has poorer engineering geological conditions and higher investment, nearly 19.9 hundred million more than Yambajan scheme CK. It also takes longer construction time.

Yambajan scheme has advantages such as shorter total length, better engineering geological conditions, less investment, and convenience for engineering construction. In particular, it avoids the disturbance in nature reserves. It also ensures the development of Yambajan geothermal resources. Thus, we recommend Yambajan scheme (CK).

6.5 Solution Taken in the Qinghai–Tibet Railway Route Selection

6.5.1 Effects of Qinghai–Tibet Railway on Ecological Environment

6.5.1.1 Positive Effects

1. Construction of railway will make it possible to transport abundant coal and petroleum resources in northwest district to Tibet by economical and convenient channels, which could meet the energy requirement of Tibet. Thus, it contributes to adjusting the energy structure of Tibet and preventing blind deforestation, which makes positive contribution to ecological environment protection and has profound significance.
2. Construction of railway will directly improve the economic development of Qinghai and Tibet, accelerate the paces of urbanization and industrialization, and encourage further adjustment of industrial structure. It will provide employment to a large group of herdsmen in the industry, service, and other sectors. Thus, the heavy load on the prairie and vegetation could be reduced. It will not only protect the ecological environment but also effectuate sustainable development, which is proved to be a successful solution that kills two birds with one stone.
3. After the Qinghai–Tibet Railway is put into operation, it will consequentially become the main way for tourism and freightage into Tibet. It will reduce the enormous transport motorcades. Thus, it will contribute to reducing the exhaust emission of motor vehicle and improving the air quality in the Qinghai–Tibet Plateau.
4. Tourism is one of the most promising and new economic departments across the world. This area has a unique natural environment and good exploitation conditions so that a series of novel and unique plateau scientific exploration tourist projects could be developed. For example, the location of "the roof of the world", original natural landscape, unique landform, rare psychrophyte, and precious plateau animals are all precious resources to attract foreign and domestic tourists.

6.5.1.2 Adverse Effects

1. Thirty water supply stations are present along the railway, which can discharge up to 5,016,420 tons of wastewater annually, and six of them have 4,204,130 tons of annual discharge quantity. The wastewater coming from water supply stations, especially the six ones, may leave slight effects on the partial aquatic ecosystem.
2. Only two gas and fuel oil boilers are present along the line in Lhasa and Golmud. The atmospheric pollutants discharged from Lhasa station and Golmud station are 647.19 and 20.84 t/a, respectively. The atmospheric pollutants may leave slight effects on the local natural ecosystem.
3. In the period from Golmud to Lhasa, the quantity of atmospheric pollutants from motor vehicles is 3978.5 t/a, which may leave slight effects on ecosystem in the strip area along the line.
4. The noises from fixed equipment in Lhasa and Golmud stations and motor vehicles may leave slight effects on the urban ecological environment.
5. Household garbage from Lhasa and Golmud stations may leave slight effect on the urban ecological environment.

6.5.2 Corresponding Measures Taken in the Qinghai–Tibet Railway Route Selection

Human activities are not isolated, and they consequentially cause changes in natural environment. When selecting route, we should consider the effects on environment from the construction and operation of railway. Thus, we take corresponding measures when we design the routes.

6.5.2.1 Wildlife Protection

Wildlife passageways are set when designing the route. As a main way to protect wildlife in nature reserves, they are widely applied along the line and play an important role.

1. Wildlife passageways setting principles

Principle-making takes ecological environment protection as a basic factor, wildlife protection in the Qinghai–Tibet Plateau as an aim, and the construction of the Qinghai–Tibet Railway as a premise. Measures are adjusted to fit local conditions, distribution is set scientifically and reasonably and wildlife passageways are applied. The principles are shown as follows:

a. insist on the same importance of the Qinghai–Tibet railway construction and wildlife protection;
b. fully consider wildlife habits;
c. make full use of surrounding topography to set passageways;
d. make full use of railway project or make little adjustment;
e. comprehensive considerations on setting passageways, which means considering gentle slope of subgrade and shape of bridge.

2. General distribution and form of wildlife passageways

From the situation of wildlife passageways set along the Qinghai–Tibet Railway, passageway locations mainly distribute in the period from south of Naji Tal (DK909+450) to Buqu third bridge (DK1396+023) in north of Tonglha Mountains. It is 487 km in length. There are 33 passageways in total (Table 6.5). The total width of the passageways is approximately 24,830 m, with 2840 m overground tunnels (11.4%), 9860 m under bridges (39.7%), and 12,130 m on the gentle slopes (48.9%). The wideness of passageways makes up 5.1% of wildlife distribution area (487 km), and accounts for 2.2% of the new built railway (1142 km).

The Qinghai–Tibet Railway, as an important project running through the Qinghai–Tibet Plateau from south to north, consequentially leaves effects on the communication of animal populations on the different side of railway. Thus, from the aspect of passageway forms, engineering designs of the Qinghai–Tibet Railway, topography, and wildlife habits are fully considered, and three forms of passageways, such as those above tunnels, under bridges, and across gentle slopes are applied.

a. In districts with high relief ridge and animals passing through, upper section of the tunnel project is used for wildlife passageways.
b. In districts with poorly dissected topography, distribution of zanjon, river, and valley and animals passing through, and passageways are mainly under the bridges combined with railway engineering.
c. In districts with gentle topography and animal passing through, across passageways are set, which means going on the railway roadbed.

3. Wildlife passageways setting

a. Alpine mountain fauna

This type of wildlife often takes steep alpine mountains as their habitats and is characteristically good at climbing but not sustained sprinting. The main animal species are bharal, argali, white-lipped deer, snow leopard, lynx, and brown bear. The districts where they live with railway passing through are Kunlun Mountain, Wulanwula Mountain, and Tonglha Mountains. These mountains are east–west trend and the railway crosses them from north to south, making influences on the east–west trend migration and communication of the fauna. The form of passageways for alpine mountain animals is above tunnels, and they could also be used by intermontane lake basin and broad valley hirst fauna. Two ridge passageways are present, in Kunlun Mountain, and Fenghuo Mountain, respectively. In the route

Table 6.5 Distribution table of wildlife passageways along the Qinghai–Tibet Railway

Number	Location	Start and terminal point	Passageway width/m	Passageway form	Interval/m
1	4 km east to pass of Wangbao Valley	DK909 +200–DK909 +700	500	Across	–
2	Sanchahe Grand Bridge	DK918 +489–DK919 +150	600.7	Bridge	8.8
3	Xiaonanchuan second Dongzhi Bridge	DK930 +346–DK930 +594	247.6	Bridge	11.2
4	Xiaonanchuan forth Grand Bridge	DK937 +232–DK937 +836	604.5	Bridge	6.7
5	1 km from Wangkunlun Sation	DK961 +300–DK961 +800	500	Across	23.5
6	Luanshigou second Grand Bridge	DK969 +531–DK970 +009	477.8	Bridge	7.7
7	Kunlun Mountain Tunnel	DK972 +200–DK974 +700	1660	Tunnel	2.4
8	Nangoukou	DK984 +000–DK984 +740	740	Across	9.9
9	First Kunnan Bridge	DK984 +740–DK985 +316	575.9	Bridge	9.9
10	South of Budongquan	DK1006 +500–DK1007 +500	1000	Across	21.2
11	4 km south to Sonam Dargye Station	DK1036 +200–DK1037 +200	1000	Across	28.7

(continued)

Table 6.5 (continued)

Number	Location	Start and terminal point	Passageway width/m	Passageway form	Interval/m
12	Chumar River 70 Dannan	DK1053 +000–DK1054 +000	1000	Across	15.8
13	Chumar River Grand Bridge	DK1062 +238–DK1062 +846	608.6	Bridge	8.2
14	Wudaoliang to middle section of Chumar River	DK1076 +000–DK1077 +000	1000	Across	13.2
15	Middle bridge of Tongxin River to Hongliang River Bridge	DK1101 +000–DK1102 +000	1000	Across	24
16	Hongliang River Grand Bridge	DK1107 +400–DK1107 +845	445.1	Bridge	5.4
17	South to Xiushui River	DK1124 +500–DK1125 +500	1000	Across	16.7
18	Grand Bridge	DK1143 +197–DK1143 +903	704.7	Bridge	17.7
19	Daxigou Grand Bridge	DK1149 +220–DK1149 +860	639.3	Bridge	5.3
20	South slope of Daxigou	DK1150 +420–DK1150 +810	390	Across	0.6
21	Fenghuoshan Tunnel	DK1154 +830–DK1156 +010	1180	Tunnel	4
22	Xiacanniqiqu Bridge	DK1189 +703–DK1190 +167	462.6	Bridge	33.7

(continued)

Table 6.5 (continued)

Number	Location	Start and terminal point	Passageway width/m	Passageway form	Interval/m
23	2 km east to 85 Daobang	DK1195 +500–DK1196 +500	1000	Across	5.3
24	3 km east to 86 Daobang	DK1203 +500–DK1204 +500	1000	Across	7
25	Nuorigouqu Bridge	DK1218 +129–DK1218 +347	217.6	Bridge	13.6
26	Tuotuo River Grand Bridge	DK1231 +208–DK1232 +594	1386.5	Bridge	12.9
27	Kaixinling Grand Bridge	DK1245 +887–DK1246 +548	660.2	Bridge	13.3
28	Tongtian River Grand Bridge	DK1275 +803–DK1276 +507	704.7	Bridge	29.3
29	South of Tongtian River	DK1293 +500–DK1294 +500	1000	Across	17
30	North of Yanshipin	DK1326 +490–DK1326 +870	379.3	Bridge	32
31	Buqushun River Grand Bridge	DK1360 +300–DK1361 +140	739.4	Bridge	33.4
32	North of Queqiao	DK1379 +500–DK1380 +500	1000	Across	18.5
33	Third Buqu Bridge	DK1395 +677–DK1396 +023	345	Bridge	15.2

design, the top of two tunnels can be used as passageways, and the principles are to go through the ridge with less disturbances to mountain form.

First, when the route passes through the stone riprap on the north slope of Kunlun Mountain, two tunnel design schemes are used: one is long tunnel scheme, which builds the Kunlun Tunnel in length of 1600 m, and the other is to build five short tunnels in aforementioned district. From the aspect of general conditions, application of long Kunlun Tunnel scheme leaves a passageway in wideness of 1.6 km for the east–west trend migration and communication of Tibetan antelope, kiang, bharal, Tibetan gazelle, and wild yak. It ensures the migration and communication of animal populations and prevents damages on tunnel environment (e.g., noises and light). Thus, it turns out to be the better scheme.

Second, Fenghuo Mountain Tunnel meets the requirement of wildlife migration and communication of kiang, bharal, and Tibetan gazelle.

b. Intermontane lake basin and broad valley hirst fauna

The animals living along the valleys are usually highly vigilant, and good at fast running in long distance. The main species are Tibetan antelope, wild yak, Tibetan gazelle, and kiang. They often find food in mountain valleys and river valleys. Excepting the hot days in summer, they rarely climb up to alpine mountains. They find food at dawn and at dusk, and sleep, and rest in mountain recessions in the evening and at noon. These animals have large foraging range, usually along the river valleys where vegetation is abundant. According to valley wildlife population distribution along the Qinghai–Tibet Railway, the areas under largest influences are Yeniugou Valley, Chumar River Valley, and Tongtian River Valley. Thus, the passageways for intermontane lake basin and broad valley hirst fauna are mainly set on gentle slope on roadbed or under bridges.

Form as across over gentle slopes: they are set in districts where animals pass frequently and the topography is gentle, 14 in total. Two passageways are present east of Wangbao Valley and Wangkun station. Wild yak, argali, and bharal pass through them back and forth, and the populations are relatively small. Thus, the passageways are set as 500 m. The rest 12 passageways are main route for migration and activities of kiang, Tibetan antelope, Tibetan gazelle, and wild yak. Their populations are relatively large, so the plan requires the passageways in wideness of around 1000 m. Considering the railway operation and safety for wildlife to pass the railway, guard bars are set on the two sides of these 12 passageways. The height of guard bars is no less than 2 m, and the length is up to practical situation.

Form as under the bridges: form as under the bridges and across over gentle slopes are supplements for each other. Many medium and large bridges serve as animal passageways in the whole route design. Of these, animals can pass through 17 bridges, in a total length of 9859.5 m. The aforementioned bridges passageways all meet the demands of migration, foraging, drinking water, and population communication of wild yak, kiang, bharal, Tibetan antelope, white-lipped deer, and brown bears. Of these, ten exceptionally large bridges span more than 500 m, making 7284.5 m in total; and seven bridges with length of 200–500 m, making 2575 m in total.

c. Wetland fauna

The Qinghai–Tibet Railway passes through seasonal buffer region of Yarlung Tsangpo Nature Reserve in middle and lower reach of Lhasa River in the district from Doilungdêgên to Lhasa. This district is home to birds such as black-necked cranes, and birds such as Tadorna ferruginea live near Damxung. The agriculture here is mostly developed all across Tibet, and the people activities are quite frequent.

The railway route goes on the west side of road in Damxung (the east side is the wetland for bird habitat). Thus, it makes no effects on the habitat. But construction scheme in this district still should be fully optimized to reduce damage on wetland and protect living and reproducing of birds here.

In the vicinity of a leather factory in urban area of Lhasa, the route passes through a winter foraging habitat of black-necked cranes, anser indicus, tadorna ferruginea, and larus brunnicephalus. Route selection has been fully optimized to ensure the shortest distance to get the other side of Lhasa River. However, the alternative scheme in Dagzêgoes goes along the Lhasa River and leaves great influences on bird habitat, which is suggested to avoid. If effective environmental protection and management measures are taken, construction time is arranged reasonably and various construction activities are standardized, the disturbance on birds could decrease. When necessary, measures should be taken to reduce train operation noises.

4. Wildlife passageways safeguard measure

a. Laws and regulations measure

Wildlife passageways have double effects. It is the important engineering facility in railway construction and the important facility in nature reserve construction. Specific location and number of each passageway should be confirmed. Regulations, systems, and items which have corresponding legal force should be established actively. Construction, management, and protection should be carried out. And they should be included in important contents of nature reserve management. For the activities and behaviors which damage the wildlife passageways or cause difficulties in migration and communication of wildlife, the legal responsibility should be pursued in accordance with the laws and regulations.

b. Organizational measure

To guarantee the construction of passageways, management should be enhanced. Main problems in passageway construction should be harmonized and solved by establishment of specific implementation measures, regulations, and requirements. It is mainly aimed to supervise the construction organization and individuals and to strictly forbid hunting, disturbing wildlife, and damaging forest and grass vegetation, blind excavation, and polluting rivers and lakes. And it encourages those people to cooperate with the local government and the nature reserve management department for supervision and inspection. Striking zone boundary sign in nature reserves should be set as a distinct mark such as the wildlife protection advertisement sign at the entrance of wildlife passageways.

c. Technical safeguard measure

Detailed engineering design should be put forward in wildlife passageway construction. It requires technical innovation and combination of engineering measures and biological measures. When applying bridge passageways, the problem of vertical height from ground and wideness between piers, increase of approach bridge length should be considered to avoid the situation that flood fills in the riverway in wet season in summer, which may affect the pass of animals. When setting bridge passageways, 200 m of protective screening should be added on both sides of approach bridge roadbed. When applying form as across gentle slopes, it should avoid damaging the surrounding original landform and the surface vegetation, and should maintain, or recover the original landscape. As for the form of tunnel, attention should be paid to make little change on the original landscape at the entrances of both sides. At least 200 m of protective screening should be set above the two-side entrances of short tunnels.

d. Other measures

Avoidance of damaging the vegetation and landform, construction of passageway tunnels, bridges, and roadbeds during construction time should ensure the ecological recovering project (vegetation and water resources). Attention should also be paid to the management of construction operation, construction materials, and transport vehicles to ensure them not to cross the experimental region of nature reserves, and not to disturb the pass of animals.

6.5.2.2 Others

1. Protect vegetation and prevent further deterioration of meadow. Vegetation is an important factor to maintain the ecosystem stabilization in the nature reserves. Thus, the existing grass and psammophyte should be first protected, and protection forest should be built. According to carrying capacity, measures should be carried out. For example, rotation grazing should be followed, livestock should be raised scientifically, rodent, and pests disasters should be prevented and further deterioration of meadow should be controlled.
2. Make reasonable use of wetland resources and prevent desertification and water and soil loss. Strictly forbid further land reclamation in lake basin area and maintain or reasonably decrease existing cultivated land and reuse some abandoned plough to plant grass artificially to partly solve the conflict of lack of winter and spring pasture. Carry out measures such as no graze or limited graze

on deteriorated meadow. Make reasonable use of land resources and prevent water and soil loss and gradual expansion of desertification area.

3. Protect and make reasonable use of water resources. This area is arid region, so the water resources should be made a good use. Clean the water which has been polluted already, increase water-processing equipment, refuse to build factories with severe pollution, and the pesticide utilized in husbandry should be replaced by nontoxic types gradually. Apply biological comprehensive treatment to replace the former high toxic residues of pesticides to prevent water resources pollution.
4. Protect and make reasonable use of wildlife resources. Because the wildlife resources have been heavily damaged, the amount of fishing should be controlled reasonably, bird nature reserves should be expanded, hunting of rare wildlife should be forbidden, and nice habitat environment should be provided.
5. Enhance whole quality-oriented education, set up environmental awareness in public, and enhance the establishment of relative laws and regulations to ensure the healthy development of ecosystem.

6.6 Summary

The ecosystem along the Qinghai–Tibet Railway is fragile, and the construction and operation of railway consequentially leave influences on it. Nature reserves act as an important component of ecosystem, protection of them shows significant environmental benefit and ecological meaning. The Qinghai–Tibet Railway route schemes fully reflect the environmental protection concept. It also protects the wildlife and plateau vegetation effectively.

At the beginning of design, environmental protection is fully considered as the important content in constructing the first-class plateau railway in the world. Harmony between construction project and environment is required. To build ecological environmental protection railway with plateau features, permafrost regions under effective protection is ensured, river pollution is prevented, adverse effects on the migration of wild animals are avoided, and damage on the landscape beside the railway is averted.

The concept of environmental protection is persisted throughout the route selection to reduce the effects on the environment from construction as less as possible. Although under extremely difficult conditions, routes still avoid the core regions of reserves and most of the experimental regions. It mainly runs in the buffer regions. In design process, several alternative schemes are put forward when route passes through Hoh Xil Nature Reserve, and the scheme which has minimum intervention on reserves and minimum disturbance on landscape is applied in the end. In TARs, Namtso—the sacred lake—and its core region are avoided. Yambajan scheme is applied to avoid Linzhou–Pengbo Black-Necked Crane Nature Reserve.

Routes avoid areas of swamp-type and lake-type wetland to the greatest extent in order not to damage the water resources and wild animal habitats. In the district where wildlife migrates, three different forms of passageways are applied such as those across gentle slopes, under bridges, and above tunnels to ensure the migration.

Routes are usually not far from the Qinghai–Tibet road to avoid double incision on whole plateau ecosystem by road and railway. To prevent the influence of interaction effect from road and railway on the stability of frozen earth, 100 m is confirmed to be the reasonable minimum distance between road and railway through the research work.

In the mountainous and hilly areas, routes avoid roadbed of slope to the greatest extent to reduce damage on stability of original mountains and artificial gelifluction and thaw slumping. Routes avoid changing and cutting the direct surface runoff. When crossing the rivers, wide angle, and large span of bridge are applied to avoid pressing the riverbed and cross section of river. On the cutbank where the mountains are steep and rivers swerve, roadbed is replaced by bridge to avoid excavation in mountains, damage on original vegetation, and effects on stability of massif.

In vertical section design, when effectiveness of roadbed engineering treatment is not good, roadbed is replaced by bridges in the districts where the ground temperature is high, or the ice content is large, or the settlement amount is large, or the lithology is unstable. In windy and sandy districts, roadbed forms and fill height are selected reasonably to protect vegetation against desertification.

On the base of the aforementioned route selection principles, the best routes of economical, reasonable, and minimal intervention are selected in the end when railway goes through the nature reserves.

References

1. Agricultural Resources Division Office, Qinghai Province. (1997). *Soil of Qinghai*. Beijing: China Agriculture Press. (in Chinese).
2. Bai, J. Q., Mei, L., & Yang, M. L. (2006). Geothermal resources and crustal thermal structure of the Qinghai–Tibet Plateau. *Journal of Geomechanics, 12*(3), 354–362. (in Chinese).
3. Cen, S. X., Gong, Y. F., & Chen, Y. Y. (2007). Climatic features of atmospheric heat source over the Qinghai–Xizang Plateau. *Journal of Chengdu University of Information Technology, 22*(3), 369–373. (in Chinese).
4. Chen, L. T. (1998). Test and application of the relationship between anomalous snow cover in interspring over the Qinghai–Xizang Plateau and the first summer rainfall in southern China. *Quarterly Journal of Applied Meteorology, S1,* 1–8. (in Chinese).
5. Chen, M. X., Wang, J. Y., & Deng, X. (1994). *Geothermal resource of China: Forming characteristics and potential assessment*. Beijing: Science Press. (in Chinese).
6. China Railway First Survey & Design Institute Group Co., Ltd. (1994). *The manual of routes* (2nd ed.). Beijing: China Railway Press. (in Chinese).
7. China Railway First Survey & Design Institute Group Co., Ltd. (2000). *Comprehensive planning at nature reserve in Nacuo*. Beijing: Internal Information. (in Chinese).
8. Chinese Geological Atlas Editorial Board. (1996). *Geological atlas of Chinese*. Beijing: Geological Press. (in Chinese).

9. Dai, J. X. (1990). *The climate of the Tibetan Plateau.* Beijing: Meteorological Press. (in Chinese).
10. Department of Natural Ecological Protection, National Environmental Protection Administration. (2006). *China nature reserve list (2005).* Beijing: China Environmental Science Press. (in Chinese).
11. Duo, J. (2003). Basic characteristics of the Yambajan geothermal field: A typical high-temperature geothermal system. *Engineer Science, 5*(1), 42–47. (in Chinese).
12. He, Y. J., Cui, G. F., Zou, D. L., Zheng, J., Dong, J. S., & Li, Y. B. (2007). Plant species diversity of main forest community types in San Jiangyuan National Nature Reserve. *Forest Research, 20*(2), 241–245. (in Chinese).
13. Hu, X. C., Sun, J. D., Yao, Z. H., et al. (2003). Research on the influence of geothermal activities and exploitation of geological environment in Tibet. *Journal of Mountain Science, 21*(12), 45–48. (in Chinese).
14. Li, W. Y. (1988). *Quaternary plants and environment in China.* Beijing: Science Press. (in Chinese).
15. Li, B. Y. (1996). *Natural environment west of Cocoa.* Beijing: Science Press. (in Chinese).
16. Li, W. H. (1998). *Ecological system and optimal utilization models in Tibet Plateau.* Guangzhou: Guangzhou Science and Technology Press. (in Chinese).
17. Liang, S. H., Chen, J., Jin, X. M., et al. (2007). Regularity of vegetation coverage changes in the Tibetan Plateau over the last 21 Years. *Advances in Earth Science, 22*(1), 33–40. (in Chinese).
18. Liu, F. G., Zhang, H. F., & Zhang, Y. L. (2005). A study on the resource use and environment policy in the Three River's Source Nature Reserve. *Journal of Qinghai Normal University (Natural Science edition), 2,* 86–91. (in Chinese).
19. Ma, T., Zhou, J. X., Zhang, X. D., et al. (2007). Preliminary studies on characteristics of vegetation distribution along the Qinghai–Tibet Railway. *Research of Soil and Water Conservation, 14*(3), 150–154. (in Chinese).
20. Qin, D. H., Ding, Y. H., Wang, S. W., et al. (2002). A study of environment change and its impacts in western China. *Earth Science Frontiers, 2*(2), 321–328. (in Chinese).
21. Shu, L., Lou, W. H., & Wang, L. J. (2003). An analysis of the ground temperature of Yambajan Tunnel. *Journal of Glaciology and Geocryology, 25*(1), 24–28. (in Chinese).
22. The China Society on Tibet Plateau. (1995). *Collected papers of seminar on the Tibetan Plateau and global change*. Beijing: Meteorological Press. (in Chinese).
23. The Expert Panels of the Qinghai–Tibet Projects. (1995). *Research on the formation and evolution environmental changes and ecosystem of the Qinghai–Tibet Plateau.* Beijing: Science Press. (in Chinese).
24. Wang, Q. Q., Qin, N. S., Tang, H. Y., et al. (2007). Study on climate change facts and their characteristics in the Qinghai Plateau in the recent 44 years. *Arid Zone Research, 24*(2), 234–239. (in Chinese).
25. Xiong, R. W. (2009). Exploration of ecological protection in nature reserves. *Yunnan Science and Technology Management, 1,* 46–47. (in Chinese).
26. Yang, Y. F., Jiang, H., Niu, F. J., & Zhao, H. Y. (2007). Space–time variation analyses of air temperature over the Qinghai Xizang Plateau in warm and cold seasons. *Plateau Meteorology, 26*(3), 496–502. (in Chinese).
27. Yi, Z. M., Zhou, J. X., & Zhang, X. D. (2006). Analysis of hydrologic conditions along the Qinghai–Tibet Railway. *Technology of Soil and Water Conservation, 4,* 14–16. (in Chinese).
28. Yuan, L. (2004). Some problems needing attention in investment in strong motion area. *Journal of Natural Disasters, 13*(2), 130–133. (in Chinese).
29. Zhou, J. X., Yi, Z. M., Li, D. X., et al. (2007). Distribution patterns of species diversity of natural vegetation along the Qinghai–Tibet Railway. *Journal of Soil and Water Conservation, 21*(3), 173–187. (in Chinese).

Chapter 7
Railroad Route Alignment in Geothermal, Aeolian, and Snowdrift Areas

Abstract Geothermal activity, dust storms, and snowdrift severely affect railway lines. Rich geothermal resources, plateau desertification, local-area development, semi-fixed and shifting sand dunes, and heavy snow are common along the Qinghai–Tibet Railway. Given these phenomena, geological line selection in the Qinghai–Tibet Plateau warrants attention.

Keywords Railway route alignment design · Geothermal resource · Dust storm Snowdrift · Scheme comparison

7.1 Road Route Alignment in Geothermal Areas

7.1.1 Distribution of Geothermal Resources Along the Qinghai–Tibet Railway

According to the temperature characteristics, geothermal resources can be classified into three groups as high-temperature (>150 °C), moderate-temperature (90–150 °C), and low-temperature (20–90 °C) geothermal resources, which exist as vapor, water–vapor mixture, and water, respectively. Low-temperature geotherms are of three types: warm water (25–40 °C), warm–hot water (40–60 °C), and hot water (60–90 °C). High-temperature geotherms generally occur at the boundary of plates with strong crustal activities, such as volcanic eruption, earthquake, and magma intrusion.

The formation and distribution of geothermal resources in China are controlled by local and global tectonics. Geothermal resources in the Tibet Plateau are distributed within the Bangong Co-Nujiang structure zone. Numerous strong magmatic activities, energy-rich high-temperature geothermal sources, and some well-known geothermal fields, such as Yambajan, are located in the geothermal belt spanning the south of Tibet. More than 200 hot springs and some hydrothermal geysers and fumaroles are distributed along the Yambajan–Nagqu rift tectonics belt, and the boiling springs and fountains here can be 90+ °C. Outside of this area, a

J. Li et al., *Geological Line Selection for the Qinghai-Tibet Railway Engineering*,
https://doi.org/10.1007/978-3-662-55572-9_7

warm spring (>50 °C) is located near Buqiangge Station of the Qinghai–Tibet Railway.

7.1.2 *Engineering Characteristics of Geothermal Fields*

The Tibet Plateau is located over the tectonic belt formed by the collision of the Indian Ocean plate and the Eurasian plate. Consequently, the local topography and geomorphology are complex and varied, with such structures as strips, fractures, and fault depressions, all of which often trigger earthquakes and volcanic eruptions. Moreover, water is abundant, with lakes feeding the geothermal systems; this, in turn, induces geo-disasters such as landslides, collapse, debris flow, earthquakes, and volcanic eruptions.

The exploitation of geothermal resources and natural activities have further changed the geological environment, which in turn accelerates, slows, or consolidates the material movement on the slope, inducing numerous geological disasters, such as changes to the structural features of geological bodies, collapse and landslides, land subsidence of dewatered aquifers, and soil salinity due to geothermal fluid erosion.

7.1.2.1 Geothermal Activity and Landslide

A landslide refers to the sliding down of a rock soil mass along a weak surface under the action of gravity. The characteristics of this sliding process are influenced by the available geothermal energy. Generally, active geothermal regions are groundwater-rich and thus groundwater easily infiltrates the sliding surfaces, triggering the landslide of rock soil masses.

Landslides triggered by geothermal activities have the following characteristics: First, rock and mineral erosion occurs at a high rate in active geothermal regions, where some high-strength rocks have eroded over time, forming weak layers along the structural surfaces and the passageways of the uplifting thermal fluid. For example, eroded granite forms the kaolinite soil layer, and the chloritization and epidotization produce weak layers in basic and ultrabasic rock mass and may be susceptible to landslides and collapse. In addition, heavy groundwater infiltrating in structural layers reduces the friction, which under the action of gravity, triggers landslides, as is the case in Tibet.

7.1.2.2 Chemical Sediment and Soil Erosion

The fixation effect and cementation of chemical sediments by geothermal activity is a surface process that is a product of the human environment. Not only can it form slopes and mountains, but it also forms the sinter landscape.

Sinter sediments are common in the active geothermal areas of Tibet. The cementation of unconsolidated sediments around these areas or the formation of a hard shell that ameliorates the inner structural stress reduces the velocity of soil erosion, stabilizes rock masses, and protects slopes.

7.1.2.3 Soil Salinization Due to Geothermal Activity

Highly mineralized water is the most common thermal fluid released in geothermal fields. Such water increases soil salinity, an important effect of geothermal activity on the local geological conditions. In Nagqu geothermal field, a famous geothermal field 1 km south of Nagqu country, the degree of mineralization of the thermal fluid is high, with an average level of 2000 mg L^{-1}. Here, the thermal fluid flows 3 km along the valley to Ciqu, resulting in the formation of a saline-alkali soil layer of width 200–400 m and area 120 hm^2; the properties of this soil are in stark contrast with those of its surroundings. In addition, such saline sedimentation and alkalinization harden the soil, eventually resulting in desertification.

7.1.2.4 Ground Subsidence

Ground subsidence is the unfavorable result of the dewatering of aquifers because of the excessive extraction of the underground hot water; examples of subsidence of up to 260 mm can be found in Yambajan geothermal field.

Recent data from Yambajan reveals that the groundwater levels have decreased by more than 10 m in all geothermal fields, and so has the head of the confined water. To prevent land subsidence, the extraction rate of groundwater resources should not exceed the recharge rate.

7.1.3 Influence of Geothermal Distribution

7.1.3.1 Influence of Regional Tectonics

Regional tectonics and the deep crustal structure strongly influence the temperature and distribution of geothermal energy. In addition, the thermophysical properties of rock as well as the presence of volcanoes, magmatism, and groundwater influence the geothermal distribution.

In addition, along the Qinghai–Tibet Railway, the tectonics are intense and seismic activities frequent because of the complex tectonics. Faults and deep-faults control the local geothermal temperature and channel the geothermal fluids. Hot springs and partially abnormal geothermal areas develop along the faults when deep hot water upwells along the fracture.

7.1.3.2 Influence of Groundwater

The distribution of geothermal energy is strongly obstructed by the groundwater activity, particularly in hilly areas. The strength of the influence of groundwater activities on geothermal energy depends on the terrain, structure, lithology, and subsurface runoff thickness of the aquifer. Occasionally, a negative geothermal gradient forms in the mountain and along the edge of the mountain area because of strong groundwater (cold water) runoff.

7.1.3.3 Influence of Terrain and Precipitation

Some studies have indicated that geothermal energy at depths exceeding 1000 m is influenced by terrain and rainfall; that is, the tectonic structure controls the influence depth, especially in mountainous areas. In addition, the rocks have survived years of strong weathering during the Qinghai–Tibet Plateau upheaval. When rocks crack, rainfall permeates directly into the soil, cooling the geothermal energy source. This cooling function decreases with soil depth. Furthermore, topographical relief produces lateral heat release, which also exerts a cooling effect on the temperature distribution.

Because of the warm climate and abundant rainfall, vast quantities of rainwater in south China permeate into the mountainous areas, resulting in strong runoffs. These runoffs affect rock cooling, especially in the karst areas. Therefore, low ground temperature regions are distributed in mountainous and hilly areas. Low ground temperature is dependent on the topography, rainfall infiltration, groundwater runoff, and tectonic fragmentation, and the effects of these factors decrease with depth, eventually disappearing at a certain depth; this depth is dependent on the characteristics of the regional geological structure.

7.1.3.4 Influence of Springs

Geothermal distribution is also affected by the distribution of hot springs. Underground hot water is often exposed on the surface of the fault structure, resulting in small abnormal-temperature regions around such surfaces; this abnormal regional temperature has no effect over large regions. Thus, hot springs exert a weak but obvious effect on the distribution of geothermal energy at depths of 1000–3000 m.

7.1.4 Influence of Geothermal Activity on the Qinghai–Tibet Railway

The engineering-geological features of geothermal regions influence those of the Qinghai–Tibet Railway as follows:

1. Adverse effects of high ground temperature

A high ground temperature adversely affects railway construction. Hot groundwater can spout out during constructing in geothermal areas. Furthermore, constructing tunnels under high temperature is difficult; these two factors put the workers at risk of injury or disease, and the mechanical equipment should be protected from the high temperature as well.

Yambajan tunnel is located at the intersection of the EW-trending fault structure zone of Mozhugongka and the NE–SW-trending fault structure zone of Nyainqentanglha; hence, the rocks here have numerous joints, such as the following:

J1 (NS 70°W, 70°E) runs through all districts. This gapping joint fissure has a slightly open structure. The spacing is 1–4 m, and the local intensive department is 0.2–0.5 m. This joint is a water-storing structure.

J2 (N 45°E, 70°N) runs through all districts. This gapping joint fissure has a slightly open structure and is confined by J1 at a spacing of 0.2–0.5 m.

J3 (N 30°W, 30°S) runs through all districts. Its structure is slightly open and belongs to gapping joint fissure at a spacing of 0.2–0.5 m.

J4 (NW 80°S) gradually develops southwards. This impermeable gapping joint fissure is also confined by J1, at a spacing of 0.4 m.

South of the fault structure zone of Nyainqentanglha, the Damxung–Yambajan fault zone has formed a tectonic rifted basin, which is an active regional scarped fault zone; hot springs have developed along the faults here. This region has a seismic-intensity classification of 9°, and fault activity here induces collapse and landslides.

A tunnel is located in Doilung Qu Valley, on the edge of Damxung–Yambajan fault zone, where earthquake activity has occurred since the late geological period. Joint J3 is long and large, with a low-angle rock mass on the left bank; it has developed bedding faults, which induces large-scale landslides. The slope here is not too steep. By contrast, the slope of the right bank is relatively steep and thus the Qinghai–Tibet Railway runs on this bank. The bedrock is severely weathered on the slope, the cleavage of the joint is exposed, and many rock masses have developed many unloading joints, producing perilous rock and collapse. Rock accumulation is common at the slope toe.

Generally, the abnormal geological phenomenon includes hydrothermal activities such as spring, fountain, and air injection in the geothermal energy area. And the main geological factors are magma action and deep circulating groundwater. The ground research in Yambajan tunnel indicates that there are no geological phenomena such as spring, heat release, and air injection in bedrock. However, fissure water freezes severely when it drains from the ground. In addition, the hot water has not been discovered under the construction of drilling and service incline in the tunnel. However, the geological phenomena of spring, heat release, and air injection are common but the air temperature has a small influence on the temperature of groundwater.

Although tectonic activity has a series of influence on the railway route alignment design, the field survey, and monitoring discover that there are not large fault belt and folds. And because of granite existence, the rock is not influenced severely by tectonic structure to develop little fault belt and joint. So, the outcrop groundwater is the shallow circulating water. The area without the condition of deep circulating water and the leakages often freezes at winter. The hot water also does not convey to the tunnel in the geothermal field because there is no large fault belt in the tunnel wall rock connecting with Yambajan tunnel. This analysis indicates that the geothermal energy has little influence on Yambajan tunnel.

High temperature has great influence on travelers and the trains when passing the tunnel.

2. Geothermal regions, with complicated geological conditions, in mostly locate in the tectonic active area and are often accompanied by active fault zone development, which causes affect the railway route design and construction.

The geological areas crossed by the Qinghai–Tibet Railway are mainly near Yambajan. Based on regional geological surveys and Yambajan geothermal-geological surveys, Yambajan lies at the intersection of the tectonic band. It includes EW-trending tectonic belt in Tibet Plateau, actuates tectonic belt in Jia-Na-Nian-Qing, the eta-type tectonic system in Tibet Plateau and SN trending tectonic belt in Bange-Kangma.

So, the region lies in the strong earthquake zone and the action is so powerful and frequent. Basing on distributing characteristics of the earthquake, the NE trending fault structure and the SN in the east of accurate tectonic belt play a dominant role in the region (Fig. 7.1).

The Qinghai–Tibet Railway extends from the structural basin of Tibet Plateau (the structural basin of Damxung–Yambajan), joined to the basin of Doilung Qu by upward steep tunnel and Lhasa River Valley by a downward steep tunnel. Yambajan tunnel lies at 9° intensity earthquake region, which is on the edge of the structural valley outcropping grandiosity in the Himalayan period. The physical regolith of granite is thicker, because the primary joints and structural joints develop under the function of thermal dilation, cool shrink, frost heave on fissure water, and ice split. In addition, the section belongs to erosion landscapes including deep river valleys, steep slope, unstable slope, and unload joint along the gradient slope in some section. Furthermore, the deep developing fissure water is mainly stored in the open and micro-open joint trenching SN. So, the slope is unstable, and easy to develop collapse landslide, collapse, rock accumulation, steep perilous rock with erosion, and rock fall.

Figure 7.2 shows that Danguoguo–Maruguo fault is located 6 km away from the Yambajan tunnel trenching SN, tending NW, and dipping 80. And, Youmu fault lies 6 km away from the tunnel tending EW and dipping 50. Both of faults are tectonic faults acting slightly in the Late Pleistocene. The outcropping granite is influenced by two faults near the tunnel, and it develops some joints trenching SN, NE, and NW and crevices. However, the fault has not been discovered during geological exploration.

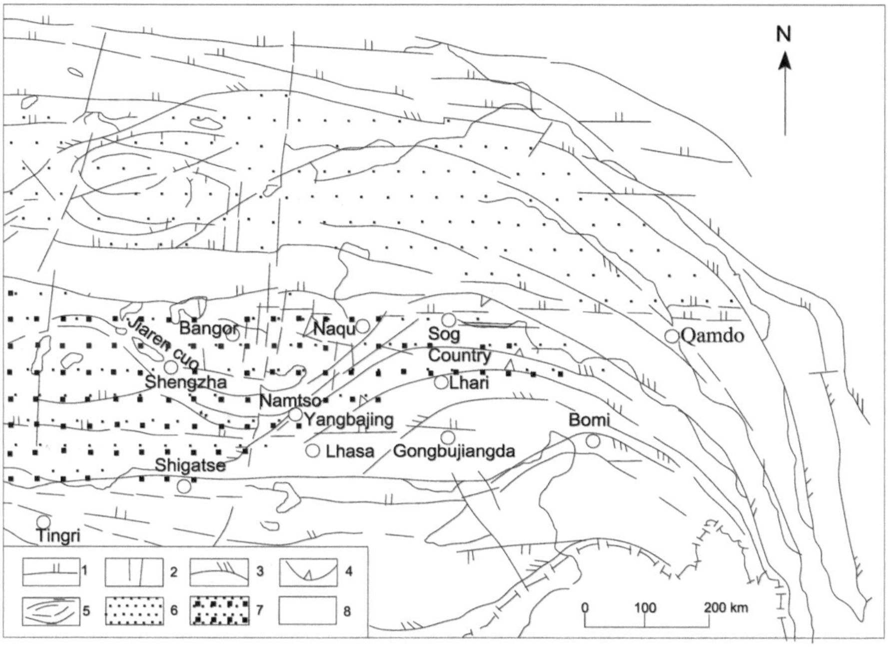

Fig. 7.1 Tectonic system in east of Tibet

7.1.5 Principles of Selecting Line in the Geothermal Areas

The principles of selecting a line in geothermal areas are as follows:

1. The dodging method is used in road route alignment design when the main location of the important structure is selected in geothermal areas. The surveys of engineering geology, hydrogeology, and geothermal activities should be carried out, based on the actual surrounding. When the dodging method is difficult to take, the road should select the edge of the geothermal areas to pass. And it also should select the narrowest area when it is difficult to avoid the geothermal area.
2. Line selection should combine other engineering geology to avoid the fault movement and the hot spring spots. For line choice at landslides, the road should avoid the large ancient landslide at first. For the line that cannot avoid such area, corresponding measures should be taken on the basis of understanding the nature and stable state.
3. Railway engineering should take subgrade, culverts, and other small engineering. And it is unfavorable to choose deep tunnels or large bridges.

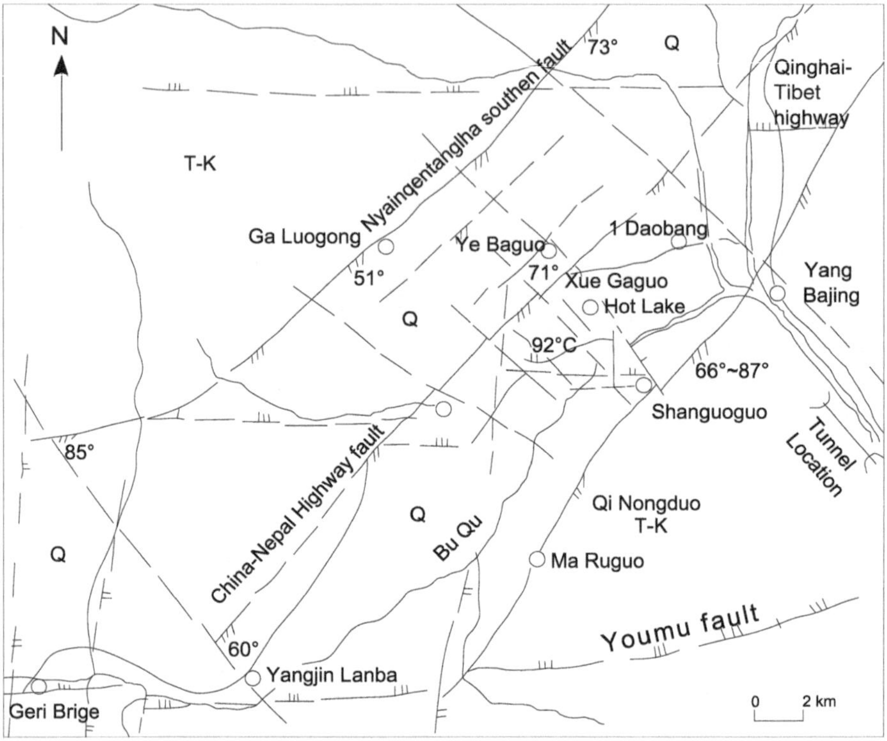

Fig. 7.2 Tectonics of Yambajan Basin

4. The hot springs are a type of important tourism resources and if the tunnel forms a larger drop leak bucket to make hot springs disappear or decrease its flowing, it affects the development and utilization of the hot spring resources.
   It is a complicated problem needing further analysis. First, tunnel construction certainly affects groundwater system. So the main method is controlling temperature on excavation, adopting reasonable lining, and setting the heat insulation. Furthermore, line selection should be selected as far as possible away from the surface with the hot springs.
5. The railway-line selection should consider the tourist resources in addition to engineering, because there are great tourism resources in the geothermal area.

7.1.6 Selection of the Qinghai–Tibet Railway in the Geothermal Area

Tonglha Mountain Bingzhan area (hot spring area) and Yambajan region are major geothermal areas along the Qinghai–Tibet Railway.

Tonglha Bingzhan area lies at the hot spring basin. It develops concealed faults. And spring water pours out of the ground, of which the temperature exceeds 50 °C. The tunnel avoids the hot springs because the population here is sparse, the elevation is more than 4800 m, and it is difficult to develop tourism resources here.

By comparison with Dagze, and other schemes, the Yambajan tunnel scheme along Tibet road is chosen finally with fully considering the geothermal tourism resources exploitation and environmental factors.

7.1.7 Summary

All types of exploiting activities should adjust measures to local conditions, especially in Tibet. For some large construction project site, the engineering construction should avoid the harmful site as far as possible where the unfavorable geological body has been produced. And the comprehensive utilization of geothermal resources should also be considered. In addition, the dodging method is used in road route alignment design, when the main location of the important structure has been selected in geothermal areas. Engineering geology, hydrogeology, and geothermal activities should be contained in field geological survey. When the dodging method is difficult to conduct, the road should select the edge of geothermal areas to pass. And when it is difficult to avoid the geothermal area, the road should select the narrowest area. Third, line selection should combine other engineering-geological condition to avoid the fault movement and the hot spring spots. For line choice at landslides, the scheme should first avoid the large ancient landslide location. For the line that cannot avoid such area, the scheme should take corresponding measures on the basis of understanding the nature and stable state. Railway engineering should take priority to subgrade, culverts, and other small engineering, choosing deep tunnels or large bridges is unfavorable. Last, the railway-line selection should consider the tourist resources in addition to the point of engineering, because there are great tourism resources in geothermal area.

7.2 Road Route Alignment Design in the Aeolian Area

Desertified land is large in the plateau, China, which is 11.95% of the total area and the vast majority distributes in Qinghai and Tibet Provinces. The main characters are: arid with scarce rainfall, high wind velocities, a large difference of temperature, severe cold, plant growing difficulty, and small, and rare vegetation. Additional characteristics are lots of sands, widely spread drift sand area, widely distributed mobile sand dune, semi-fixed sand dunes, and the tall, and complex sand dunes.

The dust storm leads to a series of severe consequences. It can cause the death of human and livestock, bury the village, farm, and pasture, damage transport, and telecommunications facilities, degrade production of soil, and deteriorate air quality.

In addition, the climate is dry, the wind is strong, and the ecological environment is fragile in the Qinghai–Tibet Plateau. The effects include strong wind, sandy resources, desertification caused by gradual global warming, glaciers, and permafrost degradation. Consequently, dust storm has been becoming more and more severe as one another key technical problem in the construction of the Qinghai–Tibet Railway.

7.2.1 Sand Causing Analysis and Spatial Distribution in Tibet Plateau

7.2.1.1 Sand Hazard Along the Qinghai–Tibet Railway

The three large natural climatic zones along the Qinghai–Tibet Railway are as follows: arid climatic zone from Golmud to the Kunlun Mountains, plateau arid and semiarid climatic zone from the Kunlun Mountains to Amdo, and semiarid climatic zone from Amdo to Lhasa.

The climate is varied along the railway and the characteristics of four seasons are indistinct, thin air, low air pressure, and larger evaporation capacity than precipitation. The major forms of precipitation are snow and hail in mountain areas, whereas, the precipitation with rain has a priority in the vast flat areas. The northwest wind and the west wind are dominant and focus on October to April the following year. Table 7.1 shows the windy weather condition along the Qinghai–Tibet Railway.

According to the survey of sand deposition, half-fixed sandy land, and semi-fixed sand dune is distributed along the Qinghai–Tibet Railway, which has a wide distribution of gravel gobi, and the sediment is significantly over a long distance. Furthermore, there are 93 severe sandstorm spots up to 88.167 km in total, and there has been a trend of to develop and expand. Effective measures should be taken to dispose of the disasters in time, otherwise, the road be sandy, the railway be buried and the traffic be disrupted inevitably and it causes severe accidents such as the derailment.

Table 7.1 Meteorological data for the wind along the Qinghai–Tibet Railway

Meteorological station	Golmud	Wudaoliang	Tuotuo River	Amdo	Nagqu	Damxung	Lhasa
Average wind velocity (m s^{-1})	2.6	4.1	3.9	4.3	4.1	2.4	2
Predominant wind direction	W	W	W	N, NE	SW	SW	E, SE
Maximum instantaneous wind velocity (M/S)	24	31	30	35	37	25	16.3
Maximum instantaneous wind direction	W	W	W, SW	W, SW	SW	W, SW	N, NE
Annual average wind days (D)	9.8	130.1	178	147.1	106	57.1	26

Table 7.2 Survey on sand deposition along the Qinghai–Tibet Railway

Zone	Sand deposition section	Sand deposition type	Thickness (m)
Golmud–Wangkun	DK918+380–DK919+930 and DK924+980–DK926+850	Semi-fixed sand land and semi-fixed sand dune	0.5–4
	Golmud to Nanshankou and DK885+840–DK887+360	Flat dunes with deepening of layer thickness	0.2–0.5
Wangkun–Tonglha Mountains	Hoh Xil Mountain, Beiluhe Basin, Tongtian River Basin, Buqu River Valley, Tuotuo River Basin and Tonglha Mountains	Semi-fixed sand land, dune, and migratory dune	0.2–3
Tonglha Mountains–Lhasa	East of Beisangqu, Sangkarigang Moutains and East Coast of Na Cuo Lake	Semi-fixed sand land and dune and crescent dune and dune chain	<5
	DK2002+330–DK2002+579	Aeolian sand	3–5

Table 7.2 shows the condition of sand deposition along the Qinghai–Tibet Railway.

7.2.1.2 Analysis of the Cause of Aeolian Disaster

1. Wind erosion climate

The terrain of Tibet Plateau is high, and most of the plateaus are with monsoon climate. The characteristics include intense sunshine, strong radiation, large daily temperature difference but small annual temperature difference, large evaporation but small annual rainfall. Furthermore, the index of annual wind erosion climatic factor is more than 50; in particular, the cold, dry and windy dry season lasts more than 6 months of which the index of wind erosion climatic factor is more than 100. It is one of the strong wind erosion force climate center in China. It provides incentives and necessary conditions for the activities of a dust storm to cause disaster weather of drought and sandstorm.

2. Severe desertification and rich sand content in soil

Table 7.3 shows that Tibet Plateau is one of the lands of severe desertification in China. The area of desertification is more than 1997.40×10^4 hm^2, which is 16.58% of the total area. The land includes three classes of desertification which is severe, moderating, and mild desertification. And there are five types desertification disaster land which include migratory dune, semi-fixed dune, fixed dune, bare, and half bare sand gravel land. The desertification mainly distributes at Nagqu, Ali and Shigatse region (accounting for 97.09% of all desertification lands).

In addition, in most of the Tibet Plateau, forest are scarce, a large area is sparse, low grassland, and the vegetation growth period is short. Most area is always bare or half bare. The basis of the surface sediment is formed by loose residual and slope deposit, diluvium, alluvium and proluvial, and lake deposits content rich sands. For example, physical gravel content of 338 soil profiles is up to 70.44% on average in Nagqu. In conclusion, windy dry weather, and loose sandy soil are a material basis for desertification. In a large area, the containing sand is exposed on the surface, which creates favorable conditions for dust emission. These make the inherent risk of natural environment desertification high in Tibet.

3. Climate drying and warming, precipitation reduction

The Tibet Plateau is sensitive to environmental change. The temperature generally rises and the temperature increasing rate reaches up to (0.10–0.30) °C/10 a. For example, Shigatse, Lhasa, and Tsetang in central Tibet, January temperature has

Table 7.3 Type and area of desertified land in Tibet (*Unit* $\times 10^4$ hm^2)

Zone	Severely desertified land	Moderative desertified land		Mild desertified land		Summation
	Migratory sand	Semi-fixed sand	Fixed sand	Naked gravel	Seminaked gravel	
Nagqu	10.08	57.63	7.46	593.34	334.16	1002.70
Ali	4.60	4.76	0.00	272.64	323.07	605.07
Shigatse	14.01	27.75	14.64	149.22	125.89	331.51
Shamnan	3.54	2.15	2.27	3.96	5.66	17.58
Qamdo	0.00	0.21	0.26	0.12	14.44	15.03
Lhasa	0.86	3.35	3.66	1.71	7.35	16.93
Nyingchi	0.56	0.95	1.03	0.05	5.99	8.58
Summation	33.65	96.80	29.32	1021.04	816.59	1997.40

increased 2.1, 1.5, and 0.8 °C, respectively. And the annual average temperature has increased by 0.4, 1.0, and 0.4 °C, respectively, compared with the 1990s and 1960s. Since the 1950s, rainfall has decreased and the precipitation decrease rate is (10–40) as mm 10 a^{-1} in most areas. In particular, the summer precipitation decreases significantly and climate warming trend is more severe. For example, the rainfall of Lhasa averaged 450 mm in the 1950s. It decreased to 386 mm in the 1990s, and the average precipitation reduced 38 mm from 1961 to 1980 compared with that from 1981 to 1994.

A series of changes have occurred in the natural environment in Tibet with the climatic warming and drying tendency. The changes are as follows: First, sandstorm activities increase sharply as the temperatures, rainfall, and evaporation, and the force of wind erosion have been changed rapidly. Second, soil wind erosion intensity has increased with the rainfall, evaporation, and the surface runoff decreases. Then the surface soil gradually becomes dry and the wind erosion resistance of soil becomes weakened. Finally, soil desertification and the wind-sand movement of the surface have been further strengthened with the change of the conditions, for instance, the climate, and soil become worse, vegetation is degradation, plant height, and coverage is decreasing.

4. Dust storm intensified by human activities

The social and economic development level in Tibet is relatively low. And the means of exploiting and utilizing resources are simple. Its living still maintains a style of "farming depend on weather", and "raising livestock depend on weather" with underdeveloped productivity, all the factors of low investment, low output, and low level of agricultural production and low income determined the main developing process as a denotative expansion in the past, such as increasing the amount of livestock, expanding cultivated land. In particular, with the increase of population, resource-exhausting development, and utilization are adopted, such as increasing livestock, and cultivated land quantity sharply and rapidly. These increasing human activities inevitably have destroyed the natural vegetation and soil. Human activity aggravates the surface wind erosion, the sand handling, and stacking effect. Particularly in basin areas, these activities have expanded the sand area and enhanced sandstorm activity intensity, which forms the artificial process of Tibetan desertification disasters, especially in larger valleys.

With the potential natural factors, the sand disaster develops in Tibet Plateau as a comprehensive processing interacted with human activities, causing the Tibet Plateau sand range expanding and the intensity strengthening. Moreover, the rapid growth of people and animals and their activities leads to the excessive exploitation and utilization of land resources. The interaction occurs between the Tibetan natural process and human process. According to research, climate change is the dominant factor for desertification in the pastoral areas with relatively less human activity and agriculture. By contrast, human activity is the main cause of desertification in farming areas.

7.2.1.3 Overview of Sand Disasters on the Qinghai–Tibet Plateau

In recent years, the study of dust weather is conducted mainly in the low-altitude area in northern China. However, there is few study on the characteristics of high winds and dust weather along the Qinghai–Tibet Railway. Because of less rainfall, harsh climate, vegetation degradation, and lots of windy days in Tibet Plateau, sections of the road subgrade construction is difficult after meeting strong wind, dust, and sand blowing weather, because dust heavily blocks the subgrade (Fig. 7.3). Through observing the date of 66 weather stations in the Tibet Plateau (taken from the national meteorological center), scholars studied the wind, dust, sand blowing weather, and the trend of the spatial distribution characteristics and changes from 1971 to 2000.

1. Characteristics of wind distribution

The Tibet Plateau windy areas are mainly gathered in the central region (showed at Fig. 7.4). In addition to the number of annual strong windy days in Qaidam Basin, the southern plateau, and the eastern edge is <40 days. The average annual days with strong wind are commonly more than 60 in the plateau. Annual average windy days of the region around Tuotuo River are over 100. The whole distribution characteristics are related to the influence of upper air jet flow in main area of plateau, while the wind in plateau border area and Qaidam Basin is affected by the ground cold air. The segment from Wudaoliang to Amdo is just located in the center of the gale area, the number of gale days in south and north part of railway is few. They are 167.8, 148.8, 135.5, 19.1, and 27.5 days in Tuotuo River, Amdo, Wudaoliang, Golmud, and Lhasa, respectively.

2. Characteristics of dust distribution

Overall the average dusty days are less in a plateau. The area with high incidence of floating dust is mainly concentrated in the Qaidam Basin and Qilian Mountain region. The center is located in Mangnai of west Qaidam, Golmud of south Qaidam, Caka of east Qaidam, and Minle of northeast Qilian Mountain. And the average dust days are 27.7 and 18.2 days, 16.9 and 21.9 days, respectively. Rich sand resources are underlying the ground surface in the western Qaidam Basin and there is less rainfall the ground. It is also one of the channels for the west cold air invasion and the way through which the dust must pass from southern Xinjiang Basin. So the dust weather is common here. However, the main area with strong wind around Tuotuo River and Wudaoliang does not have conditions for dust weather. Precipitation and rainy days are much more in the southern plateau, where vegetation condition is better, with no dust conveyed externally and less dust days. The distribution characteristics of the floating dust are most remarkable in the spring.

a. Depositing sand under Basuoqu Bridge

b. Depositing sand on subgrade of sbutment of Tuotuo Bridge

c. Depositing on left between 1476+160 to DK1477+400

d. Gravelly desert landscape in Dangla Mountains

e. Depositing sand on right subgrade of Xiushui River Bridge

f. Depositing sand on toe of subgrade of Tongtian River Bridge

Fig. 7.3 Depositing sand on railway

3. Characteristics of blowing sand distribution

Two forming conditions of blowing sand and sandstorms are the sand-underlying ground surface and the strong wind. The distribution of vulnerable areas to micrometeorology windy days are roughly the same with the distribution of strong wind, and average blowing sand days are more than 70.9 at Shiquan River Station in the west. According to the observation, blowing sand days may be more in the

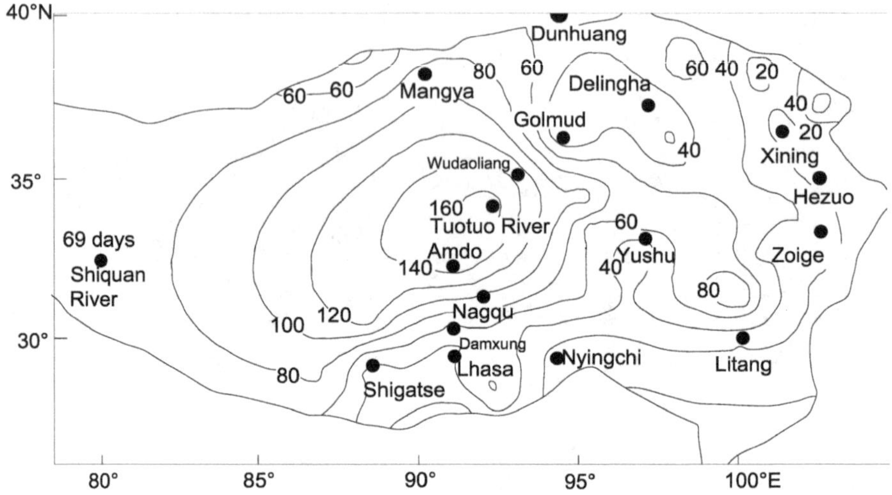

Fig. 7.4 Annual strong wind days in Tibet

northern Tibet Plateau and next south of Xainza and the surrounding area. Another area with large numerical value is located in the Qaidam Basin, which has two centers, Mangnai and Golmud, and annual average blowing sand days are 34.9 and 40.1, respectively. Much of the southeastern plateau Jiangheyuan annual blowing sand days are less five. The number of blowing sand days is relatively large in the region from Tuotuo River to Golmud, and the average annual blowing sand days are 12.6–40.1 days. But in Damxung–Amdo section, it is relatively small, only 1.9–7.1 days on average.

4. Characteristics of sandstorm distribution

The geographical distribution of sandstorm days is very similar to the geographical distribution of blowing sand days in Tibet Plateau. Its general distribution characteristics are more days in northwest but less days in the southeast. Because of the complex plateau terrain and weather climate, strong sandstorm occurs regionally. Annual average sandstorm days at each station are as follows: 19.4 days in Shiquan River in the west, 18.3 days south of Nagqu, 10.2 days in north Shannan Zêtang Station, 14.3 days in Wudaoliang, 11.2 days in Qaidam Basin Mangnai, 13.2 days in Golmud, 13.4 days in Gangcha in the eastern Qinghai and 11.1 days in Xinghai. The days are few in the southeast plateau, mostly less than a day, even without dust storm weather here. However, dust storms are more along the Qinghai–Tibet Railway, the annual average is 3.0–14.3 days from Lhasa to Wudaoliang. Among them, the Lhasa–Amdo section is 1.5–5.7 days, Tuotuo River–Golmud is 11.1–14.3 days. Therefore, this data analysis shows that the sand hazard is severe along the Qinghai–Tibet Railway.

7.2.1.4 Seasonal and Interannual Variability Characteristics of Sandstorm Disasters

1. Seasonal characteristics

The monthly and seasonal data show that the windy days with dust and sand blowing weather are more in winter and spring (more than 75% of the days all the year round), less in summer, and fall in Tibet Plateau. The days with blowing sand and sandstorms are more than those with fly ash, resulting from the influence of the jet stream winds, except for the northeast edge of Qaidam Basin and Plateau.

Annual change character of windy days in seven stations is as follows: most of the windy days are in March with the least windy days in August, except the Golmud Station, where the most of windy days are in May and the least in August. From the seasonal change analysis, strong wind often happens in spring and summer at Golmud, but at most of the other stations, it occurs in winter and spring. We use the Tuotuo River as an example to illustrate the frequency of strong wind along the railway, which is characterized by the strong wind. The windy days in spring account for nearly half of the total number of a year at the Tuotuo River. And windy days in winter account for 63% of annual strong windy days, among which windy days are more than two-thirds in March. Occasionally, windy weather occurs more than twice in a day.

Contrast to the low-altitude area of northern China, where the sandstorm mostly occurs in spring and few in autumn, at most of the south of Wudaoliang Station in the Qinghai–Tibet Plateau, sandstorm days in winter are more than those in spring, and the days in autumn are slightly more than those in summer. It suggests that the sandstorm in the plateau and the sandstorm in north China do not belong to the same system. The former is upper air jet flow, appears earlier, the days in winter are more than those in spring, and the days in autumn are more than those in summer. And the latter is the low-altitude wind, with little effect downstream because of the effect of high altitude wind diffusion.

2. Interannual variability

Annual variation trend of windy days on Tuotuo River is contradictory with Shiquan River by analyzing wind characteristics and the annual average. The gale days reflect it more clearly. The former is stepping up, but the latter is stepping down. Furthermore the time of mutations is not in sync. The time of mutations at Tuotuo River is 1969 and 1986, and the time of mutations at Lion Spring River is around 1975 and 1991, centralizing 5–6 years to the west.

There is a decreasing trend of sandstorm days in Tuotuo River. From the mid-1970s to the early 1990s, sandstorm days in Shiquan River were at the peak, but less in 1960s, and 1990s. The interannual variation trend of gale days and sand dust storm days is anti-correlated, which means that the year of more (less) windy days often corresponds to the year of less (more) sandstorm days. This feature in Shiquan River was more evident before the 1990s.

7.2.1.5 Desert Landscape Types Along the Qinghai–Tibet Railway

Desert landscape pattern is diverse. According to the shape, it can be divided into level sand ground, barchan, reticulate dunes, and pyramid dunes. According to the sand dunes of direction relations, they are divided into vertical and horizontal sand dune. The main forms are flat sand (semi-fixed and fixed sandy lands) crescent dune (mobile sand dunes), but the scale along the Qinghai–Tibet Railway is small.

7.2.1.6 Movement of Sand Flow

The critical wind speed, which makes the sand particles start to move, is called the starting velocity. Start wind speed is concerned with sand-grain size, ground surface properties, the water content of sand, and other factors. When the wind speed is greater than the starting wind speed, the wind raises the loose sand and form airflow containing sand, which is known as the sand flow. Table 7.4 shows the relationship between the grains of sand start wind speed and grain size, water content.

7.2.1.7 Form of Sand Movement

The movement of sand is concerned with the sand flow form, the strength of the wind and sand-grain size and quality. It includes three forms of creeping (slow moving), saltation, and suspending. Additionally, the saltation is the major form with an average of 78% in the total sediment. Creeping is second, accounting for 20%, and the suspending is the least. Creeping and saltation mainly occur at a height of 10 mm over the ground.

7.2.1.8 Structure of Sand Flow

The distribution of the transporting layer at different heights in the handling of the grains of sand is called sand flow structure. The sand flow structure changes with wind speed, the ground surface properties, and the number of grains of sand entering in the air. The height of sand flow decreases with the height in the vertical

Table 7.4 Relationship between start-up air speed, grade, and moisture content

Grade (mm)	Start-up air speed under different moisture content				
	Dry state	Content (%)			
		1.0	2.0	3.0	4.0
0.18–0.25	3.8	4.6	6.0	10.5	12.0
0.25–0.50	4.8	5.8	7.5	12.0	–
0.50–1.00	6.0	7.0	9.5	12.0	–
1.00–2.00	9.0	10.8	12.0	–	–

distribution of sediment concentration. And its grains of sand are mainly gathered in the airflow close to the ground. When the wind speed exceeds the start wind speed significantly, the sediment concentration in the upper airflow increases rapidly.

7.2.1.9 Wind Abrasion and Accumulation

Wind abrasion or sand flow accumulation is mainly determined by the wind, sand, obstacles, and the underlying surface. Because of the increase of wind speed or the reduction of sand, the sand flow is in an unsaturated state and thus produce wind abrasion or be moved easily. And because of the wind speed reduction or obstacles, the sand sinks down, and gets accumulation. If the sand concentration in air is roughly equal to the volume of sand sinking down in the air, neither wind abrasion nor sand accumulation occurs.

7.2.2 Sand Harm to Railway

When the sand flow is encountered with the blocking of the roadbed and the upper structure of the line of in the process of migration, the sand pile up at the line, and bury the roadbed, and the rail, which brings a series hazards in railway operation and maintenance.

7.2.2.1 Type of Sand Harm to Railway

Main hazards of sand for the railway is sand burial and wind erosion.

1. Sand burial

There are two reasons for sand burial: one is that the sand falls and accumulates when the wind speed is decreased, and thus buries the subgrade. The second is that the mobile dunes cover the roads.

There are three types of sand burial: one is sheet sand burying. It occurs in larger scale and forms more quickly and mainly occurs in the areas with sand flow activity. In the initial stage, the sand is thin, and the train can pass it. If the sand resource is rich, much sands accumulate and block the traffic. The second type is ligule sand burying. When the line crosses dunes in the flowing sand dune area, or there are obstacles on the upper side of the roadbed in the strong wind-sand flow area, ligule sand is formed. Ligules sand burying is swift and the thickness is larger. It disrupts traffic easily in case of strong winds. The third type is heap sand burying. The disasters can be predicted because the sand dunes moving direction and the speed can be measured. Once it is formed, the clear work is hard because of the large amount of sediments. The light harm is to interrupt traffic, and the heavy harm is to change the line of the road.

2. Wind erosion

Under the direct effect of the sand, the sand, or soil particles are blown away from the roadbed. The embankment bottom part is cut, collapsed, and empty, which decreases the width and height of roadbed, and affects or interrupts the traffic. The degree of wind erosion is concerned with the wind, wind direction, the form of subgrade, the packing material, and the protective measures.

Embankment wind erosion occurs on the windward slope shoulder and the upper part. Particularly, wind erosion is most severe at the high embankment. It often forms a slope with wind erosion. The wind erosion depth often reaches a dozen centimeters, the largest can reach tens of centimeters, which makes the whole shoulder eroded by the wind. And the maintenance quantity is increased and the traffic safety is affected.

The erosion in slope and the top of the slope is severest. When the dominant wind direction is parallel to the line, the slope often turns into furrows shape on both sides with wind erosion. And its depth can reach more than 20 cm. When the dominant wind direction is vertical to the line, slope forms round, or irregular shape. Fangs shape and pocket pot hole are formed after the windward slope is damaged by wind erosion. A large number of sand fall within the cutting to block line.

7.2.2.2 Sand Effect on Railway Operation and Maintenance

Wind sand has become a great burden to the normal operation and maintenance of railway, and major disasters are as follows:

1. Affect driving safety. When sand reaches the rail head on both sides, it affects the train operation, and it derails the train when much sand covers the rail surface.
2. Produce line disease and reduce the quality of the line. After accumulating lots of sands in ballast bed, grains of sand infiltrate into railway ballast by the train vibration. And they are gradually gathered at the bottom of the ballast bed to raise sleepers and rail to form a triangle pit and low joint diseases. These diseases affect track geometry condition and reduce the quality of the line.
3. Increase the line maintenance and maintenance workload. To ensure the safety of driving in wind season, the comprehensive section line patrol should be enhanced to keep the line clear in time.
4. Shorten the service life of rail and fasteners. The sand intensifies the rail abrasion. In a sand disaster area, the railway abrasion is increased by 4–5 times than that of the sand free area. Rail damages also occur here much more often than in areas without sand disaster. When the sand contains salt, rail fastener is susceptibly corroded, which shortens the service life of the rail and fastener.

Bridges and culverts are blocked and the flood draining is affected. Sand easily blocks bridges to affect flood draining. If the sand is not cleared in time, it may cause flooding.

7.2.3 Sand Region of Railway Survey and Design, the Principle of Line Selection and Prevention Measures

According to the analysis on the first phase of the Qinghai–Tibet Railway from Xining to Golmud segment, the train derailment may occur when the thickness of sands is more than 20 mm higher than the rail with an extending length of tens of meters. The ideal plan for preventing sand is to combine planting with engineering. However, because of the thin air, drought, and cold climate and the unfavorable natural conditions, the plant growth period is short along the Qinghai–Tibet Railway. Therefore, the survey and design and the principle of line selection and control measures should be combined with the characteristics of Tibet Plateau. Short-term protection should combine with long-term protection. And protection should do simultaneously with road works. It gives priority to comprehensive treatment without destruction of the ecological environment with mechanical protection.

7.2.3.1 Survey on Sand Area

The Qinghai–Tibet Railway survey should be closely combined with the local nature, sand landforms, regional environment, and other specific conditions. The main works are as follows:

a. Investigate sandstorm topography, geology, hydrology, climate, vegetation, and other natural conditions along the planned route.
b. Investigate the disease of sand along the line and put forward the governance solution.
c. Combine with the plan of path in the sand area and put forward intentional recommendations.
d. Initially set the engineering measures and materials.
e. Macroeconomic analysis to possible line scheme comparison, make a preliminary budget, and work on the necessity and feasibility of research to determine the construction, submit the preliminary feasibility study report and project proposal.

1. Collecting regional meteorological data

Meteorological data are basic for tracing sand, sand morphogenesis, and mobile rule, choosing the route plan, designing the roadbed protection. Meteorological data should include temperature, precipitation, evaporation, humidity, wind conditions (direction, speed and the changes in different seasons) the radian of the sandy wind and its frequency, and the season in which it occurs. The available wind direction condition is indicated by power chart and the sandy wind vector diagram.

2. Engineering-geological investigation

   a. Mainly investigate the distribution and forming conditions of sand areas. First of all, the sand distribution range, and the distribution of various sand landforms should be investigated to choose appropriate lines or protective measures. In addition, the location, size, and quantity of the sand should be surveyed. Finally, all types of natural conditions (e.g., material, the ground vegetation, flow direction, the local air, geomorphology, hydrology and human activity) with effects on sand formations should be found.
   b. The survey of movement characteristics of dune. The direction, mode, and speed of the mobile dunes should be mainly found out. Visiting local residents, surveying, aerial photos contrast, analysis of existing methods of wind direction and wind speed data should be applied.
   c. Other investigations. It includes physical and chemical properties of the sand, the local vegetation cover °, plant species and ecological characteristics, and groundwater depth.

3. Material survey

The foundation needs many protective materials in the sand area. Attention should paid to the investigation of subgrade protective material. Usually, it is necessary to investigate local gravel and pebbles, gravel, and clayey soil origin, reserves, and the distance. It also needs to investigate the source, reserve, and transport conditions of the material such as wheat straw, reeds, and sand sagebrush.

4. Water survey

Domestic and engineering water is inadequate in sand area. Groundwater should be found out to use as water resources. The detailed survey is applied and the relationship between groundwater and geomorphology, geology, and surface plant should be noticed.

5. Investigate local sand experience

People have struggled with sand in the long-term in desert region. They have accumulated rich experience and effective methods, which should be carefully surveyed and applied with the professional knowledge of road.

7.2.3.2 Railway-Line Selection Principles in the Aeolian Area

Through summarizing the past experience in line selection and the engineering practice in sand region in recent years, the principle of line selection and the main points of the circuit layout are as follows.

1. Handle the relationship between future and nowadays

Line selection should handle the relationship between the near future and nowadays in sand. It should combine engineering and operating conditions, the regions national economic development prospect and the overall planning to do a series comprehensive research. The engineering should take all the valuable plans into consideration, fully demonstrate the location, and requirements through the selection of major control points.

2. Reasonable route to avoid severe drift sand area

In sand line scheme comparison, apart from the construction period, and the initial construction investment, the construction, maintenance, and operation should also be considered in engineering. Danger quicksand should be avoided in route selection when the detoured routing is not too much long and the engineering cost is reasonable. So, it is economical for the construction and operation. However, routing choice still needs to depend on the situation.

3. Make full use of favorable terrain

If the road must cross through the sand area, it should make full use of all types of favorable terrain. The road should be chosen as far as possible in the area where the sand damage is relatively light such as the lake shoal, valley terraces, the ancient river bed, and the marginal zone.

4. As far as possible to cross through the favorable parts of fluid dunes

If drift sand areas cannot be avoided or the cost is not reasonable, and the line has to flow through the sand dunes, the narrowest parts of these fluid dunes, and the corrosion parts of the sand dunes on the upwind side should be selected. It also uses the lowlands between open hills as far as possible to pass through the low point in the sand dunes.

For example, the line should choose the flanking parts or the chain bealock at dunes when routing at the chain of crescent dunes and sand dune area. Whereas, the road should pass through cross-cutting site between the main girders of depressions vice beam as far as possible in the reticulate dune. And the road is designed with the alternating of dike and cutting. The main control point is the average height of vice beam on line longitudinal slope.

5. Make the line direction parallel to dominant wind direction

When selecting a line in the sand area, the direction of line should be parallel to the local dominant wind direction or intersect with a small angle to decrease wind erosion to the embankment and sand-burying hazard.

6. Choose reasonable plane curve and profile combined with terrain, wind direction

The plane curve should be used as far as possible in the desert region. Particularly, plane curve with small radius is unfavorable. If the plane curves have to be set, it should be set up in embankment. And the convex of plane curve faces the dominant wind direction. Line profile should not be just for the sake of smooth. In accordance with technical indicators, the line should comply with the natural terrain and avoid cutting terrain. The design of embankment is better, and its height should meet the needs of the longitudinal slope, but it cannot be too high.

7.2.3.3 Design and Construction of Engineering Sand Prevention

1. General content of engineering sand-prevention design

According to the seasonal changes of the sand activity, the main wind direction, especially the angle between roads and the main wind direction in the designing area, the principles are that the local conditions are adjusted by measures, the priority is given to prevention and the efficiency should be considered.

a. Sediment transport is applied in the strong wind and the unsaturated sand flow. The road is designed into arc section, and sediment section is set along the road. These methods help the sand flow pass through the road smoothly. When the wind is not strong, the wind plate is set up to increase wind pressure. The permeable chemical material is adopted to reduce ground roughness, making the sand flow go through easily. In addition, fences, and ditch are set up in upwind to intercept drift sand.
b. Rows of feathers and earth embankment are set up in the rich sand desert area and the angle between main direction and route is less than 45°, which can change the direction of sediment deposition and make use of terrain to keep away from sand disaster.
c. The resistance fixation combined with sand-prevention system is considered in the rich drift sand areas. A single type resistance sand-fixation belt can be set up in one side of upwind in a single wind direction area. The double direction protection is set up in dominating with the main wind direction and existing double or more wind direction. All directions should be treated equally in distinguishing primary and secondary wind direction.
d. Two cases are generally considered in plateau area of sand subgrade design: one is protecting the roadbed, and another is protecting engineers on both sides of roadbed.

The roadbed protection is changing the nature of soil, reducing wind speed, covering the roadbed, and isolating the effect of the wind.

The protective measures mainly include paving pebbles, clay protection, turf protection, rubber cement sand plate rubber asphalt sand plate, and spraying asphalt emulsion, or salt brine. Pebbles (crushed) protection is a main way in plateau area. Many literature have a detailed design scheme and we do not state again here.

The main protection on both sides of subgrade engineering is to take steps to cut off the contact between wind and sand surface, and to increase the surface roughness to reduce the speed of the near-surface wind. Specific practices are heavy material, asphalt emulsion, and sand-fixation and sand squared barriers with grass covered, squared sand barriers with pebbles (crushed) stone covered, squared sand barriers with clay covered.

Here is the protective method used commonly: pebbles (crushed) stone is used to set square grids in sand areas. It is a half hidden barrier. And it not only fixes in situ sand surface, but also stops the moving of the outside sand and makes it piled up around. It plays a dual role of sand-fixation and resistance of sand and its effect is remarkable. Because of the fact that more than 90% sediment discharge is within 0–20 cm high, and most of them concentrate in 0–10 cm high, squared sand barriers with pebbles covered should be 20 cm above the ground. If the main wind direction is single and no other harmful wind exists, it can be changed into determinant sand barriers which are parallel with the dominant wind direction. The row spacing, decided by the gradient of sand dune and wind, is generally 1–2 m. The top height of the barrier should be 5 cm higher than the bottom height of the previous one in level. The function of this type of barrier is same as the sandy square, but it can save materials and labors.

2. Engineering measures principle

Engineering method, also known as mechanical sand-fixation, namely physical sand-fixation, is to use the physical properties of the sand to control the harm of sand flow and to prevent the sand dunes from moving forward by setting the engineering measures. The basic approaches of engineering measures are as follows: to stop the grains of sand, to control surface wind erosion, to accelerate the sand movement, to deposit the sand flow, to change sand flow direction, and change the holistic movement to sand flow dispersion. Engineering method can be summarized as follows from engineering mechanics principle:

a. Blocking contact between the gas and solid, and inhibiting interface interaction happened on the interface of air flow and sandy surface.
b. Strengthening the cohesion of the sand, and improving the ability to resist wind erosion.
c. Increasing the vertical distance to overcome the sand flow movement and decreasing local resistance.
d. Reducing and eliminating all types of resistance in the sand dunes movement or guiding redirection accumulation of the sand flow.

The effective controlling method used frequently is seldom a single one, but a combination of several methods. The chemical sand-fixation method is that the fluid sand is sprayed by adhesive to consolidate the surface, which improves the erosion resistance on the surface of the sand body to eliminate the erosion.

3. Attention in construction

   a. The construction organization arrangement should be well done because of the unfavorable plateau weather, the sand activities, large protective material quantity, and the large transporting capacity.

b. Roadbed, road shoulder, and slope filled with fine sand should be provided a comprehensive protection. And it should be completed at once with the fill and excavation. Gravel could be chosen as protective material and the section size should comply with the design requirement. Pebble protection should be packed tight compaction and fill gaps with coarse sand. Pavement of slope should be paved from down to up. Blocks are built first and then pave stones. Gravel in block should be 2–3 cm below the edge of block to prevent slipping.
c. According to experience in construction, the main belt should be first set in the windward side, following which the vice belt is set on both sides of the subgrade. The sand barrier set in this way is stable and durable.
d. Vegetation should be protected strictly in the construction process. The pebble in surface should not be applied as barrier. This bare gravel layer to provide new sand. The roadbed and surface on both sides should be tidy and the rest of the material should be cleaned up lest the erosion and sand burial be caused.

4. Maintenance

Subgrade maintenance is hard and arduous in plateau sand. Because of the unfavorable natural conditions, the large sandy wind, and the dry climate, the following measures should be applied in the maintenance:

a. According to the local wind conditions, the structure, and location of the protective measures are adjusted at any time. The equipment damaged by wind erosion and sand burial needs to repair and change in time.
b. The work plan and the detailed rules should be worked out to clean the accumulating sand on the bridge in time. Maintenance personnel should keep contact with the meteorological stations nearby in windy seasons. They should know the wind situation and strengthen the tour road work. And a guard should be assigned in severe sand disaster.
c. The investigation should be strengthened and workers are organized regularly to fill the location, types, causes, existing protective measures, and improvement suggestions of disasters into sand resume.

7.2.3.4 Sediment Control System Design

Comply with the principle that prevention is prior to treatment and both should be combined. Semi-fixed sand and dune have a wide distribution and with an ability of long-distance sand transporting. The road easy to accumulate sand should take the engineering protection measures. Engineering protection measures include preventing wind erosion, paving pebbles, and gravels to stabilize sand and building stone squares barrier and high barrier to stabilize sand. The sand subgrade worksite is 93 locations of per 88.167 km, which is 10% of all length. 4 locations of per 1053 m silty-fine sand cuttings use gravel to cover the slope, 34 locations of per

25,153 m embankments, which use dry flaky stones to protect slope. And sand-prevention system combined with the plane and solid is set in 56 locations of per 64.5 km subgrade sides. The disposing sand covers 38,507 mu. Gravel square covers 13,666 μ and flatting gravel covers 7296 μ.

1. Subgrade body protection

Dry flaky stone layer with 0.3 m thickness is applied in road shoulder and the road shoulder should be under 1.0 m to prevent shoulder and slope from erosion. The dry flaky stones in the whole slope are used to protect area with severe wind erosion, such as the sand subgrade in Cuona Lake.

A slope direct to the top is adopted in silly sand cutting and the slope area. When the slope height is less 6.0 m, the slope rate should be set to 1:1.75, and when the slope height is more than 6.0 m, it should be 1:2.0. The platform for depositing sand is set up on cutting of 3.0 m wide, and side ditch is unnecessary. Whole section is covered by gravel with the thickness of 0.2 m. A sand platform is established in the cutting part with the width exceeding 2.0 m.

2. Surface protection

The function of paving gravel with thickness of 0.20 m can fix local sand. Coarse particle soil such as gravel soil is solid and durable and can prevent wind and rain erosion. In addition, it can keep temperature, and is helpful for plants growing. When the angle between the wind and resistance engineering is <30°, gravel can pave whole sand. Otherwise, gravel can be paved outside of sand in start-up wind speed.

The squared sand barrier can fix and resist sand. The geogrid is not used in Tibet because of the strong ultraviolet ray. And the wheat straw hood and haulm are lack. So, the sandy barrier with stone squares is designed to fix and resist sand.

The width of surface protection is determined by the angle between the dominant wind direction and the route:

a. When the angle between the dominant wind direction and route is more than 30°, the width of windward side is 200 m, and the width of lee ward side is 100 m.
b. When the angle between the dominant wind direction and route is from 6°–30, the width of windward side is 100 m, and the width of lee ward side is 50 m.
c. When the angle between the dominant wind direction and route is less than 6°, the width of both windward side and lee ward side are 50 m.

3. High vertical sand barrier

High vertical sand barrier plays a role in resisting sand. It is generally set up as a row, and two, or three rows should be set if a large volume of sand is present. The height is 1.50 m and the stack spacing is 30 m. Considering the activities of wild animals, cattle, and sheep and other livestock in Tibet, High Vertical barrier is decorated with a feather. High vertical barrier structure adopts drafty fence. Because of lack of branches, thorns, and reed, the concrete fence is applied. There are five types of structure forms including rung perforated concrete block sand fence, hanging round piece of concrete block sand fence, activity board type concrete block sand fence, and rotating plate concrete block and brick block sand fence. The rung perforated concrete block sand fence has been widely used in Cuona Lake. And the other four structures have been widely used in Wudaoliang to Yanshiping and Tonglha Mountain area. In totally, the resistance function of all type of sand fence is same, but the structural form is different.

7.2.4 The Qinghai–Tibet Railway-Line Selection in the Aeolian Area

The parts of road which are affected by sand are Golmud to the Kunlun Mountains, Wudaoliang to Tuotuo River, Tonglha Mountain to Touerjiu Mountain and Amdo to Cuona Lake. Considering the unfavorable geology, the main schemes are the scheme from Wanbaogou to Xiaonanchuan and the scheme for Xidatan flowing area. The two schemes are as follows:

7.2.4.1 Scheme Comparison from Wanbaogou to Xiaonanchuan

Belonging to the valley terrace of Kunlun River, Xiaonanchuan, and Yeniugou join in Sanchahe, which is called Kunlun River. Kunlun River passes through in mountains such as a snake in the east–west direction. The width of modern river bed ranges from 200 to 400 m and I grade terrace is higher than terrace 3–10 m. II grade terrace is mostly incomplete. III grade terrace is higher than terrace 30–50 m. The quay is steep and more complete. And the longitudinal gradient of which is 10‰. Xiaonanchuan runs south–north. And the longitudinal gradient of which is 22–25‰. The semi-fixed dunes distribute in the left bank of Kunlun River and local trend to platform margin. The sand disaster is severe in the outfall of Sanchahe. The height of the lines is controlled by the cables and the oil-transporting tubes in the left of Xiaonanchuan.

1. Introduction of line scheme

Combining with the terrain, geological conditions, and riverbed longitudinal slope, the following three schemes are designed for the bridges with different height in the right bank of Kunlun River and Sanchahe (Fig. 7.5).

CK plan: the line starts from Wanbaogou, going upward along III grade terrace in the left bank of Kunlun River. Its length is 698.2 m and the height is 54 m. It strides Kunlun River in downstream of Sanchahe Highway Bridge of 180 m to reach the high-grade terrace in the left bank of Kunlun River, westward striding Xiaonanchuan, and southward reaching Xiaonanchuan Station finally. The length of the line is 11.04 km. The bridge is five and the length of bridge is 1.7 km. The static investment is 176 million Yuan. The line passes through sand and ligules semi-fixed sand in the left bank of Kunlun Mountains internally.

CIK plan: the line starts from Wanbaogou, going west forward along I grade terrace in the left bank of Kunlun, striding route bridge, pipeline, and Kunlun River. Its length is 155 m and the height is 34 m. The next is same as CK scheme. The length of line is 10.5 km, bridge is 3, the length of bridge is 1.96 and the static investment is 170 million Yuan. Gully, washout, and soil deposit are severe in the platform on both sides of river. Sand deposition appears occasionally on left bank.

CIIK plan: the line starts from Wanbaogou, going upward along III grade terrace in the left bank of Kunlun River, striding Yeniugou in Sanchahe. Its length is 960 m and height is 50 m. It reaches final point along high-grade terrace in left bank of Xiaonanchuan. The length of route line is 10.62 km, bridge is 5, the length of bridge is 1.46 km and the static investment is 177 million. The location with wind sand is longer on left bank of Kunlun River and the length across semi-fixed sand dune by evacuation is 1 km.

2. Evaluation of schemes

CK solution: the high platform is used, so the line is higher. Slope of a line from Sanchahe to Xiaonanchuan is not sufficient. The height of Sanchahe super-large bridge is higher, but the location of the bridge is better and the length of bridge is shorter. And the length of line across semi-fixed dunes is also short and the engineering budget is less.

CIK scheme: sufficient slopes are used between Sanchahe and Xiaonanchuan Bridge to reduce the height of Sanchahe Bridge and the length of line. However, although the height of bridge is reduced, the length of bridge is increased because of the restriction of high steep slope. Comparing with CK plan, the length of route reduces 540 m, the height of Sanchahe Bridge reduces 34 m and the investment

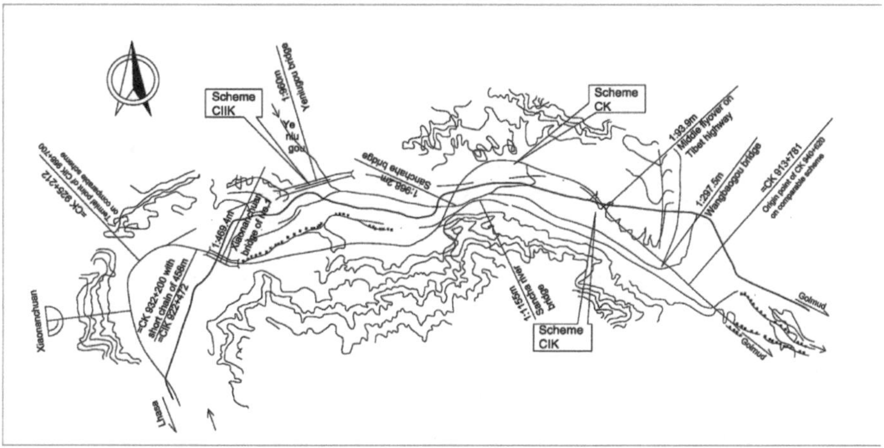

Fig. 7.5 Comparison of route alignment project between Wangbaogou and Xiaonanchuan

reduces 5.68 million Yuan. But, the bridge location is poor and oblique crossing degree is large, and the length is 460 m longer than CK scheme. And it also disturbs highway, pipeline, and optical cable. The deposit and washout is found on the edge of high bench and sand deposition is severe in some parts.

CIIK scheme: it omits a river-crossing work. And the length of route and bridge are also 480 and 40 m shorter than that of CK solution. However, the line has to be striding upstream Kunlun River. The road is high, and the length, and height of Yeniugou Bridge are long and high. The emphasis quantities increase and the evacuation need passing through semi-fixed sand land 1 km to reducing the height of Yeniugou Bridge.

To sum up, CK scheme can make full use of the terrace on both sides. Bridges is a basic orthogonal with least construction cost and relatively well geological conditions. So here CK scheme is recommended.

7.2.4.2 Route Scheme Comparison in Flowing Xidatan Area

The road in this section belongs to Xidatan Rift Valley north of the Kunlun Mountains. The valley is steep in south, the top of mountain is eternally covered with snow, and the wide pluvial fan and flood plain develop in piedmont. The north of the valley is middle and low mountains, its terrain is ups, and downs, pluvial fan and flood land distribute in front of piedmont and at the foot of clearance, and vast amounts of sands distribute in valley and gully. The width of valley from north to south is 2–4 km and the riverbed natural longitudinal slope is 15‰–18‰. Tibet Highway extends actually from east to west in the valley. Figure 7.6 presents a comparison of the schemes.

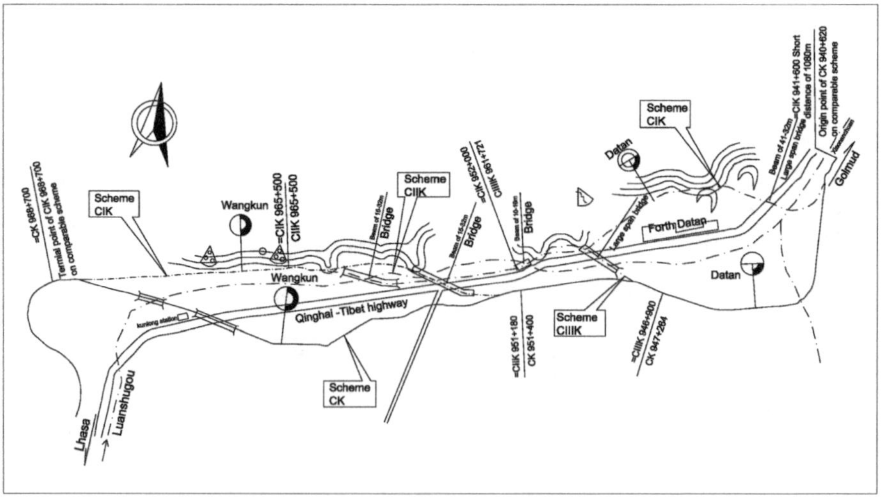

Fig. 7.6 Comparison of route alignment project in overflow area of Xidatan

1. The main engineering-geological problems affecting route plan

    a. Dust storm: it is mainly distributed in the right part of the toe of slope, depression, gully inside, and the surface of the diluvial fan. Its scale is large and the thickness is uneven. The thickness of pluvial fan surface and mountain slope foot is 0.5–2 m, and the thickness of gully and depressions is 2–10 m.
    b. High-intensity seismic region: the basic seismic-intensity is 8°. Line selection should avoid major projects and lower the height of the bridge.
    c. Overflow area: it is distributed in the north and south sides of the Tibet Highway. The surface runoff is developed into mountain glacier as a solid reservoir south of the Tibet Highway. The water melted from ice and atmospheric precipitation is a major supplement. And its damage degree is higher than that in the north of Tibet Highway.
    d. Diluvial fan: the mountain is steep on both sides of the Tibet Highway, and the gully is developed. The road should fill the circuit, for the diluvial fan is more on the line position.
    e. Permafrost: the frozen soil in this period is distributed in a shape of island, the harm to engineering are uplifting in cold season and melting down in the warm season. The length of distribution of permafrost exceeds one-third of the comparison length, mainly near the Wangkun Station.

2. Introduction of line scheme

South highway solution (CK): the route starts from the comparison starting point, going westward to Datan Station. The route passes through pluvial fan and sheet flood section with continuous mesh. Wangkun Station is set in the east of oil pump station and south of the highway. The route finally spans Tibet Highway and TNE drainage line to reach the comparison destination. The line length is 28.18 km, each 11 large and medium bridges is 1.05 km and the static investment is 339 million Yuan.

North road solution (CIK): the route extends westward from the comparison starting point. A grand bridge is applied spanning over the river and the Tibet Highway. The length and height is 1354 and 30 m. Datan Station is set at the hillside in north of Xidatan oil depot. Then the route goes westward north of Tibet Highway along the left bank slope, striding over the river, and advancing along foot on the left bank slope. Wangkun Station is set in northeast of oil pump station to reach the terminal point. The line length is 27.1 km, each 3 large and medium bridges is 2.04 km and the static investment is 368 million Yuan.

Combination scheme i (CIIK): to cancel the grand bridge engineering in north highway scheme (CIK) and avoid the large comprehensive station location near beach, the design selects the best location across a river. The line advances in overflow area south of the highway, striding over river, and highway and advancing at side slope foot north of the highway. The line length is 28.1 km, each 9 large and medium bridges is 0.95 km and the static investment is 336 million Yuan.

Combination scheme ii (CIIIK): the length of route across overflow area is shortened in the scheme on the basis of a combination plan (CIIK). The early striding road distributes north of the highway. But, bridge engineering increase. The line length is 28.26 km, each 8 large and medium bridges is 1.71 km and the static investment is 366 million Yuan.

3. Line scheme evaluation

CK plan: the height of watershed is 5300–5500 m on the Kunlun Mountains south of Xidatan Valley and snow cover has distributed on it for many years. Snowmelt water and a low precipitation (annual average rainfall is 276 mm) are converged into linear trickle to flow along the slope surface. The erosion gully is formed on a steep slope. The trickle forms runoff in the form of facet mesh and radial, because it does not have fixed filament line after leaving mountain pass. It is called sheetflood. The catchment area is large above the mountain pass at the Dongdatan–Xidatan intersection (CK944+000–CK950+880). The scale of proluvial fan is formed by sheetflow, its primary sector, and sheetflow beach are large. And the catchment area is small from shed to slope toe, because the area is icy and snowy from CK950+880 to Wangkun Station. The flow is generated by snowmelt and precipitation. The scheme is feasible after being applied with engineering measures.

CIK plan: line walks largely along compressional concealed fault in Xidatan. There are not major projects and the influence is not strong. The three large crescent flowing dunes are located near Xidatan depot. The sand dune with length of 500 m and height of 30–40 m is developing, which influences the line. Furthermore, aeolian sand damages the line in the valley and gully from Xidatan to Wangkun. When the line passes through the large rock heap, the excavation should be applied. The development of groundwater at the toe of slope has larger influence on line.

CIIK and CIIIK: the area that lines pass through has difficulty in eradicating potential geological hazards, such as sand damage, and rock pile.

In conclusion, the geological conditions in the scheme that line is distributed north of the road are poor. Some severe geological disasters such as sand damage, rock pile, and rock fall are caused. The shortcomings of CK scheme are that the line is located in overflow area, bridge engineering is concentrated, and heavy ancillary work is required. But the other conditions are better than other schemes. Although the length of line in CIK solutions is 1.08 km shorter than that in the CK, the investment is approximately 29 million Yuan higher. The investment of CIIK solutions is 3 million Yuan less than CK, but other geological problems exist. CIIIK project investment is 27 million more than CK scheme and its geological conditions are poor. Therefore, CK solution is recommended.

7.2.5 Summary

The main diseases brought by sand for railway are burying lines, eroding shoulder slope, blocking bridges. It deposit sand on track bed to make train derailment. Shoulder be evacuated and sleeper head be exposed by severe wind erosion to threat traffic safety. Rainstorm destroy the railway for the poor drainage of blocked bridges. Therefore, the sand can cause long-term effects on railway operation.

For line selection in sand area, all types of factors contributing to the sand dunes should be further found out, including the nature, qualities, and activity of sand dunes, the direction, and power of wind, sand resources, and topography. Anyway, the line should avoid drift sand area as much as possible.

If drift sand area cannot be avoided in large sand dune areas, the line should be selected on the edge of sand dunes as far as possible. It can also pass through both sides of the river, paleo-channel, and lake basin grass land in sand dunes. If line locates in area with continuous sand dunes, the line should pass through low dunes as possible to avoid being buried by drift sand. In local drift sand area with semi-fixed and fixed sand dune, the line should pass through the upper side of wind to avoid going through leeward. The may be parallel to the local dominant wind direction as possible. Otherwise, a large amount of sand would be accumulated in the upper side of the embankment to erode shoulder and sand deposits easily. Overall, the profile design shall be in the form of embankment, and keep a certain height.

7.3 Road Route Alignment Design in Snowdrift Areas

In China, blowing snowdrift region accounts for 55% of the total territory. It mainly distributes in the Tibet Plateau and the surrounding mountains, Tianshan Mountains and northern Xinjiang, Inner Mongolia, and northeast in China. The latitude of southern boundary is lower than other blowing snowdrift region in the northern hemisphere. The steady snow covering area is 420×104 km^2 in China. The area of snow covering is 230×10^4 km^2 in the Qinghai–Tibet Plateau (except the northern Tibet Plateau and the Qaidam Basin) so that the snow resources mainly distribute here and the disaster of pastoral often occurs. Livestock is endangered and the traffic is interrupted by blizzards and snow. National defense construction and people's normal life are also influenced. Some research indicates that the area of Tianshan Mountains, Altai, southeast of Tibet, north Yunnan, Sichuan–Tibet Highway, Tibet Highway, Tonglha Mountain area, the west side of Greater Khingan Mountains and northern foot of Yan Mountain are severely influenced by snowdrift hazard and different levels of snowdrift disasters occur every year.

According to snow stability, forming conditions, distribution scope and expression, snowdrift, and avalanche are the main snow disasters along the Qinghai–Tibet Railway. Snowdrift is the main snow disaster according to the scale and frequency of damage. Its formation mechanism, types, characteristics, and spatial distribution are completely controlled by local climate and terrain.

7.3.1 *Cause Analysis of Snowdrift and Spatial Distribution in the Qinghai–Tibet Plateau*

7.3.1.1 Analysis of Snowdrift Formation in the Qinghai–Tibet Plateau

Snowdrift is an untypical two-phase flow of air with snow grains. It is a weather phenomenon that the snow grains is rolled and moved by wind. It redistributes natural snow to induce and aggravate ice–snowmelt flood, avalanches, landslides and directly affects production and living with blocking traffic, interrupting the grid, and stopping work and production.

Transition and deposition are two processes for formatting snowdrift, which can be completed in the following conditions. Heavy snow, the wind to start snow grains and appropriate terrain are the essential conditions. The material source is snow and snow cover, the power is the wind, which forms wind flow, and determines the direction and motion of wind flow. Suitable terrain that reduces wind speed or separates the wind can form vortex decelerating zone to deposit snow. The several characteristics are as follows in snowdrift regions along the Qinghai–Tibet Railway:

1. Rich snow source

Snow cover and snow are material sources of snowdrift. A larger snowdrift is formed from a stable snow cover with thicker snow (more than 10 cm). The formation and development of the snowdrift are also influenced by the physical and mechanical properties of the snow, such as the density of snow, snow-grain size, and its hardness. Table 7.5 is the condition of start wind speed to various state of snow, when the temperature lowers below 6 °C.

The Tibet Plateau is known as "the roof of the world". South Asian monsoon blows northwards to the east of the plateau from the bay of Bengal and influences the Qinghai–Tibet Railway. South Asia monsoon is characterized by the humid and rainy climate, which increases rainfall along the Qinghai–Tibet Railway. For example, in the section from Tonglha Mountains to Nagqu, the annual precipitation is 200–400 mm, which is 2–3 times greater than that in Tibet Plateau inland. In addition, the height of 90% section of the Qinghai–Tibet Railway is 4000 m. The

Table 7.5 Relationship between start-up air speed and snow in different state

	Grade (mm)	Average density (g cm^{-3})	Start-up air speed (m s^{-1})
New dry snow	<1	0.06	2.0
Fine snow	<10.5	0.18	3.7–4.3
Old fine snow	0.5–1.0	0.23	6–8

temperature is low, and the average temperature is −16.8 to −12.8 °C in winter. So, the snowfall in winter is large, which accounts for 15–25% of the annual rainfall, and the depth of snow is more than a foot with a stable thickness of 60 cm. Heavy snow provides ample material source for snow disaster.

2. Strong wind

Wind is a dynamic condition to form snowdrift. The different type and motion of snowdrift are determined by levels and status of wind. According to the test of professional department, when the wind speeds are 4 and 8 m s^{-1}, respectively, in the height of 5 cm from the ground, the result is that even though the latter wind speed is two times as the former, the transmission capacity of snow of latter wind speed is 27 times as the former.

Snow-bearing wind includes drifting snow, blowing snow, and storm; Table 7.6 clarifies the relationship between these activities.

The geographical location is unique, geological conditions complex, air thin, and the climate changeful in the Qinghai–Tibet Plateau. Air speed in the plateau is greater than that in the plain. Except for Kunlun Mountains, Tonglha Mountains, and Nyainqentanglha Mountains, the terrain is flat, and the elevation difference is less 200 m without hindering for airflow. The wind is strong in winter, the surface wind speed is more than 4 m s^{-1}, and occasionally storm wind with 30 m s^{-1} speed appears. The snow is blown easily by the wind, because the particle size and water content are low. Therefore, re-carried snow flow accumulation often occurs along the Qinghai–Tibet Railway and the accumulation thickness is up to 6–7 m in some area.

The general direction of the wind is northwest in winter. The Qinghai–Tibet Railway runs northeast–southwest. The angle between them is approximate to 80° and it is the best angle to accumulate snow beside roadbed. So, the harm of snowdrift cannot be avoided along the Qinghai–Tibet Railway.

3. Inducing factors—global warming

The researchers pointed out that, as early as 1981, the snow was an important factor to control long-term climate change in Tibet Plateau. As the weather became

Table 7.6 Relationship between snow stream and wind

Type	Wind velocity range (m s^{-1})	Wind drift height (m)	Horizontal visibility (km)	State
Drifting snow	5.5–7.9	<2	10	Snow grains are blown and run with the wind
Blowing snow	8.0–10.7	>2	<10	Snow is rolled up
Snowstorm	10.8–13.8	very high	<1	The sky is indistinguishable with snowfall

warming, the snow influenced atmospheric circulation in East Asian and China to intensify flood and drought disasters.

For example, snowfall in 1985 and 1986 was the largest in the 1980s in the northern hemisphere and it was also largest in Tibet Plateau. Abnormal snow occurs north of Tonglha Mountains, Bayan Har Mountains, and A'nyemaqen. The depth of snow was 100 cm, but it was only 10–20 cm in a normal year. The snow disaster has been the severest in Tibet Plateau for 40 years. The strong snow appeared in the Himalayas in January and February 1989 and rare snowstorms occur in Aksu in January to March 1993. Continuous snow fell in southern Qinghai and blocked 4400 km road.

By 2030, if the global average temperature rises 0.5 °C again, snowfall and snow cover have increased by 26 and 18.5%, respectively, in Tibet Plateau, mountain areas, and the middle and lower reaches of Yangtze River. And the appearance of the blizzard disaster is more frequent and severer.

7.3.1.2 Overview of Temporal and Spatial Distribution of Snowdrift in Tibet Plateau

1. Snowdrift distribution in the Qinghai–Tibet Plateau

Snowdrift occurs frequently and its harm is severe in Tibet Plateau. It is mainly distributed in the northeast, southeast (including Hengduan Mountains and western Sichuan Plateau) and the surrounding hills. The strong snowdrift is distributed at the height of 5000 m in the south and west mountain area and the snowdrift is perennial. The development of snowdrift is affected by the dry and cold climate and the little rainfall.

The main snow disaster is snowdrift along the Qinghai–Tibet Railway. And the length of snowdrift is nearly 90% of all snow disasters. However, avalanches only occur in the Kunlun Mountains, Tonglha Mountains, and Nyainqentanglha Mountains, and its amount, and scale is far less than the wind snow. The snow disaster is mainly distributed in the region between Kunlun Mountain Pass and Nagqu, and the disaster is more severe in the region between Tonglha Mountains and Nagqu. The length affected by the snow disaster is 300–450 km and 27–40% of the total length of the Qinghai–Tibet Railway. So, the snow disaster governance should be focused on.

2. Snowdrift season change process in the Qinghai–Tibet Plateau

According to research, the seasonal distribution, and duration of snow are controlled by annual changes in rainfall and temperature. The detailed conditions are as follows:

1. The snow begins in mid-September, expanding rapidly before winter, the peak appears in January and the melting process lasts long continuing from February to June. The storing volume of snow reaches the largest in winter (from

December to February), then reduce in spring (from March to May), and drops down to the smallest in autumn (from September to November), which account for 45.2, 28.0 and 21.2% respectively. The annual variation of peak range is striking, and the differing amount of water equivalent is 30 billion steres between abundant snow and less snow of years.

2. The amount of snow is decreases with subzero temperatures, its peak occurs in alternating seasons and does not appear in the coldest month. The snowing period is long in Tibet, generally from September to May of the next year. Snowdrift disaster is produced as long as strong wind occurs. The actual hazard days of snowdrift are 30–60 days in a year. Most of them concentrated on February–April, which accounts for 50–70% of all snowdrifts.

3. Snow annual variability in the Qinghai–Tibet Plateau

The annual variability of snow is presented as a perceptible annual fluctuation process superimposed with a continuous and slow increasing trend in Tibet Plateau. Since the mid-1980s, the trend has been more obvious and the annual amplitude has increased significantly. The annual fluctuation of snow depth is remarkable in the Qinghai–Tibet Plateau. The average depth is 15.0 cm along the railway when snow is heavy. The depth of snow was 21.3 cm in 1985 and 1986, and the depth was only 10.4 cm in 1984 and 1985. The spatial difference of annual amplitude of snow depth is extraordinarily large. The significant annual fluctuation is highly concentrated in the eastern plateau (E90°–100°), whose area is a quarter of the plateau. The difference with average is 50–110 cm in Tonglha Mountains, Bayan Har Mountains, and A'nyemaqen Mountain. The value is largest in the world, which is most significant in Eurasia. It is worth noting that the max difference comparing with the average does not occur in the area with most abundant snow and ice. For example, the average depth of snow is 10–20 cm in Tonglha Mountains to Bayan Har Mountains and A'nyemaqen Mountain.

7.3.2 *Influence of Snowdrift*

The movement of the snowdrift is affected by many factors, such as the physical and mechanical properties of particle density of snow, snow, snow depth, and hardness, and some natural conditions of solar radiation, air temperature, ground temperature, and surface roughness.

7.3.2.1 Influence of Snow Depth

The stable snow layer with a thickness of more than 20 cm forms snow-driving flow easily. That is to say, snow obstacle may be produced at a certain section of state road when the depth of stable snow is more than 20 cm. The depth of local

Table 7.7 Relationship between snowdrift strength and snow source

Observation point number	1	2	3	4	5	6
Start-up wind speed on height of 1 m (m s^{-1})	5.9–6.0	6.6–6.8				7.5
Nature of snow	Drifting snow	Drifting snow	Drifting snow	Drifting snow	Drifting snow	Drifting snow
Wind direction	NW	SE	NW	NW	MW	SE
Snow source state	Between unrich and rich	Very rich	Unrich	Far from snow source	Unrich	Rich
Strength of moving snow (g m^{-1} s^{-1})	13.92–3.96	32.94	5.65	12.2	6.46	27.30–25.94

snow is 30–60 cm, which offers the material condition for snowdrift along the Qinghai–Tibet Railway.

The intensity of the moving snow is used to represent the amount of snow contained and the characteristics of snow grains transmission. The intensity of moving snow refers to the amount of snow passing through a unit perpendicular area to the wind direction in a time unit. The length of the time duration is deepened by the abundant snow resources and the mountainous topography. These factors result in the inhomogenous snow distribution in the horizontal plane and form the difference in snow storm flow duration, and the snow flow content and the density decrease with time lasting. If snow source is rich, the difference is small. And if the speed is same or close, the differences of the intensity of moving snow are also large (Table 7.7).

7.3.2.2 Influence of Temperature

The nature of the snow is changed easily by the difference between the temperatures in air and in snow surface. If the snow density changes little, with temperature rising, the water contained in the snow increases, the viscous force between snow grains enhances and the start wind speed increased significantly.

7.3.2.3 Influence of Snow-Grain Size

Under the low temperature, the starting wind speed is determined by snow-grain size. It is observed that when the snow-grain size is less 2 mm, the starting wind speed is increased with the snow grains becoming large and the acceleration is slower.

7.3.2.4 Snow Density Effect on Strength Start Wind Speed and Snowdrift

The relationship of density, the intensity of snow grains, and starting wind speed is complicated. Under normal circumstances, the start-up wind speed, and moving snow intensity are increased with the density of snow.

7.3.2.5 Influence of Surface Roughness on Start Wind Speed

The start-up wind speed increases with the surface roughness of general naturally underlying surface. Because the grass and shrubs block the air movement near Earth. The snow grains on the ground is not easy to start and the moving snow grains are stopped to form snow cover.

The start wind speed is decreased with roughness on the snow surface. If the snow surface is uneven, the vertical movement speed of airflow is relatively larger. In addition, the contacting surface between snow grains and the air is enlarged in the snow concave and convex part. So, the snow grains start easier than those in the smooth snow surface. Table 7.8 shows the wind speeds at 10 cm high in various types of snow started of different roughness cases in field.

7.3.3 Snowdrifts Influence on the Railway

Snowdrift generally occurs in the snowing or after snowing, mostly after snowing. When the wind direction is suitable, heavy snowdrift deposition on excavation section, cuttings, and half cutting sections on a slope and lee side area bury the lines and block the traffic. In addition, the deposited snow on railway produce water lubrication between the rail and wheel, which impedes the train operation, it is difficult to restart the train and difficult to move in the long upward slope, consequently, it cause an accident of parking or delay.

7.3.3.1 Influence of Snowdrift for Cutting

Observational studies show that numerous factors affect the wind speed near ground atmosphere. It mainly includes the roughness of the ground, the obstacles, and local

Table 7.8 Start-up air speed under different roughness

Z_0 (cm)	3.1000	1.0000	0.5930	0.2450	0.1400	0.0210	0.0026
V_t (m s^{-1})	1.5	1.8	2.6	2.7	3.1	3.8	3.9
State of moving snow	Mounded snow ripple	Undulating snow ripple		New drift snow board		Smooth windscreen	

terrain changes. So, the shape of subgrade cross-section plays a crucial role in snowdrift disaster if the other conditions are same. When the wind flow is vertical to the cutting, wind speed is reduced sharply because of the rapidly widening cross-section, and lee slope blocking action. The wind speed in the windward slope toe and the cutting centers are 10 and 26% of that in the sloping top in a depth of 2.0 m light cutting. So, a great of snow grains accumulate in the cutting.

Whether the snow stops the car in deep cutting mainly depends on the length of cutting windward slope and the quantity of location snow. When the snow cover is less than the critical storing snow volume of cutting at windward side slope, traffic problem here not occur. When it exceeds the critical storing volume, transportation become difficult.

Wind tunnel simulation experimental research on appraise of snow and wind tunnel flow also proves that when the angle between the wind direction and the zero-point subgrade, low or high embankment, and leeward, or windward half cutting sections is small, the snow disaster is light and does not occur at tailwind. However, cutting and tunnel is different. Whatever the angle is, a snowdrift accumulation be formed at cutting. But the location, scope, and severity are different. The snow covers near the entrance and the exit of the wind cutting, and then extends gradually thickening at both ends, extending to the middle of cutting and becoming more severe. When the direction increases but the angle is <45°, there are snows on both ends. The snow on the entrance is more than that on the exit. The snow on both ends spreads to the central cutting and the snow extension at the entrance is faster than that at the exit. An s-shaped long snow embankment is formed in cutting finally. When the direction is more than 45° but <75°, in addition to a small amount of snow covering the exits, snowdrift is mainly accumulated in the upper hand on the cut slope to gradually form a snow tongue or snow eaves to block the road. When cut is perpendicular to the wind flow, most of the wind snow accumulate in the windward cut slope to quickly form a cornice to bury the cutting as the snow cornice moves forward.

The length of cutting before embankment may also affect the accumulation of snowdrift. If the length of cutting before embankment is short, the thickness of snow be larger, if the length of cutting before embankment is long, the thickness of snow be small. In addition, when the width of cutting is invariant, the depth of snow in cutting increases with the depth of cutting.

The covering snow in tunnel is mainly located in the entrance and the exit of the cutting. Snow changes with the tunnel trends and the angle with wind direction. If the angle is less than 45°, snow mainly accumulate inside the cutting before the tunnel entrance. With the angle increasing, the snow on both ends of the tunnel cutting approaches and extends to the entrance and the exit. While the wind snow enters into tunnel, its direction is parallel, or vertical to the wind flow.

7.3.3.2 Influence of Snowdrift on the Embankment

The movements of snowdrift and the snow deposition are influenced greatly by embankment height. For example, for an embankment of 1.5 m high, when the wind blows snow from a long distance, the wind speed change is not large when passing through a flat area, but the wind speed is reduced at the windward slope foot and reaches minimum at the slope toe to unload heavy snow grains deposition. Then the speed of unloading snowdrift increases gradually and arrives at a maximum speed at the road shoulder. And then the speed is also close to the maximum on the whole road without forming snow covering on the road. The speed of snow decreases rapidly when the snow goes through lee side and the snow grains over the pavement parts deposit at lee slope near the slope toe. Since then, the snowdrift blows into the flat area, and the wind speed returns to be normal gradually.

When the snowdrift blows over the embankment of 15.1 m high, the snow deposition condition is totally different from the 1.5 m high embankment. The motion of snowdrift is similar, but there is a significant difference on the road. When the height of embankment is 15.1 m, the speed reaches to maximum at the road shoulder windwards. Then, the speed decreases, and appears minimum in the center of the road. Air velocity is increased again to reach the submaximum in the lee side of the road. It indicates that when the embankment is higher, although the airflow speed increases along the slope, the boundary layer separation is severe on the road. Therefore, a small amount of snow appear on the high embankment.

The direction of the Qinghai–Tibet Railway is warp. Severe snowdrift and strong storm threaten the Qinghai–Tibet Railway section south of Kunlun Mountains in Tonglha Mountain Pass, Nagqu, Damxung. In particular, a severe hole jam phenomenon be formed at the cutting near the entrance and the exit of the tunnel section, which should be taken into full consideration.

7.3.4 *Avalanche Effects on Railway*

The snow cover slides down by gravity in certain conditions on the hillside, which causes a chain reaction on the slopes and in the snow to cause numerous snow body collapse called avalanche or “snow landslide”, “snow shifting sand”, and “snow mountains” in some areas. Huge avalanche often happens accompanied with a thunderous noise, and produces a huge blast and mushroom cloud. It produces a great effect and can reach more than 50 t m^{-3}, and the collapse of the snow can reach more than 107 m^3. It is a severe natural disaster in snow mountain area.

However, avalanches have not occurred in all the mountains. The necessary condition of avalanche formation and occurring is a certain depth of snow and a range of slope. Slope plays an important role. And vegetation types, orientation, and roughness are causes that cannot be ignored. Research shows that slope avalanches occur roughly between 20° and 50° and the optimal gradient slope is 30°–45°.

Studies for many years have shown that the terrain cutting coefficient exceeds 150° in the avalanche growth topography. When the terrain cutting coefficient is <150, even though the solid precipitation is large, the avalanche is less.

The Qinghai–Tibet Railway passes across the hinterland of the Qinghai–Tibet Plateau. The height of plateau datum plane is at around 4000 m. The elevation difference is relatively small between the plateau mountain (Kunlun Mountains, Tonglha Mountains, Nyainqentanglha Mountains) and the plateau. The region of Qumar, Tuotuo River, Lhasa river are upstream location. The depth of cutting topography is shallow and the terrain cutting coefficients are not too large, a little more than avalanche minimum limit of 150. The cutting coefficient is more than 150 at Kunlun Mountains, Tonglha Mountain Pass, as well as Nyainqentanglha Mountains, with the avalanche of terrain condition. So, the avalanche hazard sections along the Qinghai–Tibet Railway are distributed mainly in the north slope of the Kunlun Mountain Pass, the north slope of Tonglha Mountain Pass, the north and the south slope of Nyainqentanglha Mountain Pass and the east slope of Nyainqentanglha Mountains from Damxung to Yambajan.

7.3.5 Railway Location Principle and Control Measures of Snowdrift Areas

Snowdrift and avalanche are two large snow disasters along the Qinghai–Tibet Railway, and the former is dominant. How to prevent snow disaster is the first problem need to solve in railway location choice and subgrade design. Therefore, apart from the political, and economic significance, the properties, and standard, relationship between present, and prospect should also be considered. In addition, wind, and snow weather, hydrology, and geology, landform, especially the snow flow operation, accumulation rule along the line should be also investigated.

7.3.5.1 Investigation and Collecting Information

To design the railway location well, the datum should be collected below:

1. Gather and analyze the temporal and spatial distribution rule and characteristics of rain fall, snows, wind, and temperature from observatory, so as to realize the snow condition.
2. Because the observatory stations are set nearby towns, dates differ from the actual condition of line. Semilocating station shall be established to observe rain fall, snow, and temperature, wind direction, and speed to compare with meteorological factors collected by observatory stations. Snowfall, mount distribution, continuous snowfall, and their amplitude of variation, extremism of the snow season, especially in the stable period of snow accumulation, should be made clear to analyze the influence of snow-bearing wind.

3. The angle between wind direction and oriented railways, and the wind speed in the main section of the line should be measured by mobile observation.
4. Snow covering snow should be measured by snow gage observation, nature snow canal observation, and snow fence resistance method. It is important to design snow embankment. If the information is difficult to collect, the maximum snow depth should be calculated by the deepness of snow and refer to the information nearby meteorological station.

7.3.5.2 Railway Selection Principle

The formation mechanism, type traits, temporal, and spatial distribution of snow disaster are subject to the local climate and terrain conditions completely. The terrain is flat, elevation is almost same, and the type of snow disaster is snowdrift along the Qinghai–Tibet Railway. The length suffering snowdrift disaster is almost 90% of all snow disasters.

The harm of snowdrift to the railway is related to climate, elevation, lines, geophysics, geology, geomorphology the profile form, subgrade, and cross-section. In general, some region with extremely severe snow disaster should be avoided. The line should not pass across it to save investment and driving time. If the region cannot be avoided, the best way is tunnel and shed. If the road direction does not deviate the general trend, the road length is not too long and the construction cost is limited, the engineer should conduct as followings:

1. Angle between the line direction and wind direction

The line should be parallel to the prevailing wind flow (except tunnel), especially for the line close to mountains. In general, the smaller angle between the line and the prevailing wind flow, the lighter the snow disaster be in the similar cases. If it is hard to reduce the angle, its angle should be enlarged to be nearly vertical. So, even though the snow disaster is severe, it is easier to set up snow prevention projects than in the sections with a small angle, and the effect is better too.

2. Line location

If the altitude of the line is low, the snow accumulates late and melt early, and the time affected by the snow disaster be relatively shorter than in the area with high altitudes. So, the mountains with low altitude is advisable for line crossing.

The sunlight in sunny slope is more than that in shaded slope. The solar radiation there is strong and snow melts quickly. So, the lines should be set in sunny slope generally. If the topography is broken and the cost of roads maintenance is high, the line should be located in shaded slope. It is appropriate for the line to go through the lee of forest, and keep a certain distance from forest. If lines should go through the forest, where trees are cut down should not be too large, only with a capacity for storing the snow swept away from the line. For example, if the depth of snow is 1 m, the range of cutting down trees is about 10 m on both sides of the line.

The local terrain should be made full use. For example, the open valley is better than the narrow one and the valley terrain with small changes is superior to the one with large changes. The ridge line is better than mountainside line. Generally, the line does not pass through narrow areas between two mountains, particularly avoiding the mountain under steep slopes.

The snowstorms in leeward slope are severer than those in windward slope. Thus lines should be set up on the windward slope instead of the leeward slope.

The snow in hairpin is usually heavier, especially in the leeward mountain bend. So, the hairpin should not be set up and embankment should be used as far as possible in the areas with rich snow source.

3. Roadbed design

In the areas with rich snow source, no matter how the angle between line trend and the prevailing wind direction, severe snow disaster happen. The cost of snow disaster prevention is higher than other forms of subgrade cross-section especially in shallow cutting and long cutting. So, embankment should be widely used in roadbed design and cutting should be tried to avoid. If the cutting is difficult to avoid, widening, and deepening ditches in the upper side, raising roadbed to make the cutting in the form of embankment, setting store ski resorts, snow ditch, or slowing down slope and open cutting should be applied in design.

Lee road, the wind road with half filling and half digging can be appropriate to add some excavation based on the actual conditions. For example, some engineering can design ski resorts, enlarge the borrow pit, broaden ditches to accommodate a large number of snow or slow down slope to prevent from snow disaster in the windward of lee road, lee windward road.

Snow is easily accumulated in the hairpin road. The line should be designed with a large radius and the prominent body in the middle back bend should be dug out. Striping the inside part of the mountain body near the mountain bend or designing a completely open return bend, both are good ways to eliminate the snow caused by vertical and horizontal vortex.

If the length of line is not too long but the snow disaster is extremely severe at mountain pass sections, structures such as tunnel, open-cut tunnel, and shed-tunnel could be used. Although the construction cost is high, the effect is good.

In the area where snowfall is affluent, attention should be paid to drainage design. When the climate becomes warmer, snowmelt runoff increase. It is useful to construct some culverts on the valley to ensure that the snowmelt runoff is drained and keep the stability of roadbed.

7.3.5.3 The Qinghai–Tibet Railway Engineering Prevention Measures in Snowdrift Area

Many types of prevention and controlling measures are applied in snowdrift area, which are commonly the methods of mechanical removal, engineering protect snow, build shelter forest, heat treatment, and chemical reagent. The priority should be given to protection, combined with engineering management.

Snowdrift is mainly influenced by wind. The main methods to prevent and control the wind snow are adjusting the wind flow, changing wind speed, and wind movement characteristics to achieve the purpose of protection management.

According to the measures and characteristics, the prevention of snow flow disaster can be summarized as the four basic types: guide, resistance, change, and clearing.

Hazard prevention and controlling of the wind snow should be based on the regulation of windy snow flow motions and the snowdrift damage, combined with the local terrain, wind speed, wind direction to take measures in local conditions. In terms of road, we should choose the best line location first to solve the hazard problem from snowdrift. The open and high terrain with less fluctuation is the most beneficial area for the snow flow passing through. In addition, the smaller angle between the wind and the line is, the less damage be. When the terrain is restricted, the line is designed with the shortest route and minimal harm in the area of snow disaster. If the line cannot pass by the area of snowdrift, the engineering can select comprehensive prevention and controlling measures combined with resistance, solidation, and transportation. Lastly, because the snowdrift is complicated and snow protection facilities can only reduce disasters, but cannot completely eliminate disasters. So, mechanical snow removal operation is necessary.

Using snow break to prevent snowdrift disaster is the main method to prevent the snow from piling up in heaps, and reduce the transporting amount of snow on the road, and improve visibility. Because of the high altitude, tall plants cannot be grown on the line along the Qinghai–Tibet Railway, so the snow protection measures should be adopted engineering measures.

Snow fence is generally set on the upper side of the road, which can make snowdrift piled up before or behind it and then decrease transporting amount of snow on the road. Resistance snowbank is a soil dam in situ tightly, whose function is similar to the fence for storing snow and resisting snow.

Snow fence can decrease wind speed near the fence on the windward side near ground surface, which forms vortex to make the jumping snow grains static and accumulated. Snow fence can be made of all types of materials of available aluminum alloy, reinforced concrete, concrete, brick, stone, clod, and fences, the shape and size of covering snow near the fence have a close relationship with the density and height of the fence, as well as the space between the fence and the ground surface.

7.3.6 Summary

Snowstorms disaster is mainly the snowdrift along the Qinghai–Tibet Railway. Avalanche only appears in the Kunlun Mountain area, Tonglha Mountains, and Nyainqentanglha Mountains. Its amount and scale are far less than snowdrift. The snow disaster is mainly distributed between the Kunlun Mountain Pass and Nagqu. Notably, snow disaster is severest in the segment from Tonglha Mountains to Nagqu.

The line comparison is conducted on the main road passing through mountains for selecting a line of the Qinghai–Tibet Railway. For example, a long tunnel scheme has been adopted in the Kunlun Mountains. It avoids not only the influence of snow on the hillside, but also the adverse geological effects of steep slope location.

The snow fence is set up near Cuona Lake to prevent snowdrift, which is proved to be effective.

References

1. Bai, J. Q., Mei, L., & Yang, M. L. (2006). Geothermal resources and crustal thermal structure of the Qinghai–Tibet Plateau. *Journal of geomechanics, 12*(3), 354–362. (in Chinese).
2. Cao, D. Y., Lei, Y. D., Deng, X. F., et al. (2007). General engineering geologic conditions along the Naij Tal–Tuotuo River section of the Qinghai–Tibet Railway. *Qinghai Environment, 17*(2), 78–80. (in Chinese).
3. Cen, S. X., Gong, Y. F., & Chen, Y. Y. (2007). Climatic feature of atmospheric heat source over the Qinghai–Xizang Plateau. *Journal of Chengdu university of information technology, 22*(3), 369–373. (in Chinese).
4. Chen, L. T. (1998). Test and application of the relationship between anomalous snow cover in interspring over the Qinghai–Xizang Plateau and the first summer rainfall in southern China. *Quarterly Journal of Applied Meteorology, S1,* 1–8. (in Chinese).
5. Chen, R. H., & Wang, Z. T. (2003). Engineering-geological route selection of the gypsum section from Buqu to Wenquan of the Qinghai–Tibet Railway. *Journal of Glaciology and Geocryology, 25*(1), 14–16. (in Chinese).
6. Chen, M. X., Wang, J. Y., & Deng, X. (1994). *Geothermal resource of China: Forming characteristics and potential assessment.* Beijing: Science Press. (in Chinese).
7. China Railway First Survey & Design Institute Group Co., Ltd. (1999). *Railway engineering geology manual.* Beijing: China Railway Press (in Chinese).
8. China Railway First Survey & Design Institute Group Co., Ltd. (2000). *Comprehensive planning of Namtso Nature Reserve*. Internal information (in Chinese).
9. China Railway First Survey & Design Institute Group Co., Ltd. (2001). *Feasibility study report of the Qinghai–Tibet Railway*. Internal information (in Chinese).
10. China Railway First Survey & Design Institute Group Co., Ltd. (2001). *Line survey and instruction of the Qinghai–Tibet Railway*. Internal information (in Chinese).
11. China Railway First Survey & Design Institute Group Co., Ltd. (2001). *The railway engineering geology reconnaissance specification*. Beijing: China Railway Press (in Chinese).
12. China Railway First Survey & Design Institute Group Co., Ltd. (2001). *Geotechnical investigation code for areas unfavorable for railway engineering (TB10027-2011)*. Beijing: China Railway Publishing House (in Chinese).
13. China Railway First Survey & Design Institute Group Co., Ltd. (2001). *Provisional regulations of geotechnical investigation in the permafrost regions of the Qinghai–Tibet Railway.* Internal information (in Chinese).
14. China Railway First Survey & Design Institute Group Co., Ltd. (2001). *Special report for nature reserve and passage for wild animals along the Qinghai–Tibet Raliway.* Internal information (in Chinese).

15. China Railway First Survey & Design Institute Group Co., Ltd. (2002). *Geotechnical investigation of Qinshui River experiment section.* Internal information (in Chinese).
16. China Railway First Survey & Design Institute Group Co., Ltd. (2002). *Preliminary design of the Tonglha–Lhasa section of the Qinghai–Tibet Railway (Geology).* Internal information (in Chinese).
17. China Railway First Survey & Design Institute Group Co., Ltd. (2005). *Interim report for the results of the research project of the Qinghai–Tibet Railway test project.* Internal information (in Chinese).
18. China Railway First Survey & Design Institute Group Co., Ltd. (2006). *Code for design of railway subgrade (TB1001-2005).* Beijing: China Railway Publishing House (in Chinese).
19. Dai, J. X. (1990). *The climate of the Tibetan Plateau.* Beijing: Meteorological Press. (in Chinese).
20. Duo, J. (2003). Basic characteristics of the Yambajan geothermal field: A typical high-temperature geothermal system. *Engineer Science, 5*(1), 42–47. (in Chinese).
21. Editorial Board of Engineering Geology Manual. (2007). *Engineering geology manual* (4th ed.). Beijing: China Architecture & Building Press (in Chinese).
22. First Highway Survey and Design Institute, Ministry of Communications. (1996). *Compilation of research and design documents for the renovation project of the Qinghai–Tibet Highway.* Internal information (in Chinese).
23. Geologic Institute of State Seismological Bureau. (1991). *The active faults in central Tibet.* Beijing: Earthquake Press. (in Chinese).
24. Hu, X. C., Sun, J. D., Yao, Z. H., et al. (2003). Research on the influence of geothermal activities and exploitation on the geological environment in Tibet. *Journal of Mountain Science, 21*(12), 45–48. (in Chinese).
25. Jiang, Z. F. (1996). Geological hazards, forming conditions, and control along the Sichuan–Xizang Highway. *Chengdu Hydrogeological and Engineering-Geological Team, 3*(16), 243–249. (in Chinese).
26. Jin, X. Z., & Gongjue, D. Z. (2007). Climatic characteristics of hail weather in Nagqu. *Tibets Science and Technology, 3,* 61–63. (in Chinese).
27. Li, W. Y. (1988). *Quaternary plants and environment in China.* Beijing: Science Press. (in Chinese).
28. Li, W. H. (1998). *Ecological system and optimal utilization models in Tibet Plateau.* Guangzhou: Guangzhou Science and Technology Press. (in Chinese).
29. Liu, Z. P. (2008). Research on the technology of engineering geology investigation on permafrost along the Qinghai–Tibet Railway. *Journal of Railway Engineering Society, 8,* 1–5. (in Chinese).
30. Liu, Z. Q., Xu, X., Pan, G. T., et al. (1990). *Geotecture and evolution of the Tibetan Plateau.* Beijing: Geological Press. (in Chinese).
31. Liu, J. K., Tong, C. J., & Fang, J. H. (2005). *Introduction of geotechnical engineering in cold regions.* Beijing: China Railway Publishing House. (in Chinese).
32. Ma, T., Zhou, J. X., Zhang, X. D., et al. (2007). Preliminary studies on characteristics of vegetation distribution along the line of the Qinghai–Tibet Railway. *Research of Soil and Water Conservation, 14*(3), 150–154. (in Chinese).
33. Meng, X. L. (2006). Geological prospecting for permafrost engineering in the Qinghai–Tibet Railway. *Railway Investigation and Surveying, 3,* 28–31. (in Chinese).
34. Meng, L. S., Gao, R., Zhou, F. X., et al. (1990). Interpretation of the crustal structure in Yadong–Golmud area using gravity anomalies. *Acta Geophysica Sinci, 11*(2), 149–161. (in Chinese).
35. Meng, H., Zhang, Y. Q., & Yang, N. (2004). Analysis of the hot springtial distribution of geohazards along the middle segment of the eastern margin of the Qinghai–Tibet Plateau. *Journal of China Geology, 2*(31), 218–224. (in Chinese).
36. Ministry of Railways. (2006). *Code for design of railway line (GB50090-2006).* Beijing: China Plan Press. (in Chinese).

37. Ministry of Railways. (2006). *Code for seismic design of railway engineering (GB50111-2006)*. Beijing: China Plan Press. (in Chinese).
38. Ministry of Railways. (2006). *Basic specification for design of railway bridge and culvert (TB10002.1-2005)*. Beijing: China Railway Publishing House (in Chinese).
39. Natural Ecosystem Protection Office, Ministry of Environmental Protection of the People's Republic of China. (2006). *China nature reserve list (2005)*. Beijing: China Environmental Science Press (in Chinese).
40. Qin, D. H., Ding, Y. H., Wang, S. W., et al. (2002). A study of environment change and its impacts in western China. *Earth Science Frontiers, 2*(2), 321–328. (in Chinese).
41. Papers and reports on permafrost in the Qinghai–Tibet Plateau. *Annual Report of State Key Laboratory of Frozen Soil Engineering*. (in Chinese).
42. Shen, W. T., et al. (2005). *Prediction and evaluation of the ecological impact of the Qinghai–Tibet Railway*. Beijing: China Environmental Science Press. (in Chinese).
43. Shu, L., Lou, W. H., & Wang, L. J. (2003). An analysis on the ground temperature of Yangbajing Tunnel. *Journal of Glaciology and Geocryology, 25*(1), 24–28. (in Chinese).
44. State Forestry Administration. (2002). *Geotechnical investigation code in permafrost regions (GB50324-2001)*. Beijing: China Plan Press. (in Chinese).
45. The China Society on Tibet Plateau. (1995). *Collected papers of seminar on the Tibetan Plateau and global change*. Beijing: Meteorological Press. (in Chinese).
46. The Chinese–British Joint Geological Expedition for the Tibetan Plateau. (1985). *Geological evolution of the Qinghai–Tibet Plateau*. Beijing: Science Press. (in Chinese).
47. The Expert Panels of the Qinghai–Tibet Projects. (1995). *Research on the formation and evolution, environmental change, and ecologic of the Qinghai–Xizang (Tibet) Plateau*. Beijing: Science Press. (in Chinese).
48. The Fourth Survey and Design Institute, Ministry of Railways. (2006). *Code for design of special railway subgrade (TB10035-2006)*. Beijing: China Railway Publishing House (in Chinese).
49. Wang, Z. H. (2007). Geological environment and disasters along the railway line in the Qinghai–Tibet Plateau. *Earth Science Frontiers, 14*(6), 031–037. (in Chinese).
50. Wang, Q. Q., Qin, N. S., Tang, H. Y., et al. (2007). Study on climate change facts and their characteristics in the Qinghai Plateau in the recent 44 years. *Arid Zone Research, 24*(2), 234–239. (in Chinese).
51. Wei, G. J. (2003). Geological line selection of the Xiaonanchuan–Wangkun unfavorable geological section of the Qinghai–Tibet Railway. *Geotechinical Engineering World, 6*(7), 41–45. (in Chinese).
52. Wu, Y. S., & Yi, M. C. (2002). Active tectonics and geological hazards of the Xidatan–Lahsa section along the Qinghai–Tibet Railway and an assessment of its engineering geology. *Journal of Geomechanics, 8*(2), 97–135. (in Chinese).
53. Wu, Z. H., Ye, P. S., Wu, Z. H., et al. (2003). Hazard effects of active faulting along the Golmud–Lhasa Railway across the Tibetan Plateau. *Modern Geological, 17*(1), 1–7. (in Chinese).
54. Wu, H. Z., Wu, Z. H., Hu, D. G., et al. (2005). *Active faults and geological disasters along the Qinghai–Tibet Railway*. Beijing: The Earthquake Publishing House. (in Chinese).
55. Wu, Z. H., Hu, D. G., Wu, Z. H., et al. (2006). Pressure ridges and their ages of the Xidatan Strike-Slip Fault in south Kunlun Mountains. *Geological Review, 52*(1), 15–25. (in Chinese).
56. Wu, Z. H., Wu, Z. H., Hu, D. G., et al. (2006). Holocene seismogenic faults along the Tanggula–Lhasa section of the Qinghai–Tibet Railway. *Geological Bulletin of China, 25*(15), 1387–1401. (in Chinese).
57. Wu, Z. H., Hu, D. G., Wu, Z. H., et al. (2006). Migrating pingos and their hazard effects in the vicinity of the railway across the northern Tibetan Plateau. *Geological Bulletin of China, 25* (1–2), 233–243. (in Chinese).
58. Yang, W. F., & Zhou, T. (2007). Research on the geological disaster in the sideslope of the tunnel at the complex geological conditions. *Shanxi Architecture, 33*(4), 318–319. (in Chinese).

59. Yang, Y. F., Jiang, H., Niu, F. J., et al. (2007). Hot spring time-variation analyses of air temperature over the Qinghai–Xizang Plateau in warm and cold seasons. *Plateau Meteorology, 26*(3), 496–502. (in Chinese).
60. Ye, D. Z., Gao, Y. X., Shen, Z. B., et al. (1979). *The meteorology of the Qinghai–Tibet Plateau*. Beijing: Science Press. (in Chinese).
61. Yi, M. C., Wu, Z. H., Hu, D. G., et al. (2003). N–S trending active structures along the Qinghai–Tibet Railway and their influences on railway bed engineering. *Journal of Geomechanics, 10*(3), 343–354. (in Chinese).
62. Yi, Z. M., Zhou, J. X., & Zhang, X. D. (2006). Analysis of hydrologic conditions along the Qinghai–Tibet Railway. *Technology of Soil and Water Conservation, 4,* 14–16. (in Chinese).
63. Yuan, D. Y., Liu, X. L., Zhang, P. Z., et al. (2003). Characteristics of the modern activity of the Reshui–Riyueshan fault zone in Qinghai Province. *Seismology and Geology, 25*(1), 155–164. (in Chinese).
64. Zang, E. M., Wu, Z. W., et al. (1999). *Permafrost degradation and road engineering*. Lanzhou: Lanzhou University Press. (in Chinese).
65. Zhao, J. C., Liu, S. Z., & Ji, S. W. (2001). Engineering-geological problems in railway engineering construction. *The Chinese Journal of Geological Hazard and Control, 12*(1), 7–9. (in Chinese).